AF361806

Plant Oxidative Stress: Biology, Physiology and Mitigation

Plant Oxidative Stress: Biology, Physiology and Mitigation

Editors

Mirza Hasanuzzaman
Masayuki Fujita

MDPI • Basel • Beijing • Wuhan • Barcelona • Belgrade • Manchester • Tokyo • Cluj • Tianjin

Editors
Mirza Hasanuzzaman
Department of Agronomy,
Faculty of Agriculture
Sher-e-Bangla Agricultural
University
Dhaka
Bangladesh

Masayuki Fujita
Laboratory of Plant Stress
Responses, Faculty of
Agriculture
Kagawa University
Dhaka
Bangladesh

Editorial Office
MDPI
St. Alban-Anlage 66
4052 Basel, Switzerland

This is a reprint of articles from the Special Issue published online in the open access journal *Plants* (ISSN 2223-7747) (available at: www.mdpi.com/journal/plants/special_issues/oxidative_stress_plant_biology).

For citation purposes, cite each article independently as indicated on the article page online and as indicated below:

LastName, A.A.; LastName, B.B.; LastName, C.C. Article Title. *Journal Name* **Year**, *Volume Number*, Page Range.

ISBN 978-3-0365-4142-6 (Hbk)
ISBN 978-3-0365-4141-9 (PDF)

Contents

About the Editors

Mirza Hasanuzzaman

Dr. Mirza Hasanuzzaman is a Professor of Agronomy at Sher-e-Bangla Agricultural University, Dhaka, Bangladesh. He received his Ph.D. with a dissertation on 'Plant Stress Physiology and Antioxidant Metabolism'from the United Graduate School of Agricultural Sciences, Ehime University, Japan. Later, he completed his postdoctoral research at the University of the Ryukyus, Okinawa, Japan. Subsequently, he became an Adjunct Senior Researcher at the University of Tasmania with an Australian Government's Endeavour Research Fellowship. Prof. Hasanuzzaman has over 200 publications in the Web of Science. He has edited over 20 books and written over 30 book chapters. His publications are cited over 11000 times as per Scopus with an h-index of 54 (as of April 2022). He is an Editor and a reviewer for more than 50 peer-reviewed international journals and the recipient of 'Publons Peer Review Award 2017, 2018, and 2019'. He is acting Associate Editor of {Plant Signaling & Behavior} (Taylor and Francis), {Frontiers in Plant Science} (Switzerland), {Biocell} (Tech Science Press, USA), {Phyton} (Tech Science Press, USA); Academic Editor of {PeerJ} (USA), {Plants} (MDPI), {Vegetos} (Springer); and Member of the Editorial Board of {Plant Physiology and Biochemistry} (Elsevier), {Annals of Agricultural Sciences} (Elsevier), etc. He is an active member of 40 professional societies and is the acting Research and Publication Secretary of the Bangladesh JSPS Alumni Association. He received the World Academy of Science (TWAS) Young Scientist Award 2014, University Grants Commission (UGC) Gold Medal 2018, Global Network of Bangladeshi Biotechnologists (GNOBB) Award 2021, and Distinguished Scientists Award (2020) from AETDS, India. He is a fellow of the Bangladesh Academy of Sciences (BAS), Royal Society of Biology, and a Foreign fellow of The Society for Science of Climate Change and Sustainable Environment. He is named a Highly Cited Researcher by Clarivate AG.

Masayuki Fujita

Dr. Masayuki Fujita is Professor in the Laboratory of Plant Stress Responses, Faculty of Agriculture, Kagawa University, Kagawa, Japan. He received his B.Sc. in Chemistry from Shizuoka University, Shizuoka, and M.Agr. and Ph.D. in plant biochemistry from Nagoya University, Nagoya, Japan. His research interests include physiological, biochemical, and molecular biological responses based on secondary metabolism in plants under various abiotic and biotic stresses; phytoalexin, cytochrome P450, glutathione S-transferase, and phytochelatin; and redox reaction and antioxidants. In the last decade, his works were focused on oxidative stress and antioxidant defense in plants under environmental stress. His group investigates the role of different exogenous protectants in enhancing antioxidant defense and methylglyoxal detoxification systems in plants. He has supervised 4 M.S. students and 13 PhD students as the main supervisor. He has about 150 publications in journals and books, and has edited 10 books.

Editorial

Plant Oxidative Stress: Biology, Physiology and Mitigation

Mirza Hasanuzzaman [1] and **Masayuki Fujita** [2,*]

1   Department of Agronomy, Faculty of Agriculture, Sher-e-Bangla Agricultural University, Dhaka 1207, Bangladesh; mhzsauag@yahoo.com
2   Laboratory of Plant Stress Responses, Faculty of Agriculture, Kagawa University, Miki-cho 761-0795, Kita-gun, Kagawa, Japan
*   Correspondence: fujita.masayuki@kagawa-u.ac.jp

Citation: Hasanuzzaman, M.; Fujita, M. Plant Oxidative Stress: Biology, Physiology and Mitigation. *Plants* **2022**, *11*, 0. https://doi.org/

Received: 7 April 2022
Accepted: 25 April 2022
Published: 28 April 2022

Publisher's Note: MDPI stays neutral with regard to jurisdictional claims in published maps and institutional affiliations.

Due to climate change plants are frequently exposed to abiotic and biotic stresses, and these stresses pose serious threats to plant growth and productivity. A key sign of such stresses at the molecular level is the accelerated production of reactive oxygen species (ROS), which cause oxidative stress. Abiotic stresses result in the reduction of photosynthetic activities of the plants and accelerate the accumulation of ROS—which are oxygen radicals and their derivatives—such as hydrogen peroxide (H_2O_2) singlet oxygen ($^1O_2$), superoxide radicals ($O_2^{\bullet-}$), and hydroxyl radical ($OH^\bullet$). These are highly reactive and usually toxic. The ROS can induce cellular injuries through protein oxidation, lipid peroxidation, and DNA damage, which finally may result in a plant's cellular death [1–3]. Recently, ROS have emerged as major regulatory molecules in plants, and their role in early signaling events initiated by normal cellular metabolic function and environmental stress is now well established. Under normal circumstances, there is a balance between the generation and the elimination of ROS. However, this balance can be hampered by different biotic and abiotic stresses, resulting in the generation of a large number of ROS that should be counteracted by the antioxidant machinery in cells. Finding ways to enhance the antioxidant defense systems in plants is a very important task for plant biologists. Recent progress in plant molecular biology and biotechnology has been targeting the development of approaches that can be used to enhance the antioxidant defense systems in plants. New knowledge acquired through research on oxidative stress, abiotic stress, and biotic stress tolerance in plants will help us to apply stress-responsive determinants and to engineer plants with enhanced tolerance to stress.

Considering the deleterious effects of ROS, their detoxification is very important in order to ensure plant production. Enhanced activities of antioxidant enzymes play a crucial role in upregulating the adaptation mechanisms of plants under oxidative stress [4]. Non-enzymatic antioxidants—such as ascorbate (AsA), glutathione (GSH), tocopherols (Toc), flavonoids, and carotenoids—and enzymatic antioxidants—such as superoxide dismutase (SOD), monodehydroascorbate reductase (MDHAR), dehydroascorbate reductase (DHAR), ascorbate peroxidase (APX), catalase (CAT), glutathione reductase (GR), glutathione S-transferase (GST), glutathione peroxidase (GPX), and peroxidase (POD)—take part in scavenging ROS and protect the plants from oxidative damage [1,2].

Advanced molecular tools and genomics tools have shown that antioxidant defense is crucial for defending the plant against oxidative stress [5]. Many researchers have revealed that modern genomics studies have advanced our capabilities of improving crop genetics, especially those traits relevant to abiotic stress management. Billah et al. [5] reviewed the literature for updated and comprehensive studies concerning all possible combinations of advanced genomics tools and the gene regulatory network of reactive oxygen species homeostasis for the appropriate planning of future breeding programs, which will assist sustainable crop production under salinity and drought conditions [5].

From several studies, it is well established that the enhancement of antioxidant defense is highly correlated with stress tolerance. Seleiman et al. [6] reported that sequential application of proline (Pro), AsA, and/or GSH rectifies ion imbalance and strengthens antioxidant systems in salt-stressed cucumber by improving growth characteristics, photosynthetic efficiency, relative water content (RWC), and membrane stability index (MSI). Additional improvements were seen in AsA, Pro, and GSH; enzymatic activity; leaf and root K^+ and Ca^{2+} contents; and their ratios to Na^+, while the same sequential applications significantly reduced leaf and root Cd^{2+} and Na^+ contents [6].

Some of extremophile and endemic plants were found to have ample capacity for antioxidant defense, and those plants provide enhanced tolerance to oxidative stress. Accumulation of secondary metabolites is a defense mechanism, as reported by Hashim et al. [7]. They found that endemic endangered species viz. *Nepeta septemcrenata*, *Origanum syriacum* subsp. *Sinaicum*, *Phlomis aurea*, *Rosa arabica*, and *Silene schimperiana*) showed elevated phenols, AsA, Pro, flavonoids, and tannins content in response to different altitudes. Secondary metabolites progressively increased in the studied species that were associated with a significant decrease in the levels of antioxidant enzyme activity [7].

ROS-induced oxidative damage can be alleviated by exogenous application of different chemical substances, such as amino acids and their derivatives, sugars, polyamines and vitamins, plant growth regulators to seeds (as seed priming), roots (as irrigation or soil incorporation), or leaves (as foliar application), at low concentrations. The systemic action of these chemicals improved various abiotic stress tolerances, including salinity, by increasing the antioxidant defense system and decreasing oxidative injuries at the cellular level [8].

Del Pino et al. [9] investigated the persistence of the effects of selenium (Se)-fertilization and found that Se biofortification increased the nutritional and qualitative values of foods in Se-deficient regions and increased the tolerance of oxidative stress in olive trees. This result indicated that trace elements have important functions in the adaptability of plants [9].

Selenium is a well-recognized trace element that has shown multifarious positive effects on plants. This element alone or in combination with other elements could provide increased plant oxidative stress tolerance [10]. The study conducted by Rahman et al. [11] revealed that supplementation of Se, boron (B), and Se + B enhanced the activities of APX, MDHAR, DHAR, GR, CAT, GPX, GST, POD, Gly I, and Gly II. This supplementation consequently diminished the H_2O_2 content and MDA content under salt stress and improved the growth parameters. The results reflected that exogenous Se, B, and Se + B enhanced the enzymatic activity of the antioxidant defense system and the glyoxalase systems under different levels of salt stress. This ultimately alleviated the salt-induced oxidative stress. Se+B supplementation was more effective than a single treatment [11].

Phytohormones are also important stress elicitors and there are many plant studies that show that exogenous application of phytohormones can mitigate oxidative stress in plants. Al-Harthi et al. [12] performed seed priming with gibberellic acid (GA) and jasmonic acid (JA) and, subsequently, the plants (summer squash) were grown in saline media. They observed that GA and JA resulted in a reduction of the concentration of Na^+, Cl^-, and the chlorophyll (Chl) *a*/*b* ratio. Increasing the activity of SOD, CAT, and APX; the quantities of K^+ and Mg^{2+}; the K^+/Na^+ ratio; and the quantities of RNA, DNA, Chl *b*, and carotene, ameliorated the growth of salinized plants [12].

Similarly, salicylic acid (SA) was also found to be very protective against oxidative stress due to its metabolic functions. This hormone was found to enhance antioxidant defense and osmolyte metabolism, which was the main cause of oxidative stress tolerance in watermelons exposed to B toxicity [13]. Exogenously applied SA promoted photosynthesis and, consequently, biomass production in watermelon seedlings treated with a high level of B by reducing B accumulation, lipid peroxidation, and the generation of H_2O_2, while significantly increasing levels of the most reactive ROS, $OH^\bullet$ [13].

The interactive effects of methyl jasmonate and SA were found to mitigate drought-induced oxidative damages in *Phaseolus vulgaris* as observed by Mohi-Ud-Din et al. [14]. Combined application of these phytohormones remarkably enhanced the drought tolerance of plants by improving the physiological activities and antioxidant defense system (SOD,

CAT, POD, GPX, and GST as well as the enzymes of the AsA–GSH cycle). Phytohormones lowered the generation of $O_2^{\bullet-}$ and H_2O_2 and the malondialdehyde (MDA) content [14].

Some of the non-enzymatic antioxidants, such as AsA, GSH, and Toc, were reported to be very effective in scavenging and detoxifying ROS. In okra, foliar spray of α-Toc showed an antioxidant potential, which promoted salt tolerance [15]. Foliar application of α-Toc significantly improved the yield in tested okra varieties by increasing the activity of antioxidants (CAT, GPX, SOD, and AsA), accumulation of glycine betaine, and total free Pro in fruit tissues under saline and non-saline conditions. However, these effects were dose-dependent [15].

Manipulating production practices or growing techniques can also help in mitigating oxidative stress in plants. Using a resistant cultivar or rootstock is one possible approach. Tao et al. [16] reported that the use of heat-resistant rootstock grafting is a viable technique that is practiced globally in order to improve plant resistance to abiotic stresses. Bitter melon (*Momordica charantia* L.)-grafted cucumber seedlings showed significantly improved heat-induced growth inhibition and photoinhibition, maintained better photosynthesis activity, and accumulated greater biomass than self-grafted seedlings [16].

Many amino acids and peptones are involved in the stimulation of physiological and metabolic functions of plants [17,18]. Peptone-induced cadmium (Cd) tolerance in *Spinacia oleracea* through modulation of antioxidant metabolism was reported by Emanuil et al. [18]. The application of peptone decreased Cd uptake and decreased levels of MDA, H_2O_2, and electrolyte leakage in spinach by increasing the activity of antioxidant enzymes. It indicated that peptone is a promising plant growth regulator that represents an efficient approach for the phytoremediation of Cd-polluted soils and enhancement of spinach growth, yield, and tolerance under a Cd-dominant environment [18].

Recently, nanoparticles (NPs) have shown increasing attention in research focusing on enhancing the growth progress and development of plants [19]. NPs have recently become more commonly employed in commercial products and industrial applications [20]. There are many studies where NPs are commonly used as growth regulators and stress elicitors [21]. In their study, Ahmad et al. [22] examined the effect of zinc oxide NPs in As-stressed soybean plants. Application of zinc oxide NPs to the As-stressed plants showed enhanced activities of the enzymes involved in the AsA–GSH cycle as well as SOD and CAT [22].

This Special Issue, "Plant Oxidative Stress: Biology, Physiology, and Mitigation" published 11 original research works and 1 review article that discussed the various aspects of ROS Biology, metabolism, and the physiological mechanisms and approaches to mitigating oxidative stress. These types of research studies show further directions for the development of crop plants that are tolerant to abiotic stress in the era of climate change.

Author Contributions: Conceptualization, M.H. and M.F.; writing—original draft preparation, M.H. and M.F.; writing—review and editing, M.H.; All authors have read and agreed to the published version of the manuscript.

Funding: This work received no external funding.

Acknowledgments: We would like to thank all contributors for their submission to this Special Issue and the reviewers for their valuable input in improving the articles. We also thank the editorial office for their helpful support during the compilation of this Special Issue. The authors acknowledge Farzana Nowroz, Ayesha Siddika, Salma Binte Salam, Shamima Sultana, and Md. Rakib Hossain Raihan for their help during the organization of the studies and necessary formatting.

Conflicts of Interest: The authors declare no conflict of interest.

References

1. Hasanuzzaman, M.; Bhuyan, M.H.M.B.; Anee, T.I.; Parvin, K.; Nahar, K.; Mahmud, J.A.; Fujita, M. Regulation of ascorbate-glutathione pathway in mitigating oxidative damage in plants under abiotic stress. *Antioxidants* **2019**, *8*, 384. [CrossRef] [PubMed]
2. Hasanuzzaman, M.; Bhuyan, M.H.M.B.; Zulfiqar, F.; Raza, A.; Mohsin, S.M.; Mahmud, J.A.; Fujita, M.; Fotopoulos, V. Reactive oxygen species and antioxidant defense in plants under abiotic stress: Revisiting the crucial role of a universal defense regulator. *Antioxidants* **2020**, *9*, 681. [CrossRef] [PubMed]
3. Hasanuzzaman, M.; Raihan, M.R.H.; Masud, A.A.C.; Rahman, K.; Nowroz, F.; Rahman, M.; Nahar, K.; Fujita, M. Regulation of reactive oxygen species and antioxidant defense in plants under salinity. *Int. J. Mol. Sci.* **2021**, *22*, 9326. [CrossRef] [PubMed]
4. Miller, G.; Suzuki, N.; Ciftci-Yilmaz, S.; Mittler, R. Reactive oxygen species homeostasis and signaling during drought and salinity stresses. *Plant Cell Environ.* **2010**, *33*, 453–467. [CrossRef] [PubMed]
5. Billah, M.; Aktar, S.; Brestic, M.; Zivcak, M.; Khaldun, A.B.M.; Uddin, M.S.; Bagum, S.A.; Yang, X.; Skalicky, M.; Mehari, T.G.; et al. Progressive genomic approaches to explore drought- and salt-induced oxidative stress responses in plants under changing climate. *Plants* **2021**, *10*, 1910. [CrossRef] [PubMed]
6. Seleiman, M.F.; Semida, W.M.; Rady, M.M.; Mohamed, G.F.; Hemida, K.A.; Alhammad, B.A.; Hassan, M.M.; Shami, A. Sequential application of antioxidants rectifies ion imbalance and strengthens antioxidant systems in salt-stressed cucumber. *Plants* **2020**, *9*, 1783. [CrossRef] [PubMed]
7. Hashim, A.M.; Alharbi, B.M.; Abdulmajeed, A.M.; Elkelish, A.; Hozzein, W.N.; Hassan, H.M. Oxidative stress responses of some endemic plants to high altitudes by intensifying antioxidants and secondary metabolites content. *Plants* **2020**, *9*, 869. [CrossRef] [PubMed]
8. Costa, S.F.; Martins, D.; Agacka-Mołdoch, M.; Czubacka, A.; de Sousa Araújo, S. Strategies to alleviate salinity stress in plants. In *Salinity Responses and Tolerance in Plants; Volume 1: Targeting Sensory, Transport and Signaling Mechanisms*; Kumar, V., Wani, S.H., Suprasanna, P., Tran, L.-S.P., Eds.; Springer: Cham, Switzerland, 2018; pp. 307–337.
9. Del Pino, A.M.; Regni, L.; D'Amato, R.; Di Michele, A.; Proietti, P.; Palmerini, C.A. Persistence of the effects of Se-fertilization in olive trees over time, monitored with the cytosolic Ca^{2+} and with the germination of pollen. *Plants* **2021**, *10*, 2290. [CrossRef] [PubMed]
10. Hasanuzzaman, M.; Bhuyan, M.B.; Raza, A.; Hawrylak-Nowak, B.; Matraszek-Gawron, R.; Al Mahmud, J.; Nahar, K.; Fujita, M. Selenium in plants: Boon or bane? *Environ. Exp. Bot.* **2020**, *178*, 104170. [CrossRef]
11. Rahman, M.; Rahman, K.; Sathi, K.S.; Alam, M.M.; Nahar, K.; Fujita, M.; Hasanuzzaman, M. Supplemental selenium and boron mitigate salt-induced oxidative damages in *Glycine max* L. *Plants* **2021**, *10*, 2224. [CrossRef] [PubMed]
12. Al-harthi, M.M.; Bafeel, S.O.; El-Zohri, M. Gibberellic acid and jasmonic acid improve salt tolerance in summer squash by modulating some physiological parameters symptomatic for oxidative stress and mineral nutrition. *Plants* **2021**, *10*, 2768. [CrossRef] [PubMed]
13. Moustafa-Farag, M.; Mohamed, H.I.; Mahmoud, A.; Elkelish, A.; Misra, A.N.; Guy, K.M.; Kamran, M.; Ai, S.; Zhang, M. Salicylic acid stimulates antioxidant defense and osmolyte metabolism to alleviate oxidative stress in watermelons under excess boron. *Plants* **2020**, *9*, 724. [CrossRef] [PubMed]
14. Mohi-Ud-Din, M.; Talukder, D.; Rohman, M.; Ahmed, J.U.; Jagadish, S.V.K.; Islam, T.; Hasanuzzaman, M. Exogenous application of methyl jasmonate and salicylic acid mitigates drought-induced oxidative damages in french bean (*Phaseolus vulgaris* L.). *Plants* **2021**, *10*, 2066. [CrossRef] [PubMed]
15. Naqve, M.; Wang, X.; Shahbaz, M.; Fiaz, S.; Naqvi, W.; Naseer, M.; Mahmood, A.; Ali, H. Foliar spray of alpha-tocopherol modulates antioxidant potential of okra fruit under salt stress. *Plants* **2021**, *10*, 1382. [CrossRef] [PubMed]
16. Tao, M.-Q.; Jahan, M.S.; Hou, K.; Shu, S.; Wang, Y.; Sun, J.; Guo, S.-R. Bitter melon (*Momordica charantia* L.) rootstock improves the heat tolerance of cucumber by regulating photosynthetic and antioxidant defense pathways. *Plants* **2020**, *9*, 692. [CrossRef] [PubMed]
17. Soad, M.M.; Lobna, S.T.; Farahat, M.M. Influence of foliar application of pepton on growth, flowering and chemical composition of *Helichrysum bracteatum* L. plants under different irrigation intervals. *Ozean J. Appl. Sci.* **2010**, *3*, 143–155.
18. Emanuil, N.; Akram, M.S.; Ali, S.; El-Esawi, M.A.; Iqbal, M.; Alyemeni, M.N. Peptone-induced physio-biochemical modulations reduce cadmium toxicity and accumulation in spinach (*Spinacia oleracea* L.). *Plants* **2020**, *9*, 1806. [CrossRef]
19. Do Espirito Santo Pereira, A.; Oliveira, H.C.; Fraceto, L.F.; Santaella, C. Nanotechnology potential in seed priming for sustainable agriculture. *Nanomaterials* **2021**, *11*, 267. [CrossRef]
20. Yan, A.; Chen, Z. Detection methods of nanoparticles in plant tissues. In *New Visions in Plant Science*; Çelik, Ö., Ed.; IntechOpen: London, UK, 2018; pp. 99–119. [CrossRef]
21. Mittal, D.; Kaur, G.; Singh, P.; Yadav, K.; Ali, S.A. Nanoparticle-based sustainable agriculture and food science: Recent advances and future outlook. *Front. Nanotechnol.* **2020**, *2*, 10. [CrossRef]
22. Ahmad, P.; Alyemeni, M.N.; Al-Huqail, A.A.; Alqahtani, M.A.; Wijaya, L.; Ashraf, M.; Kaya, C.; Bajguz, A. Zinc oxide nanoparticles application alleviates arsenic (As) toxicity in soybean plants by restricting the uptake of as and modulating key biochemical attributes, antioxidant enzymes, ascorbate-glutathione cycle and glyoxalase system. *Plants* **2020**, *9*, 825. [CrossRef]

Review

Progressive Genomic Approaches to Explore Drought- and Salt-Induced Oxidative Stress Responses in Plants under Changing Climate

Masum Billah [1], Shirin Aktar [2], Marian Brestic [3,4,*], Marek Zivcak [3], Abul Bashar Mohammad Khaldun [5], Md. Shalim Uddin [5], Shamim Ara Bagum [5], Xinghong Yang [6], Milan Skalicky [4], Teame Gereziher Mehari [1], Sagar Maitra [7] and Akbar Hossain [8,*]

[1] Institute of Cotton Research, Chinese Academy of Agricultural Sciences, Anyang 455000, China; kazimasum.agpstu20@gmail.com (M.B.); fiamieta21@gmail.com (T.G.M.)

[2] Institute of Tea Research, Chinese Academy of Agricultural Sciences, South Meiling Road, Hangzhou 310008, China; aktershirin1992@gmail.com

[3] Department of Plant Physiology, Slovak University of Agriculture, Nitra, Tr. A. Hlinku 2, 949 01 Nitra, Slovakia; marek.zivcak@uniag.sk

[4] Department of Botany and Plant Physiology, Faculty of Agrobiology, Food, and Natural Resources, Czech University of Life Sciences Prague, Kamycka 129, 165 00 Prague, Czech Republic; skalicky@af.czu.cz

[5] Bangladesh Agricultural Research Institute, Gazipur 1701, Bangladesh; abkhaldun@gmail.com (A.B.M.K.); shalimuddin40@gmail.com (M.S.U.); happyshalim@yahoo.com (S.A.B.)

[6] State Key Laboratory of Crop Biology, College of Life Sciences, Shandong Agricultural University, 61 Daizong St., Tai'an 271000, China; xhyang@sdau.edu.cn

[7] Department of Agronomy, Centurion University of Technology and Management, Village Alluri Nagar, R.Sitapur 761211, Odisha, India; sagar.maitra@cutm.ac.in

[8] Department of Agronomy, Bangladesh Wheat and Maize Research Institute, Dinajpur 5200, Bangladesh

* Correspondence: marian.brestic@uniag.sk (M.B.); akbarhossainwrc@gmail.com (A.H.)

Citation: Billah, M.; Aktar, S.; Brestic, M.; Zivcak, M.; Khaldun, A.B.M.; Uddin, M.S.; Bagum, S.A.; Yang, X.; Skalicky, M.; Mehari, T.G.; et al. Progressive Genomic Approaches to Explore Drought- and Salt-Induced Oxidative Stress Responses in Plants under Changing Climate. *Plants* **2021**, *10*, 1910. https://doi.org/10.3390/plants10091910

Academic Editor: Yukio Kurihara

Received: 9 August 2021
Accepted: 11 September 2021
Published: 14 September 2021

Publisher's Note: MDPI stays neutral with regard to jurisdictional claims in published maps and institutional affiliations.

Abstract: Drought and salinity are the major environmental abiotic stresses that negatively impact crop development and yield. To improve yields under abiotic stress conditions, drought- and salinity-tolerant crops are key to support world crop production and mitigate the demand of the growing world population. Nevertheless, plant responses to abiotic stresses are highly complex and controlled by networks of genetic and ecological factors that are the main targets of crop breeding programs. Several genomics strategies are employed to improve crop productivity under abiotic stress conditions, but traditional techniques are not sufficient to prevent stress-related losses in productivity. Within the last decade, modern genomics studies have advanced our capabilities of improving crop genetics, especially those traits relevant to abiotic stress management. This review provided updated and comprehensive knowledge concerning all possible combinations of advanced genomics tools and the gene regulatory network of reactive oxygen species homeostasis for the appropriate planning of future breeding programs, which will assist sustainable crop production under salinity and drought conditions.

Keywords: salt; drought; plants; ROS; genomics; approaches; integration

1. Introduction

Global crop productivity is restricted due to abiotic stresses such as drought, salinity, flooding, nutrient deficiency, and environmental toxicity. Among these abiotic stresses, salinity and drought are the most severe constraints for sustainable agriculture on a global scale. Nearly 7% of terrestrial land is affected by salinity [1], while drought is widespread and increasingly common in recent years due to climate change. Altogether, salinity- and drought-affected lands cover approximately 10.5 and 60 million km^2, respectively [2]. Furthermore, climatic changes worsen the frequency and intensity of water shortages in subtropical areas of Asia and Africa. As stated by the UN climatic report [http://www.solcomhouse.com/drought.htm; accessed date on 12 July 2021], rising temperatures are melting the Himalayan glaciers that feed Asia's largest rivers (Indus, Ganges, Brahmaputra, Yangtze, Mekong, Salween, and Yellow), and those glaciers may disappear by 2035. Additionally, long-term trends indicate that the progressive proliferation of salinity has caused the dilapidation of arable land [3]. For instance, in California, over the last century, 4.5 out of 8.6 million hectares of wetted agricultural land have become salt-affected [4]. Currently, it has become a pertinent problem for crop production [5], mostly in arid and semiarid areas.

Based on numerous estimations, the world population will increase to over 9.7 billion by 2050, which will continue to exacerbate current global food insecurity issues. It is estimated that, over the past 50 years, improved crop productivity has brought about an increase in world food production by up to 20% per capita and decreased the proportion of food-insecure people existing in developing countries from 57% to 27% of the world population [6]. As a result, crops will need to cope with abiotic stresses such as drought and salinity and double productivity to further diminish food insecurity and support the growing human population in more ecologically sustainable ways.

Both drought and salinity stresses induce cellular dehydration, which causes osmotic stress, removal of water from the cytoplasm into the apoplast, and eventually evaporation into the atmosphere [2]. Moreover, early responses to salt stress and drought are comparable in plants. For example, plant cells prevent water loss by increasing the ionic constituents and decreasing the osmotic potential in stressed cells. Due to the similar mechanisms of the stress response in plants, it appears that drought and salinity tolerance mechanisms might be functionally interchangeable [7]. It is well known that stress response mechanisms involve several particular physiological and biochemical pathways that allow plants to adapt to unfavorable conditions. A number of abiotic stress factors, such as salinity, drought, high temperatures, and osmotic stresses, lead to the overproduction of reactive oxygen species (ROS), which cause serious cellular damage and hamper photosynthesis. To protect or repair these injuries, plant cells use an intricate defense system, including a number of antioxidative stress-related defense genes that, in turn, prompt changes in the biochemical plant machinery [8]. ROS production and antioxidant regulation all occur in a synergistic, additive, or antagonistic way and are associated with the control of oxidative stress.

Nevertheless, plant stress response mechanisms are controlled by convoluted networks that are determined by environmental and genetic factors that are often difficult to untangle, thereby impeding traditional breeding approaches [9]. Considering that the conventional breeding strategies for crop improvement are largely aimed at improving yield to meet the demands of an ever-growing world population, breeders have to implement innovative approaches in agriculture to combine high-yield and abiotic stress-tolerant traits in crops [10]. Recent scientific advances and the abovementioned challenges in agriculture have directed the development of high-throughput techniques to pursue and take advantage of plant genome research for the improvement of stress-tolerant crops. Thus, these genomics approaches focus on the entire genome, involving genic and intergenic positions, to attain new insights into the functional and molecular responses of plants, which will sequentially offer specific techniques for crop plant improvement. Recently, many scientists have revealed promising outcomes toward understanding the molecular mechanisms of abiotic stress tolerance in prospective crops using progressive molecular biology practices [10–21]. The mechanisms involved in crop salt and drought stress responses are discussed in Figure 1. In this review, we described in detail how to mine the functional genes involved in drought and salt response in plants, using methods such as traditional QTL, transcriptomic analysis, and GWAS. Then, we explored approaches such as epigenetic regulators, gain-of-function, RNAi, TALENs, ZFNs, CRISPR, base editing, and primer editing for functional verification with an example of target genes generated by the aforementioned approaches. Finally, we summarized the methods for generating salt and drought-tolerant crops. The review aimed to offer comprehensive knowledge for improving salt- and drought-tolerant crops using modern genomics strategies regarding

ROS regulatory networks. The overview of earlier studies on the advancement of genomics approaches will help in the investigation of upcoming research instructions for improving salt- and drought-tolerant crops.

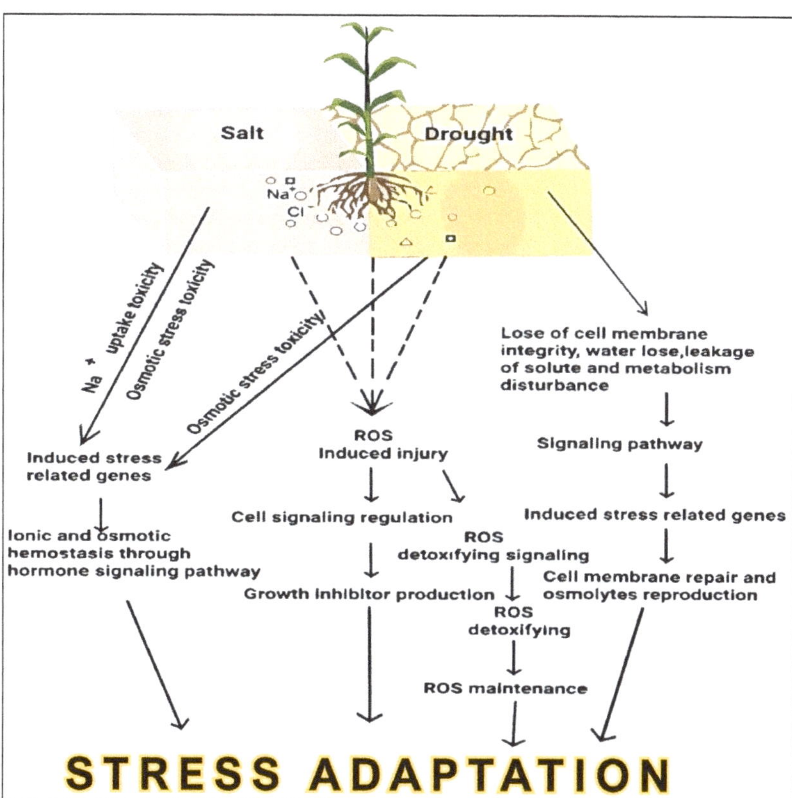

Figure 1. The mechanisms involved in crop salt and drought stress responses.

2. Mining Approaches for Salt and Drought Stress Response Genes

To improve drought and salinity stress tolerance in crops, we first require comprehensive knowledge concerning the complex mechanisms of plants that respond to stresses. Detecting the genes/markers/QTL regions associated with drought and salinity stress responses is the first crucial step toward reaching the required understanding for breeding drought and salinity stress-tolerant crop varieties. For the discovery of a gene, various strategies are available in both model and nonmodel crops; here, some of the most advanced are discussed briefly.

2.1. Quantitative Trait Loci (QTL) Analysis

A quantitative trait locus (QTL) is a gene or a region of DNA that is associated with the variation of a quantitative/phenotypic trait that must be polymorphic to affect the biological population. QTL mapping has been a powerful tool for dissecting genetic variants underlying quantitative traits in numerous biological studies and breeding programs. There are two primary concerns when using QTL mapping. One is the power for QTL identification under a controlled false-positive rate, and the other is the accuracy of QTL localization [22]. QTL mapping has been applied as a technique for identifying genomic regions significantly correlated with grain output and various genetically intricate characteristics in cereal crops. This technique is particularly powerful when genetic variation is studied concerning numerous complex traits, where it is possible to identify and differentiate genomic regions that contribute to different characteristics of interest. The data relevant to QTL mapping can be conducive to enhancing the genetic potential of crops via marker-assisted breeding [23]. Currently, scientists can link the molecular mechanisms of genes found in QTLs to demonstrate the genetic and physiological basis of traits such as grain yield. A nice example of this cooperation of QTL mapping, trait scoring, and breeding can be found in using green coloration as a metric of drought resistance in sorghum. The genetic dissection of molecular QTLs associated with green coloration during drought lends convenience to demonstrate the basic mechanisms of physiology and investigate the molecular causes of drought tolerance in sorghum and different grasses [24]. Reducing the genomic sizes of QTLs facilitates enhanced targeting of pertinent genomic regions. Improving the fine mapping of QTLs improves the efficiency with which breeders can understand the significance and mechanisms of QTLs relevant to their traits of interest [24]. Enhanced QTL mapping is particularly relevant when deconvolving complex genomic regions. For example, hypostasis of alleles within QTLs, QTL-QTL genetic interactions, context-dependent activities of QTLs, and the QTL marker position itself impact the articulation of a complex trait such as the yield of grain under drought stress [25]. Interestingly, fine mapping of QTLs revealed that an individual main QTL controlling membrane potential vastly improved marker-assisted selection for salinity-stressed barley [26]. Thus, fine QTL mapping is required for marker-aided QTL pyramiding to improve drought tolerance [27]. Identification of QTLs for abiotic stress tolerance suggests augmentations that can be used for further genomics studies toward the detection of noble genes of salt and drought tolerance to develop a new variety [28]. Several examples of QTLs (quantitative trait loci) for improving crop plant production under salinity and drought stresses are discussed in Table 1.

Table 1. Known QTLs (quantitative trait loci) for improving crop plant production under salinity and drought stresses.

Stresses	Crops	Major Effect/Finding	References
Drought stress	Cowpea	Detected QTL relevant to salt-tolerant and sensitive varieties	[29]
Drought stress	Wheat	Detected genetic loci to major morpho-physiological traits, components of yield, and grain yield	[30]
Salinity	Barley	Detected QTLs related to stomatal and photosynthetic traits associated with salinity tolerance	[31]
Drought	Sorghum	Identified QTLs associated with flowering and drought resistance	[24]
Drought	Rice	Improved crop yield under drought tolerance	[25]
Drought and submergence tolerance	Rice (TDK1)	Drought and submergence tolerance and yield stability	[25]
Drought and flood	Rice	Detected drought and salinity tolerance varieties based on developmental and physiological traits	[32]
Salinity	Rice (Pusa Basmati 1121)	Detected two QTLs for drought and one QTL for salt stress	[32]
Drought	Upland rice	Identified QTLs relevant to leaf rolling, leaf drying, leaf relative water content, and relative growth rate under water stress	[33]

2.2. Forward Genetics and the Candidate Gene Strategy

Crop plants with stress tolerance have been generated by the transference of genes/loci from definite donor parents, either through a forwarding genetics method that includes the determination of a gene function linked with a phenotype or the identification of novel stress-tolerant donor lines created by the use of mutagenesis. In contrast, reverse genetic breeding approaches could offer an understanding of gene functions and structure/sequence information to predict traits for adapting stress-tolerant cultivars using transgenic and advanced breeding tools. In genomic studies, researchers have implemented these approaches for the genetic improvement of various model and nonmodel species toward salt and drought stress tolerance [34–41].

2.3. Transcriptomics Analysis

Transcriptome analysis refers to the study of the transcriptome of the entire set of RNA transcripts that are generated by a genome, under given times and circumstances or in a specific cell. Transcriptomic analysis techniques play a crucial role in the identification of candidate gene functions and pathways that respond to specific environments [42]. In the last decade, universal transcriptome analysis approaches have been particularly advantageous for functional genomic studies that offer comprehensive molecular mechanisms of certain phenomena. Primarily, a global transcriptome study was initiated with suppression subtractive hybridization (SSH) and cDNA-AFLP and acquired a quantum dive to RNA-seq with the progression of NGS platforms [10]. The information content of an organism is held in its genome and articulated through transcription. The basic purposes of transcriptomics are to record the transcription of all species, including mRNAs, noncoding RNAs, and small RNAs; to determine the transcriptional configuration of genes in terms of their start sites, $5'$ and $3'$ ends, splice variants, and other posttranscriptional modifications; and to calculate the varying expression patterns of every transcript throughout development and under diverse conditions [40,41]. Currently, transcriptome profiling has progressed into nearly all organisms and represents how information attained from sequence data can be converted into a wide knowledge of gene functionality [42]. Plant stress-response mechanisms frequently employ the use of transcription factors (TFs). A TF is a protein that targets, typically, multiple genes that comprise a regulon and influences their expression patterns. Thus, TFs are a powerful tool for the genetic regulation of many downstream genes and processes, including abiotic stress responses [43]. In the case of salt and drought stress, transcripts related to the upregulation of vital biochemical pathways required for cellular osmotic balance, abscisic acid, and cellular water uptake are controlled by TFs [44]. The role and example of various transcription families through transcriptome analysis relevant to salt and drought stress tolerance are discussed in Table 2.

Table 2. Identification of different TF families through transcriptome analysis relevant to salt and drought stress tolerance.

SL No.	TF Family	Gene ID	Crop Variety	Target Stresses	References
1	AP2/EREBP	*TaERF3*	*Triticum aestivum*	Drought, Salt	[45]
2	bZIP	*GmbZIP1,*	*Glycine max*	Drought, Salinity,	[46]
3	MYB/MYC	*StMYB1R-1*	*Solanum tuberosum*	Drought	[47]
4	NAC	*OsNAC5,* *GmNAC20*	*Oryza sativa,* *Glycine max*	Drought	[48,49]
5	WRKY	*TaWRKY44*	*Triticum aestivum*	Drought, Salt	[50]
6	*AREB/ABF*	*AREB1,* *AREB2/ABF4* and *ABF3*	*Arabidopsis thaliana,*	Drought	[51]

2.4. Association Mapping

In genetics studies, association mapping (also well known as linkage disequilibrium mapping) refers to the regular genome-wide distribution of several genes together with other measurable loci (markers) in predicting marker-trait relatives [52] applied in various crops, including rice, barley, maize, sorghum, and wheat, to identify the significant genes or markers that confer a given trait [53]. Numerous genes have been indicated as being connected with abiotic resistance by applying association mapping [54]. It has also been applied to inquire about the limitations within focused parameters and molecular markers in different crops [55]. Association mapping is extremely efficient in experimental varieties with complex or unknown genotypes or those that have a large regeneration time.

Association mapping of drought-related varieties in barley was applied to terminate a conventional biparental system of QTL mapping [56]. Furthermore, association mapping has been used to progress the development of QTL maps [57]. A detailed discussion on the association mapping for the sustainability of crop production under salinity and drought stresses is available in Table 3.

Table 3. Association mapping for improving crop production under salinity and drought stresses.

Stresses	Crops	Target Gene	Major Findings	References
Salinity	Cowpea (*Vigna unguiculata* (L.)		Association mapping for salt tolerance at germination and seedling stages and the identification of SNP markers associated with salt tolerance in cowpea	[58]
Drought	Wheat	*RM223*	Demonstrated a strong power of joint association analysis and linkage mapping for the identification of important drought response genes in wheat	[59]
Salinity	Cotton (*Gossypium hirsutum* L.)		Provided reference data for the use of MAS for salt tolerance in cotton	[55]
Salinity	Cotton (*Gossypium arboretum*)	(*Cotton_A_37775* and *Cotton_A_35901*)	Provided fundamental information to produce novel salt-tolerant cultivars	[54]
Drought	*Pearl Millet*	*PMiGAP*	Development of high-yielding drought- and submergence-tolerant rice varieties using marker-assisted introgression	[60]

2.5. Genome-Wide Association

GWAS (genome-wide association study) is a potent presumption-free method used to identify and dissect the genetic regions associated with a certain trait. Typically, GWAS is performed by scoring the phenotypes and sequencing many individuals to link genotype to phenotype, thereby linking genetic variants to a given trait (Figure 2) [61].

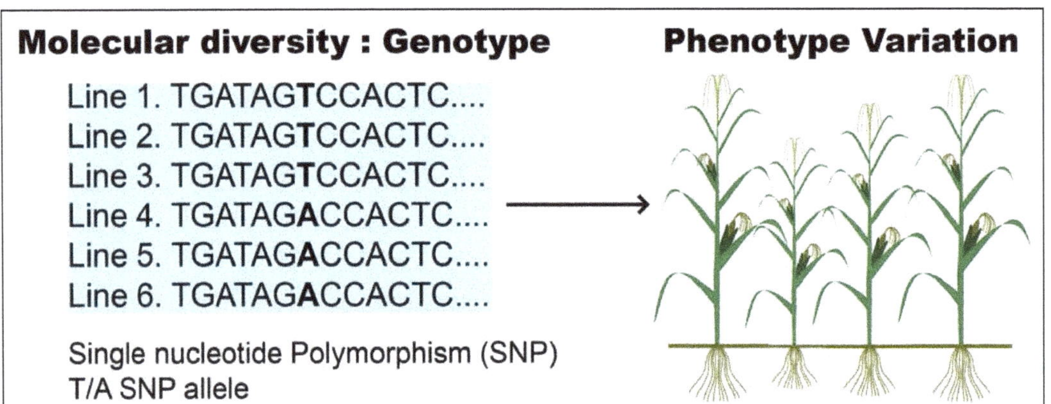

Figure 2. Connection of genotype to phenotype variation.

GWAS applies large markers and several populations of non-cross-executed lines to provide larger mapping exploration than traditional QTL mapping based on a cross-evolved segregating population, leading to the detection of unknown or unexpected genes. It has been applied to separate complicated genetic parameters in leading crops such as rice and wheat under salt and drought stress. Additionally, GWAS has been effectively conducted to designate QTLs for particular characteristics in wheat (e.g., grain yield, morphology relevant to leaf rust disease, and end-usage quality), thereby applying various systems of molecular markers to bolster breeding resources [62,63]. GWAS has detected more than 2000 loci for simple human diseases to date [64]. Therefore, compared with QTL mapping, GWAS delivers an in-depth, cost-efficient mode of gene investigation and detection of molecular markers.

GWAS focused on the flowering period of saline-treated rice identified 11 loci bearing 22 important SNPs linked to stress responses. The potential genetic determinant of germination was identified on chromosome one, close to the saline conditional QTL regulating Na^+ and K^+ levels. Approximately 1200 candidate genes linking development to sodium and potassium ion allowances were detected [65]. Thus, GWAS offered an informed list of candidates for saline tolerance-connected gene cloning and uncovered responsive genetic elements relevant to salt stress [66]. GWAS is also important to perceive the genetic architecture of complex characteristics to improve drought tolerance [67]. Recently, "No-Genome-Required-GWAS" approaches have provided easy and efficient identification of genetic variants underlying phenotypic variation in plants [68]. Details on genome-wide association mapping for identifying QTLs under salinity and drought stresses are discussed in Table 4.

Table 4. Genome-wide association mapping for identifying QTLs under salinity and drought stresses.

Stresses	Crop Variety	Major Effect/Finding	References
Heat prone	*Spring wheat*	Yield stability	[69]
Drought	Rice (indica and japonica)	Identified QTL containing promising candidate genes related to drought tolerance by osmotic stress adjustment	[70]
Salt stress	*Arabidopsis thaliana*	Provided a comprehensive view of AS under salt stress and revealed novel insights into the potential roles of AS in plant response to salt stress	[71]
Salinity	*Rice*	Candidate genes can be identified by QTL	[65]
Drought	Barley (*Hordeum spontaneum*)	Exploring the genomic basis of reproductive success under stress in wild progenitors with expected ecological and economic applications	[72]
Drought	Willow (paper-mulberry)	A core set of candidate genes encoding proteins with a putative function in drought response was identified	[73]
Salinity	Wild barley	Across many traits, QTLs that increased phenotypic values were identified	[74].
Salinity	Rice	Unveiled genomic regions/candidate genes regulating salinity stress tolerance in rice	[75]
Drought	Alfalfa (*Medicago sativa* L.)	Improved alfalfa cultivars with enhanced resistance to drought and salt stresses	[76]
Drought	Rice	Drought-induced alterations to DNA methylation that may influence epigenetics	[77]
Drought	Wheat	Thirty-seven of the significant marker-traits were detected under the drought-stressed condition	[67]
Drought	Wheat	Identified a QL on chromosome 4H	[78]

2.6. Next-Generation Sequencing

Sequencing technologies include several techniques that generally consist of template preparation, sequencing and imaging, and data analysis [79]. Next-generation sequencing (NGS) integrates technologies that inexpensively and efficiently produce millions of short DNA sequence reads mainly in the range of 25 to 700 bp in length [80]. These technologies have made it possible for scientists to investigate crops at the genomic and transcriptomic levels to assist diversity analysis and marker-assisted breeding [80]. The relevance of NGS appears to be endless, permitting quick presses forward in numerous fields associated with the biological sciences. NGS has also afforded a wealth of knowledge for biology studies via end-to-end whole-genome sequencing of a broad diversity of organisms [81]. Whole-genome sequence studies have focused particularly on detailed information on genomics criteria, including regulatory sequences, coding and noncoding genes, GC content, and repetitive elements, which would be utilized in functional characterization, such as microarray or tiling arrays. Additionally, NGS can be used to address many remaining biological questions by means of resequencing targeted areas of concern or whole genomes (as is being performed for the human genome [82]), *de novo* assemblies of bacterial and lower eukaryotic genomes, cataloging the transcriptomes of cells, tissues, and organisms (RNA sequencing), genome-wide profiling of epigenetic markers and chromatin structure using additional seq-based methods (ChIP-seq, methyl-seq, and DNase-seq), and species classification and/or gene discovery by metagenomics studies [50].

3. Functional Genomics Approaches

After identifying a QTL/allele/gene, the next sensible step is to characterize the gene before incorporation into a cultivar by studying several physiological, molecular, and biochemical pathways of genes. Thus, functional genomics approaches were extensively implemented to determine the gene functions and the connections between genes in a regulatory network that would be utilized to produce improved crop varieties. Consequently, there have been multiple tools developed for the characterization of gene function; some of the most exploited are described briefly.

3.1. Epigenetic Regulators

In wider definitions, the term 'epigenetics' frequently refers to a type of overall nongenetic (unrelated to DNA sequence *per se*) heredity at various levels. That is, epigenetics illustrates a number of dissimilar methods of genetic regulation whose temporal and heritable constituents have not in all cases been decided [83]. For example, methylation of DNA generally interferes with gene expression by way of gene silencing [84]. The reduction of methylation in resistance-associated genes activates chromatin and the expression of genes, which offers long-term or enduring resistance under stress conditions [85]. Epigenetics sustains the identity of stress memory in plants, which helps pre-exposed plants fight comparable stress throughout subsequent exposures. Histone modifications, DNA methylation and demethylation, and ATP-dependent chromatin remodeling are some of the epigenetic changes performed by plants during drought stress [86]. Epigenetic responses to drought stress have been studied in numerous plants, particularly the stress memory and gene activation marker *H3K4me3*, which has been used to carry out genome-wide ChIPseq analyses in Arabidopsis [87]. Furthermore, the HAT genes in rice (*OsHAC703*, *OsHAG703*, *OsHAF701*, and *OsHAM70*) [88] and the *HvMYST* and *HvELP3* genes in barley were also shown to be involved in epigenetic regulation in drought responses [89]. DNA methylation and histone modifications may have a similar result on stress-inducible genes, as salinity stress influences the expression of a range of transcripts in soybean [48]. Work in rice underlined that hypomethylation in reaction to salt stress may be associated with changes in the expression of DNA demethylases [90]. The transcriptional adaptor ADA2b (a modulator of histone acetyltransferase activity) is responsible for hypersensitivity to salt stress in *Arabidopsis thaliana* [91].

3.2. Gain-of-Function Lines

Gain-of-function methods have been extensively used for the study of gene function in plants and are considered among the most useful tools for gain-of-function phenotypes. Gain-of-function lines are generated through the arbitrary activation of endogenous genes by transcriptional enhancers and the regular expression of individual transgenes by transformation [9]. This method employs the phenotype of gain-of-function lines that overexpress a selected gene family and can be executed without meddling from other gene family members that allow the categorization of functionally unwanted genes [10]. Alternatively, the overexpression of a mutant gene can be expressed due to the presence of higher levels of nonfunctional protein causing a superseding negative interface with the wild-type protein. To overcome this event, a mutant type could be used to compare the wild-type protein allies, resulting in a mutant phenotype. Conversely, heterologous expression of a gene in the yeast-hybrid system is an alternative way to characterize genes. In the first gain-of-function approach, a strong promoter or enhancer element is arbitrarily inserted into the plant genome with the help of T-DNA [11], which stimulates a gene near the site of the harbor. Other gain-of-function approaches involve cDNA overexpression and open reading frame (ORF) overexpression, whereas full-length cDNAs or ORFs have been cloned into a strong promoter downstream. Under the switch of the CaMV35S promoter, various abiotic stress response genes have been characterized by the use of ectopic overexpression of cDNAs [11–14].

3.3. Gene Silencing and RNA Interference Techniques for Salinity and Drought Stress

Suppression of a gene is referred to as gene silencing in plants and fungi and interference RNA (RNAi) in animals and is generally thought of as a controlling mechanism of gene expression mostly in eukaryotic cells [92]. RNA interference (RNAi) has been considered one of the most crucial discoveries in molecular genetics during the last several years for posttranscriptional gene silencing (PTGS) cosuppression [93]. RNA silencing hints at a nucleotide sequence-specific procedure that prompts mRNA degradation or translation termination at the posttranscriptional level in plants arbitrated by small RNAs (sRNAs), which are divided into two classes: microRNAs (miRNAs) and small interfering RNAs (siRNAs). However, RNAi was properly adapted into antisense-stranded RNA as an operative technique to constrain gene expression [94]. Silencing a gene through transgenic expression of sRNAs has been extensively implemented for abiotic stress-related gene function functional efforts. Currently, the virus-induced gene silencing (VIGS) technique for posttranscriptional gene silencing is extensively used for rapid and efficient gene function studies related to salt and drought stresses [95–98]. It can also be used for both forward and reverse genetic studies. Target gene silencing techniques for improving crops under salinity and drought stresses are discussed in Table 5.

Table 5. Target gene-silencing techniques for improving crop variety under salinity and drought stresses.

Stresses	Crops	Silencing Gene	Major Findings	References
Cold, drought, salt stress	Rice	*OsNAC5*	RNAi lines became less tolerant of these stresses than control plants	[58]
Salinity	Arabidopsis	*sos1*	*thsos1*-RNAi lines of *Thellungiella* were highly salt-sensitive	[99]
Salinity	pepper	*CaATG8c*	The silencing of *CaATG8c* made pepper seedlings more sensitive to salt stress	[100]
Salinity	*Alternanthera philoxeroides*	*ApSI1*	Significantly decreased tolerance to salinity	[101]
Drought	*Alternanthera philoxeroides*	*ApDRI15*	Plants were more sensitive to drought stress than the control plants	[101]
Drought	Tomato	*SpMPK1, SpMPK2, and SpMPK3*	Reduced drought tolerance in tomato plants	[102]
Drought	wheat	*Era1* and *Sal1*	Played imperative roles in conferring drought tolerance	[103]
Drought, salt stress	Cotton	*GH3.17*	Enhanced drought and salt stress	[104]
Salinity	Cotton	*GhWRKY6*	Downregulation of *GhWRKY6* increased salt tolerance	[105]

3.4. Genome Engineering (TALENs, ZFNs, CRISPR/Cas)

Recently, several functional genomics-based strategies have been developed for genetic engineering. To improve crops for sustainable food production, targeted genome engineering has become a substitute for conventional plant breeding and transgenic (GMO) strategies, including transcription regulators, epigenetic modifiers, DNA integrators, TAL effector nucleases (TALENs), zinc-finger nucleases (ZFNs), clustered regularly interspaced short palindromic repeats (CRISPR)/Cas (CRISPR-associated proteins), and base editors and prime editors. Until recently, the existing methods have been considered to be unwieldy. Both TALENs and ZFNs could be used to mutagenize genomes at exact loci. However, the problem is that these systems need two altered DNA-binding proteins flanking a sequence of interest, each with a C-end FokI nuclease unit [106]. For plant research, these techniques have not been extensively implemented. Recently, a technique based on the bacterial clustered regularly interspaced short palindromic repeats (CRISPR)/Cas (CRISPR-associated proteins) type-2 prokaryotic adaptive invulnerable system has been developed as an alternate process for genome engineering [106]. The CRISPR/Cas (clustered regularly interspaced short palindromic repeats/CRISPR-associated proteins) system was first identified in bacteria and archaea and can cleave exogenous DNA substrates [107]. CRISPR/Cas has since been modified to be used as a gene-editing technology. However, CRISPR/Cas9 has largely overtaken the other aforementioned gene editing practices. Investigators express similar stories: a few years ago, they started working on projects using both TALENS and CRISPR/Cas9 side-by-side but rapidly established CRISPR systems [108]. Graphical presentations of the CRISPR/Cas9 techniques are available in Figure 3.

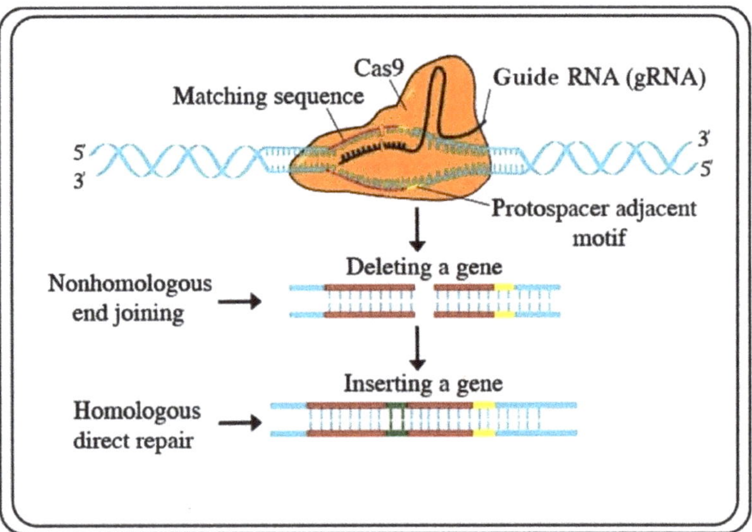

Figure 3. CRISPR/Cas9 is a powerful tool for genome editing of Cas9 to a guide RNA that directs the complex to a place on the DNA double helix and contains the code for the addition of a new DNA sequence at the double-stranded break. Source: adapted and modified from [www.stockadobe.com; accessed date on 12 July 2021].

The beginning of CRISPR has made it conceivable to rewrite host DNA by introducing some major amendments. These modifications include gene replacement, deletions, inversion, knockouts, and translocations [109]. Using CRISPR/Cas9 tools, several genes, such as OsERF922, OsPDS, OsERF922, ERFs, OsHAK1, Badh2, OsRR22, and TMS5, were knocked out, and a predictable phenotype was attained [110–116]. More promising are the potential forecasts of this technique for producing plants with specifically tailored purposes, i.e., biofuel production, synthetic biology, disease resistance, phytoremediation, etc. [117]. This technique also offers a new method for abiotic stress breeding programs [113]. Several examples of CRISPR/Cas9 technology-mediated improvements to plant tolerance to abiotic stress are discussed in Table 6.

Table 6. CRISPR/Cas9 technology-mediated improvements to plant tolerance to abiotic stress.

Target Genes	Crops	Target Stresses	References
TaDREB2 and *TaERF3*	Wheat	Abiotic stress response	[118]
ScNsLTP	Sugarcane	Drought and chilling resistance	[119]
MaAPS1 and *MaAPL3*	Banana	Cold and salt	[120]
MeKUP	Cassava	Salt, osmosis, cold, and drought resistance	[121]
MeMAPKK	Cassava	Drought resistance	[122]
GhPIN1–3 and *GhPIN2*	Cotton	Drought resistance	[123]
GhRDL1	Cotton	Drought resistance	[124]
CpDreb2	Papaya	Drought, heat, and cold resistance	[125]
OsDST	Indica mega rice cultivar	Salt and Drought	[126]
SlNPR1	Tomato	Drought	[127]
Leaf1,2	Rice	Drought	[128]

3.5. CRISPR-Mediated Base Editing and Prime Genome Editor

It is well known that CRISPR is a powerful genome-editing technique. CRISPR can change genes and edit DNA sequences by producing double-strand breaks in double-helical DNA, leaving the cell to repair the breakage (Figure 4).

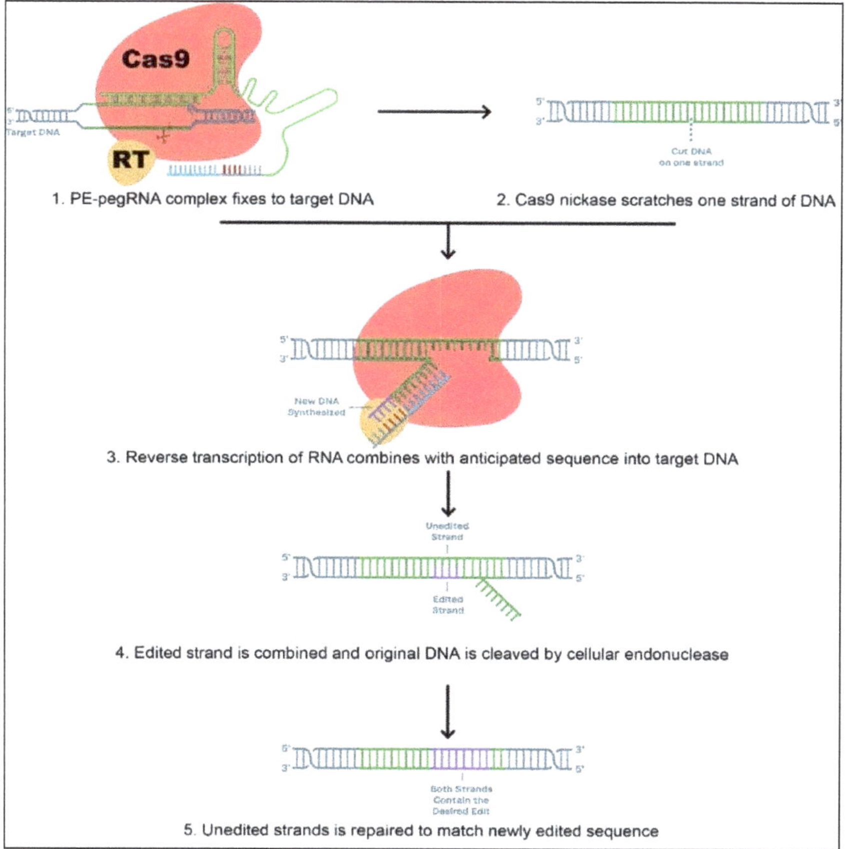

Figure 4. A new tool (prime editor) of DNA manipulation that couples two enzymes, Cas9 (brown) and reverse transcriptase (yellow), to a guide RNA (red) that directs the complex to an exact place on the DNA double helix and contains the code for the addition of a new DNA sequence at the double-stranded break. [Figure modified from: https://www.synthego.com/guide/crispr-methods/prime-editing; accessed date on 12 July 2021].

The control mechanisms over the repair process are the main limitations in basic research and plant sciences. However, several groups recently reported the "base editing" system, a new approach for site-directed mutagenesis of genomic DNA. Base editing tools are highly efficient, reduce the rate of off-target effects, and do not require DNA double-strand cleavage or donor template repair. These methods make use of a Cas9 nickase fused to various deaminases. Specific C-to-T or A-to-G transitions in genomic DNA are catalyzed by these fusion proteins. The base editor and Target-AID (target-activation-induced cytidine deaminase) systems are two representative architectures of cytidine base [127,128]. Therefore, engineering of single-plasmid CRISPR-mediated base editing tools for *S. meliloti* that included adenosine base editors (ABEs), cytidine base editors (CBEs), and glycosylase base editors (GBEs) is capable of achieving both base transitions (A-to-G, C-to-T) and transversions (C-to-G) [129]. Base editing has become a widely applicable tool for gene disruption in a variety of bacteria [17,22,28,130]. Nevertheless, the new invention "prime editor" makes the successful addition or deletion of exact sequences within the genome possible with minimum off-target effects [129].

The creators claim that their tools can precisely target approximately 89% of recognized pathogenic human genetic variants. Prime editing may have fewer bystander mutations than base editing, especially when multiple Cs or As are present in the editing activity window [131]. It is also less constrained by the availability of protospacer adjacent motif (PAM) than other methods such as homology directed repair (HDR), non-homologous end joining (NHEJ), or base editing, because the PAM-to-edit distance can be greater than 30 bp on average [26]. Nevertheless, there is a large suite of base editors that have been developed with improved efficiency, product purity, and DNA specificity, as well as broad applicability [25]. Although prime editing has the potential to replace base editors, the technology is still in its early stages and is typically less efficient than current generation base-editing systems with superior on and off-target DNA editing profiles [20]. Consequently, a suitable editing strategy for specific applications must be chosen based on various criteria for gene-editing, such as the desired edit, the availability of PAMs, the efficiency of editing, and off-target/bystander mutations.

4. The Development of Salt- and Drought-Tolerant Crops with High Yielding Capacity

The generation of crop varieties with a high level of tolerance to salinity and drought is vital for creating full yield potential and sustainable production. Generally, there are two methods to integrate enhanced traits such as drought and salinity stresses in plants: genetic engineering and breeding programs.

4.1. Genetics Engineering

The advent of modern genetic engineering strategies offers the generation of plants with rising abiotic stress tolerance. Under abiotic stress conditions, several genes of crop plants in different pathways lead to upregulation of expression. Stress-responsive genes and their controlling genes can be transferred and expressed in different species using an Agrobacterium-mediated transformation system involving molecular, biochemical, and physiological changes that direct an increase in plant growth, development, and yield under stress environments [16]. Currently, the use of stress-inducible promoters for the expression of stress response genes has confirmed a time-specific and optimal level of expression. Salinity and drought are major environmental stresses that adversely affect the growth and development of crops; thus, a number of genes encoding proteins involved in the biosynthesis of stress defensive elements, including glycine betaine, mannitol, and heat shock proteins, have been used for abiotic stress tolerance, as well as several transcription factors, such as MAPK, bZIP, AP2/EREBP, WRKY, and DREB1 [17,18]. However, over-expressed transgenes can function as positive regulators of tolerance to a single stress or multiple stresses, such as salinity, drought or both. Therefore, the newly developed transgenic plant might have to be tolerant to single or multiple stresses, have high yields, and be devoid of harmful pleiotropic traits. Posttranslational modifications, orthologous gene expression of effectors from wild relatives or halophytes, gene expression by regulating miRNA activity, osmoprotectants, gene pyramiding, engineering of transcription factors, chaperones, late embryogenesis, metabolic pathways, abundant proteins, epigenetics, and even chaperones have been implemented to produce a new generation of transgenic plants [130]. Successful salinity- and drought-tolerant transgenic crops were produced and approved for cultivation as food and feed [23,27–29].

4.2. Gene Introgression

Introgressiomics is designated as an extensive systematic improvement of plant genomes and populations through bearing introgressions of genomic fragments from wild crop relatives relative to the genetic background of established crops to develop new cultivars with promising traits [24]. Through introgression, greater genomic plasticity can be attained in a crop using exotic genetic material that was previously nonexistent within the genome [104]. For crop improvement, genetic engineering strategies are relatively faster than traditional breeding programs, as well as cloning of genes responsible for imperative traits and introgression into plants [104]. To develop salt- and drought-tolerant varieties, a particular breeding program can be established through an understanding of the physiological and genetic mechanisms of these stresses. MAS improves the speed and efficacy of breeding because genetic markers are unaffected by the environment, are efficient to use in early generations [105], and can be useful for the introgression of target genes. Successful stories of introgression in various crops for many traits, including both abiotic and biotic stress tolerance/resistance, have been implicated from wild relatives in cultivation without affecting yield and quality [24,106,107].

4.3. Marker-Assisted Breeding and Transference of Genes

Marker-assisted breeding is a process that permits breeders to track traits over generations of breeding using genetic markers associated with a given trait. In marker-assisted breeding, DNA markers associated with desirable traits are used to identify and choose plants containing the genetic locus that confers the desirable trait. DNA markers have a high probability of increasing the capability and accuracy of traditional plant breeding via marker-assisted selection (MAS). MAS allows for quicker and more efficient selection of desired crops, as cultivators can reliably test for the presence of a genetic marker associated with a trait rather than waiting to assess the trait itself. The most efficient and extensively applied method for MAS is marker-assisted backcrossing [132]. Marker-assisted breeding is in contrast to the direct addition of a gene or multiple genes to enhance a trait, such as genetic modification. Using genetic markers in breeding depends on the phenological acclimatization of the acceptor genotype, and the introduction of a new marker or allele may be necessary to increase the yield. With the advent of molecular markers and MAS technology, numerous studies have capitalized on such technology to identify genes or QTLs affecting sequence tagging in different plant species during different developmental stages, to identify genes or QTLs that were introduced into different plant varieties, and to gain an overall deeper and more efficient understanding of QTLs that contribute to complex traits [133]. Marker-assisted breeding for improving crop quality under salinity and drought stresses is discussed in Table 7.

Table 7. Marker-assisted breeding results for improving crop quality under salinity and drought stresses.

Stresses	Crops	Target Genes	Major Effect/Finding	References
Drought and Salt	Cotton		Significant associations between polymorphic markers and drought and salt tolerant traits were observed using the general linear model (GLM)	[48]
Salinity	Rice	*RM223*	Transferring genes from one variety to another and their use in MAS	[134]
Drought	Rice		Developed high-yielding rice cultivars suitable for water-limited environments through marker-assisted breeding	[135]
Salinity	Rice	*NAL1*	High yield through optimizing transportation efficiency of photosynthetic products by marker-assisted selection	[136]
Drought and flood	Rice		Developed high-yielding drought- and submergence-tolerant rice varieties using marker-assisted introgression	[25]
Drought	Rice		Provided a higher yield advantage	[137]
Drought	maize		Improved grain yield under drought stress conditions	[138]
Drought and salt	Wheat	TaCRT-D	Increased plant stress tolerance and the functional markers of TaCRT-D for marker-assisted selection in wheat breeding	[139]
Salinity	Rice		Developed new salt-tolerant rice germplasm using speed-breeding	[140]
Drought	Rice		Stimulated 10–36% higher yield among different inbred lines	[141]

Current advances in genomics and genome sequencing in rice have made it feasible to locate and precisely map a certain number of genes via linkage to DNA markers. MAS can be applied to control the presence or absence of genes and has also been applied to assess the contributions of such genes conferring traits that have been introduced into extensively developed varieties [26]. Coupling genomic resources with the utility of MAS, breeders can now gain unprecedented insight into the genetic regulation of complex traits. MAS is a large advantage for developing new crop varieties because crops with ineligible gene aggregations can be dispelled from the selection process. This offers breeders the opportunity to focus on a reduced number of candidate lines for breeding targets in successive generations [131]. It has been shown that association mapping along with population formation and screening of cotton germplasm can improve QTL assignment and MAS [131]. Combining MAS and GS (genomic selection) with adequate genetic variety, databases, analytical instruments, and well-established climate and soil data is a powerful way to produce modern varieties with high drought resistance that can be readily inaugurated into appropriate agricultural programs [27]. These methods could produce a high number of lines of a crop appropriate for propagating crops in a range of drought and salinity stress ecosystems. Furthermore, incorporating these data can lead to the creation of varieties that can be further optimized to control largely heritable principal secondary characteristics. MAS delivers precise, rapid, and profitable progress toward the development of crop varieties that can be applied to abiotic stress tolerance [26]. A graphical presentation of the development of a new crop variety by marker-assisted selection is available in Figure 5.

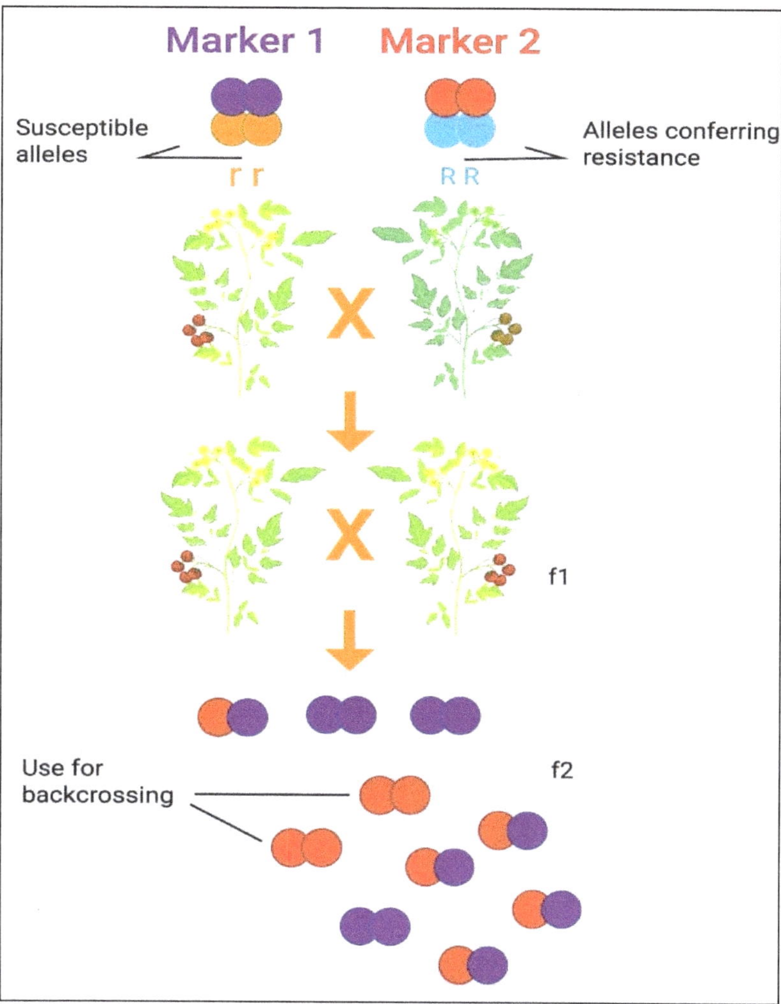

Figure 5. Development of a new crop variety by marker-assisted selection. Source: modified from: [http://b4fa.org/bioscience-in-brief/plantbreeding/how-do-you-develop-a-new-crop-variety-by-marker-assisted-selection-mas/; accessed date on 12 July 2021]. Note: Marker 1 and Marker 2 confer susceptible and resistance alleles, respectively; f1 and f2 indicate the first and second filial generations of offspring, respectively.

5. Involvement of Genes in the Regulation of ROS in Abiotic Stress Tolerance

Reactive oxygen species (ROS) are assumed to play roles in many noteworthy signaling reactions in plant metabolism. Under drought and salinity environments, interrupting photosynthesis and increasing photorespiration intermittently alter the regular homeostasis of cells and influence the production of ROS in mitochondria, chloroplasts, and peroxisomes (Figure 6) [142,143].

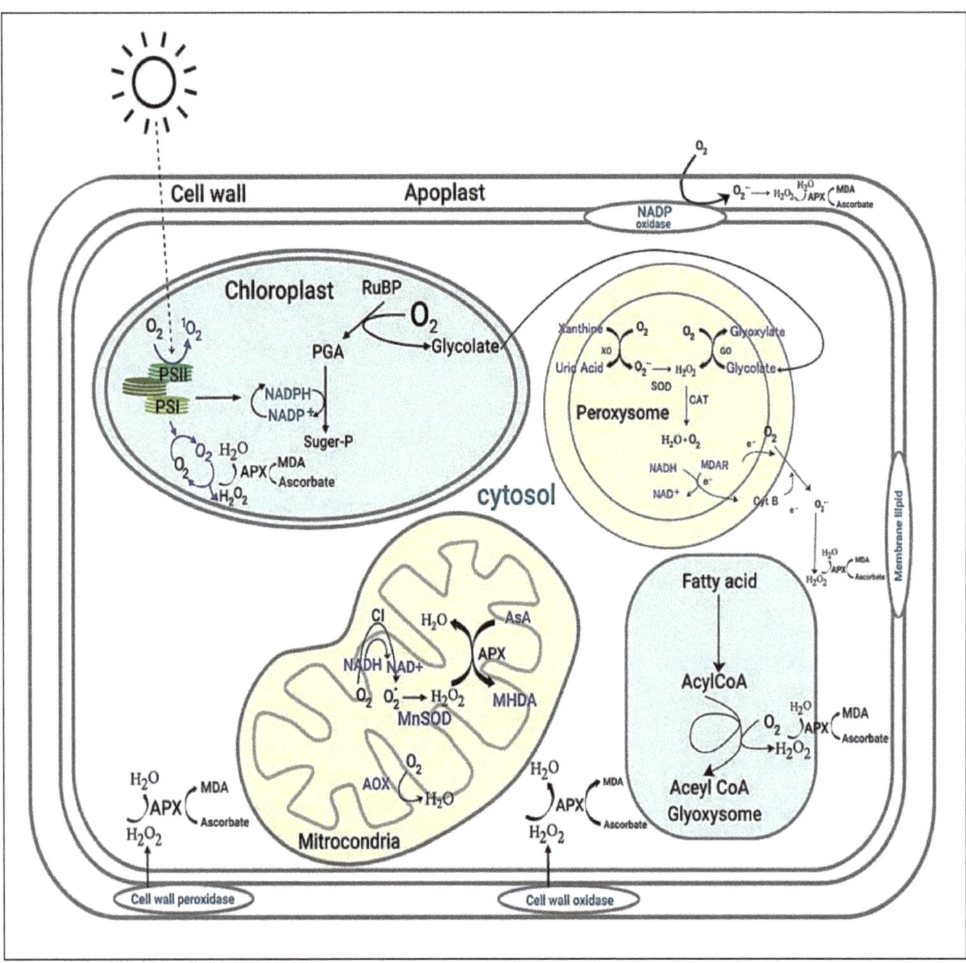

Figure 6. Sites and typical regulation of ROS in plant cells. PSII; Photosystem II, PSI; Photosystem I, MDA; Malondialdehyde, APX; Ascorbate Peroxidase, PGA; 3-Phosphoglyceric acid, H_2O_2; Hydrogen peroxide, SOD; Superoxide dismutases, CAT; Catalase, MDAR; Monodehydroascorbate reductase, NADH; Nicotinamide adenine dinucleotide, Alternative oxidase.

In addition to organelles, the plasma membrane together with the apoplast is the main site for ROS production in response to endogenous signals and exogenous environmental stimuli [144]. Overproduction of ROS in plant cells is extremely reactive and noxious to proteins, lipids, and nucleic acids, which finally results in cellular damage and death initiated by stressful environments [142]. ROS-scavenging enzymatic antioxidants (SOD, APX, CAT, GPX, MDHAR, DHAR, GR, GST, and PRX) and nonenzymatic antioxidants (GSH, AsA, carotenoids, tocopherols, and flavonoids) are located in different sites of plant cells, and they directly or indirectly play a key role in ROS homeostasis via different unique pathways to avoid oxidative damage. In addition, soluble sugars as well as disaccharides, raffinose family oligosaccharides, and fructans play a dual role in ROS maintenance [145]. Consequently, crop plants have executed several interrelated signaling pathways to operate different groups of genes (Figure 7), which are induced under stress conditions to generate different classes of proteins, for example, protein kinases, enzymes, transcription factors, molecular chaperones, and other efficient proteins, subsequent to various physiological and metabolic reactions to improve tolerance to multiple environmental stresses.

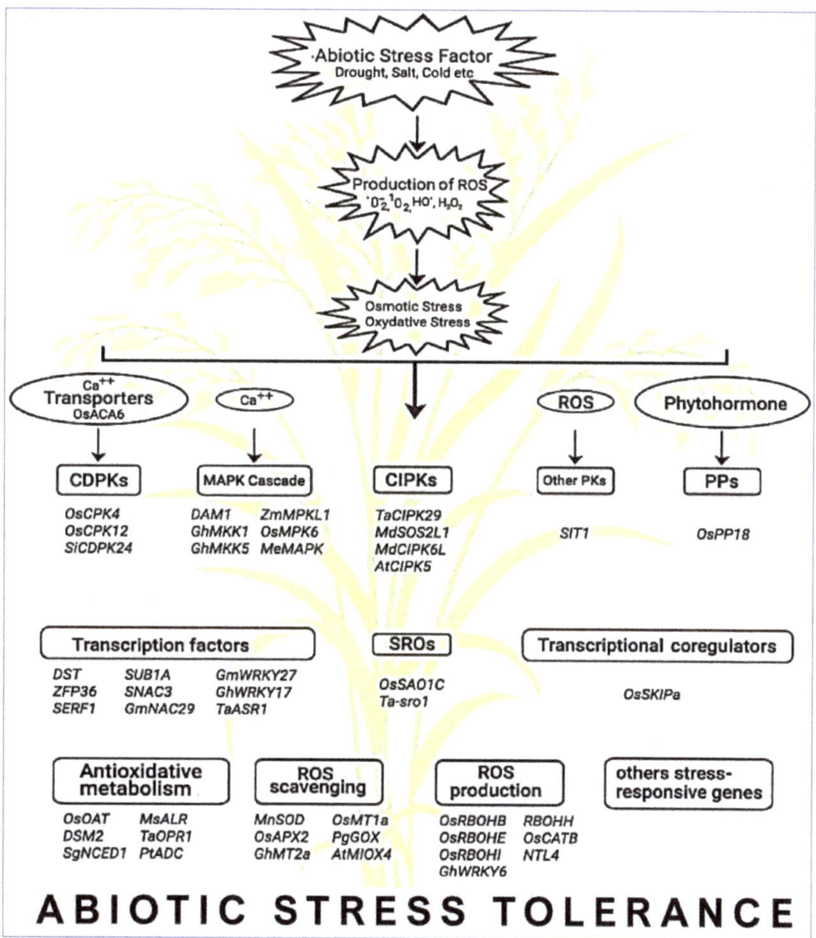

Figure 7. A general view of major genes that are intricate in abiotic stress resistance through ROS maintenance in crops. MAPK, mitogen-activated protein kinase; CDPK, calcium-dependent protein kinase; CIPK, calcineurin B-like protein-interacting protein kinase; PK, protein kinase; PP, protein phosphatase; SRO, similar to RCD.

It is well known that antioxidants stimulate gene expression linked with responses to various environmental signals to exploit protection through the regulation of cellular ROS levels and redox state [146]. The characteristics and roles of selected genes and their processes under salinity and drought stresses are discussed in detail in Tables 8 and 9.

Table 8. Characterized genes involved in abiotic stress tolerance through ROS regulation in crops.

Genes	Origin	Transformation Receptor	Protein Function	Major Functions	Signaling Hormone	Approaches Used	References
GhMKK1	*G. hirsutum*	*N. benthamiana*	MAPKK	Influences oxidative, ROS scavenging, salt and drought tolerance	Abscisic acid (ABA)	Reverse genetics	[147]
DSM1	*O. sativa*	*O. sativa*	MAPKKK	Influences oxidative, ROS scavenging, drought tolerance	ABA	RNA interference, Reverse genetics	[148]
DSM2	*O. sativa*	*O. sativa*	MAPKKK	Influences oxidative, ROS scavenging, drought tolerance	ABA	RNA interference, Reverse genetics	[149]
MEKK1	*Arabidopsis*	*Arabidopsis*	MAPKKK	Influences oxidative, ROS scavenging, abiotic stress tolerance	ABA	Reverse genetics	[150]
GhMAPKKK49	*G. hirsutum*	*G. hirsutum*	MAPKKK	ROS scavenging, salt, drought, and wounding stresses	ABA, gibberellins (GB), methyl jasmonate (JA), salicylic acid (SA), 6-benzyl amino purine, a-naphthyl acetic acid, and ethylene (ET)	Transcriptome	[151]
MKK1, MKK2, MKK6	*Arabidopsis*		MAPKK	Stimulate oxidative, ROS scavenging, abiotic stresses	SA	RNA interference	[150,152]
OsCPK4	*O. sativa*	*O. sativa*	Calcium-dependent protein kinase	ROS scavenging, drought, and salt stress	SA	Reverse genetics	[153]
OsCPK12	*O. sativa*	*O. sativa*	Calcium-dependent protein kinase	ROS scavenging, influences oxidative salt stress	ABA	RNA interference, Reverse genetics	[154]
SiCDPK24	*Setaria italica*	*Arabidopsis*	Calcium-dependent protein kinase	ROS scavenging, drought stress	ABA	Reverse genetics	[155]
TaCIPK29	*T. aestivum*	*N. benthamiana*	CBL-interacting protein	ROS scavenging, salt stress	ABA and ET	Reverse genetics	[156]
MdCIPK6L	Apple	*Arabidopsis*	CBL-interacting protein kinase	ROS scavenging, salt, osmotic/drought and chilling stresses	ABA	Reverse genetics	[157]

Table 8. *Cont.*

Genes	Origin	Transformation Receptor	Protein Function	Major Functions	Signaling Hormone	Approaches Used	References
MdSOS2L1	Apple	tomato	CBL-interacting protein kinase	ROS scavenging, salt stresses	ABA	Reverse genetics	[158]
AtCIPK5	Arachis diogoi	Arabidopsis	CBL-interacting protein kinase	Salt and osmotic stress tolerance	NA	Reverse genetics	[159]
SIT1	O. sativa	O. sativa	Lectin receptor-like kinase	ROS production, salt sensitivity	ET	Reverse genetics	[160]
OsMPK6	O. sativa	O. sativa	MAPK	ROS scavenging, salt stresses	SA	RNA interference	[161]
ZmMPKL1	Zea mays	Zea mays	MAPK	ROS production, drought sensitivity	ABA	CRISPR/Cas9, Reverse genetics	[162]
MeMAPK	Cassava	NA	MAPK	osmotic, salt, cold, oxidative stressors	ABA	Transcriptome	[163]
ZmMKK3	Zea mays	N. benthamiana	MAPK	ROS scavenging, osmotic tolerance	ABA	Reverse genetics	[164]
OsPP18	O. sativa	O. sativa	Protein phosphatase 2C	ROS scavenging, drought and oxidative stress	ABA	RNA interference, Reverse genetics	[165]
DST	O. sativa	O. sativa	zinc finger C2H2	ROS scavenging, drought and salt stress	Cytokinins	QTL, RNA interference, Reverse genetics	[166,167]
ZFP36	O. sativa	O. sativa	zinc finger C2H2	ROS scavenging, stress and oxidative stress	ABA	RNA interference, Reverse genetics	[168]
OsTZF1	O. sativa	O. sativa	Zinc Finger Protein CCCH	ROS scavenging, drought, high-salt stress	ABA	RNA interference	[169]
OsWRKY30	O. sativa	O. sativa	WRKY	ROS scavenging, drought tolerance	SA	Reverse genetics	[170]
GhWRKY6	G. hirsutum	Arabidopsis	WRKY	ROS production, drought and salt stress	ABA	Transcriptome, VIGS, Reverse genetics	[105]
EcNAC1	Helianthus annus	Helianthus annus	NAC	ROS scavenging, salt stress	ABA	Reverse genetics	[171]
NTL4	Arabidopsis	Arabidopsis	NAC	ROS production, drought stress	ABA	RNA interference, Reverse genetics	[172]
EcbHLH57	Eleusine coracana	N. benthamiana	bHLH	ROS scavenging, salt, oxidative and drought stress	ABA	Reverse genetics	[173]

Table 8. *Cont.*

Genes	Origin	Transformation Receptor	Protein Function	Major Functions	Signaling Hormone	Approaches Used	References
JERF3	*O. sativa*	*O. sativa*	Ethylene response factor (ERF)	Drought and osmotic stress	ET	Reverse genetics	[174]
MnSOD	*N. plumbaginifolia*	*M. sativa*	MnSOD	ROS scavenging drought stress	NA	Reverse genetics	[175]
OsAPX2	*Medicago sativa*	*Medicago sativa*	APX	ROS scavenging, salt tolerance	ABA	Reverse genetics	[176]
PgGPX	*Pennisetum glaucum*	*O. sativa*	GPX	ROS scavenging, salinity and drought stress	SA	Reverse genetics	[177]
MsALR	*M. sativa*	*N. benthamiana*	NADPH-dependent aldose/aldehyde reductase	Antioxidative metabolism, drought and oxidative stress	NA	Reverse genetics	[178]
AtMIOX4	*Arabidopsis*	*Arabidopsis*	MIOX	ROS scavenging, salt tolerance	ABA	Reverse genetics	[179]
MtPP2C	*Medicago truncatula*	NA	PP2C	ROS scavenging, drought and cold stress responses	ABA	Transcriptome	[180]
OsAHL1	*O. sativa*	*O. sativa*	AHL	ROS scavenging, drought resistance	ABA, SA	GWAS, Reverse genetics	[181]
OsHK3	*O. sativa*	NA	HK	ROS scavenging, salinity and drought stress	ABA	RNA interference	[182]
IcSRO1	*Ipomoea cairica*	*Arabidopsis*	SRO	ROS scavenging, salt and drought tolerance	ABA	Transcriptome, Reverse genetics	[183]
OsCATB	*O. sativa*	*O. sativa*	CATB	ROS production, drought stress	ABA	Transcriptome	[184]
RBOHH	*O. sativa*	*O. sativa*	NADPH Oxidase	ROS production, drought stress	ET	CRISPR/Cas9, Reverse genetics	[185]

Table 9. A summary of identified genes and their processes under salinity and drought stresses.

Functional Category	List of Genes	Type of Stress	Biological Function and Signaling Pathway	Tools Used	References
		Protein kinase			
MAPKKK	*MEKK1, MEKK2, MEKK3, MEKK4, MAPKKK18, GhMAP3K40, OsMAPKKK63, GhMAPKKK49 DSM1, DSM2*	Influences oxidative, abiotic, and biotic stress.	Growth and development; ABA	RNA interference, reverse genetics	[150,186,187]
MAPKK	*MKK1, MKK2, MKK6, GhMKK1,*	Influences oxidative, salt and drought	Growth and development; SA	Transcriptome, reverse genetics	
	MKK3, GhMKK3	Influences oxidative, salt, and drought stresses	Growth and development; SA	RNA interference, reverse genetics	[150,188]
	MKK4, MKK5 GhMKK4, GhMKK5,	Influences oxidative, drought	Growth and development; JA	RNA interference, reverse genetics	
	MKK7, MKK8, MKK9, MKK10, RhMKK9, GhMKK9, ZmMKK10	Salt and/or drought	Growth and development; ET	Reverse genetics	
	VvMKK2, VvMKK4	Influences oxidative, salt, and drought	Growth and development; SA	Reverse genetics	[177]
MAPK	*MPK3, MPK6, MPK10 OsMPK6, ZmMPK3, RhMPK6, ZmMPK6-2, OsMPK3, ZmMPK3*	Influences oxidative, abiotic, and biotic stresses	Cell cycle regulation, cell division; JA and ET	RNA interference, reverse genetics	[150,189]
	MPK4, MPK5, MPK11, MPK12, MPK13, OsMPK4ZmMPK4-1, OsMPK5, OsMPK5, ZmMPK5	Influences oxidative, salt, and/or drought	Cell cycle regulation; SA	RNA interference, reverse genetics	[150]
	MPK1, MPK2, MPK7, MPK14, ZmMPK7, OsMPK2AtMPK7, OsMPK7, GhMPK7	Influences oxidative, salt, drought	Circadian-rhythm-regulated; JA, SA	RNA interference, reverse genetics	[150]
	MPK8, MPK9, MPK15/16/17/18/19/20 GhMPK17, ZmMPK17	Influences oxidative, salt, drought	Cell cycle regulation; JA	RNA interference, reverse genetics	[161]
CDPK	*OsCPK4 OsCPK12 SiCDPK24 FaCDPK4, FaCDPK11 StCDPK3, StCDPK23*	Influences oxidative, salt, drought	Responses to developmental and environmental cues; SA, ABA	Transcriptome, RNA interference, reverse genetics	[190]
CIPK	*TaCIPK29 MdCIPK6L MdSOS2L1 AtCIPK5*	ROS scavenging, salt and osmotic stress tolerance	tissue and organ development; ABA	Reverse genetics	[167–170]

Table 9. *Cont.*

Functional Category	List of Genes	Type of Stress	Biological Function and Signaling Pathway	Tools Used	References
Transcription factor					
bZIP	*ABF3, BF4* *ABF3, ABF4* *FtbZIP5, PtrABF* *OsbZIP23,* *OsbZIP12,* *OsbZIP71, OsbZIP46* *OsbZIP72, ZmbZIP4* *OsbZIP62, TabZIP*	Salt, drought	Light signaling, seed maturation, flower development; ABA	Transcriptome, RNA interference reverse genetics	[191–200]
bHLH	*MYC2, AtbHLH17, AtbHLH68, AtbHLH122,* *FtbHLH2, FtbHLH3, PebHLH35, OsbHLH148*	Salt, drought	Growth, development, response to various stresses; JA, ABA	Transcriptome, RNA interference, reverse genetics	[201–208]
NAC	*ANAC019, ANAC055,* *ANAC072, ANAC042,* *TaNAC29, OsNAC6,* *OsNAC5, OsNAC9,* *OsNAC10, TaRNAC1,* *GmNAC109,* *CaNAC035*	Salt, drought	Plant growth and development range from the formation of shoot apical meristem, floral organ development, reproduction, lateral shoot development; ABA	Transcriptome, RNA interference, reverse genetics	[209–219]
AP2/ERF	*CBF1, CBF2, CBF3,* *AtERF53, AtERF74,* *AhDREB1, OsDREB1, OsEREBP1, OsERF7,* *GmERF3, ZmDREB2A,* *SlERF5*	Salt, drought	Regulation of plant growth and development; ABA	Transcriptome, RNA interference reverse genetics	[220–230]
MYB	*AtMYB44, AtMYB96,* *AtMYB20, OsMYB4,* *OsMYB6, OsMYB48-1,* *OsMYB91, GmMYB76,* *GmMYB92, GmMYB177*	Abiotic stresses	Circadian rhythm, regulation of primary and secondary metabolism; ABA, JA	Transcriptome, RNA interference reverse genetics	[231–239]
WRKY	*OsWRKY11, OsWRKY45, TaWRKY1,* *TaWRKY33,* *cWRKY023, ZmWRKY33, VvWRKY2*	Salt, drought	Growth and development; ABA	Transcriptome, RNA interference reverse genetics	[240–245]
ROS-scavenging					
SOD	*FSD1, FSD2, FSD3* *CSD1, CSD2, CSD3* *MSD1*	Salt, drought	Antioxidant defense against oxidative stress; ABA	RNA interference, Reverse genetics	[246]
	CmSOD	Oxidative stress	ABA	reverse genetics	[247]
	CsSOD	Drought	JA and gibberellin (GA3)	Transcriptome	[248]

Table 9. *Cont.*

Functional Category	List of Genes	Type of Stress	Biological Function and Signaling Pathway	Tools Used	References
APX	*APX1-APX7*	Salt and or drought	Growth regulation; ABA	Transcriptome	[246]
	OsAPX1, OsAPX2	Oxidative, Salt, drought	ABA	Transcriptome	[249]
	OsAPX3, OsAPX4	Salt and drought	ABA	Transcriptome	[249]
	OsAPX5, OsAPX6 and OsAPX7	salinity	ABA	Transcriptome	[250]
	AgAPX1	Drought	NA	Reverse genetics	[251]
	TbAPX	Salt	ABA	Reverse genetics	[252]
	CytAPX	Salt	ABA	Reverse genetics	[253]
CAT	*CAT2, CAT3, ScCAT1*	Salt and/or drought	ABA	Transcriptome	[254]
	HuCAT3	Salt and drought	NA	Transcriptome	[255]
	VsCat	Salt	Salt	CRISPR/C as9	[198]
	CsCAT3	Tolerance to heat, cold, salinity and osmotic condition	ABA	Transcriptome, Reverse genetics	[256]
GPX	*GPX1, GPX2, GPX5, GPX6 and GPX7*	Abiotic stress	Plant development, multiple signaling pathways	Transcriptome	[257,258]
	PgGPx	Salinity and Drought	NA	Reverse genetics	[177]
	ClGPX	Salinity and Drought	ABA	Transcriptome	[258]
	NnGPX	Salt	NA	Reverse genetics	[259]
	OsGPX5	Salt	ABA	Transcriptome, RNA interference	[260]
MDHAR	*MDAR2-4*	Salt	Stress protection; ABA	Transcriptome	[256]
	AeMDHAR	Salt	NA	Reverse genetics	[261]
	AtMDAR1	Ozone, salt and drought stress	ABA	Reverse genetics	[262]
	TrMDHAR	salt	ABA	Transcriptome	[261]
DHAR	*SlDHAR1 and SlDHAR2*	salt	Stress protection; NA	Transcriptome	[263]
	DHAR1 and DHAR3	Salt	ABA	Transcriptome	[256]
	LcDHAR	Salt and drought	NA	Transcriptome, Reverse genetics	[264]
	TrDHAR	Salt	ABA	Transcriptome	[265]

6. Conclusions

The adverse effects of climatic change and an increasing population pose a momentous challenge to crop production and food security, particularly in developing countries. Thus, it is a prerequisite to understand plant response mechanisms to abiotic stresses, namely, salinity and drought, at the molecular level to improve crop productivity. To overcome these circumstances, conventional breeding systems are no longer appropriate avenues to bolster crop production. In this review, we mainly discussed advanced molecular genomics tools focusing on plant genes in response to abiotic stress mechanisms to update our knowledge on the rapid development of high-yielding crop varieties under salt and drought stresses. Moreover, we summarized the recent studies of plant genes and differentiated them according to their molecular functions in response to salt and drought and reported recent advances in these stress-response mechanisms. Finally, the integration of any two or all three genomics approaches would be used to generate salinity- and drought-tolerant crops.

Author Contributions: Conceptualization, M.B. (Masum Billah); writing—original draft, M.B. (Masum Billah), S.A. and T.G.M.; review and editing, M.S.U., S.A.B., A.B.M.K., T.G.M., M.B. (Marian Brestic), M.Z., X.Y., M.S., S.M. and A.H.; funding, M.B. (Marian Brestic), M.Z., X.Y., M.S. and A.H. All authors have read and agreed to the published version of the manuscript.

Funding: This research was funded by the 'Slovak University of Agriculture', Nitra, Tr. A. Hlinku 2,949 01 Nitra, Slovak Republic under the projects 'APVV-1465 and EPPN2020-OPVaI-VA-ITMS31301 1T813'.

Institutional Review Board Statement: Not applicable.

Informed Consent Statement: Not applicable.

Data Availability Statement: Most of the recorded data are available in the tables and figures of the manuscript.

Acknowledgments: The authors extend their appreciation to the editor and anonymous reviewers for their valuable comments, which have allowed for considerable improvement of the manuscript.

Conflicts of Interest: The authors declare no conflict of interest.

References

1. Rozema, J.; Flowers, T. Crops for a Salinized World. *Science* **2008**, *322*, 1478–1480. [CrossRef] [PubMed]
2. Zhang, X.; Lu, G.; Long, W.; Zou, X.; Li, F.; Nishio, T. Recent progress in drought and salt tolerance studies in *Brassica* crops. *Breed. Sci.* **2014**, *64*, 60–73. [CrossRef] [PubMed]
3. Athar, H.R.; Ashraf, M. Strategies for Crop Improvement against Salinity and Drought Stress: An Overview. In *Salinity and Water Stress*; Springer: Berlin/Heidelberg, Germany, 2009; pp. 1–16.
4. Catlin, P.; Hoffman, G.; Mead, R.; Johnson, R. Long-term response of mature plum trees to salinity. *Irrig. Sci.* **1993**, *13*, 171–176. [CrossRef]
5. Munns, R.; Tester, M. Mechanisms of Salinity Tolerance. *Annu. Rev. Plant Biol.* **2008**, *59*, 651–681. [CrossRef]
6. FAO. Unlocking the Water Potential of Agriculture. 2003. Available online: https://agris.fao.org/agris-search/search.do?recordID=XF2006444621 (accessed on 7 April 2006).
7. EL Sabagh, A.; Islam, M.S.; Skalicky, M.; Raza, M.A.; Singh, K.; Hossain, M.A.; Hossain, A.; Mahboob, W.; Iqbal, M.A.; Ratnasekera, D.; et al. Adaptation and Management Strategies of Wheat (*Triticum aestivum* L.) Against Salinity Stress to Increase Yield and Quality. *Front. Agron.* **2021**, *3*, 43. [CrossRef]
8. Kumari, V.V.; Roy, A.; Vijayan, R.; Banerjee, P.; Verma, V.; Nalia, A.; Pramanik, M.; Mukherjee, B.; Ghosh, A.; Reja, H.; et al. Drought and Heat Stress in Cool-Season Food Legumes in Sub-Tropical Regions: Consequences, Adaptation, and Mitigation Strategies. *Plants* **2021**, *10*, 1038. [CrossRef]
9. Sinclair, T.R. Challenges in breeding for yield increase for drought. *Trends Plant Sci.* **2011**, *16*, 289–293. [CrossRef]
10. Abbai, R.; Subramaniyam, S.; Mathiyalagan, R.; Yang, D.C. Functional genomic approaches in plant research. In *Plant Bioinformatics*; Springer: Berlin/Heidelberg, Germany, 2017; pp. 215–239.
11. Kartseva, T.; Dobrikova, A.; Kocheva, K.; Alexandrov, V.; Georgiev, G.; Brestič, M.; Misheva, S. Optimal Nitrogen Supply Ameliorates the Performance of Wheat Seedlings under Osmotic Stress in Genotype-Specific Manner. *Plants* **2021**, *10*, 493. [CrossRef] [PubMed]
12. Zhang, D.Y.; Tong, J.; He, X.; Xu, Z.; Xu, L.; Wei, P.; Huang, Y.; Ebrestic, M.; Ema, H.; Eshao, H.-B. A Novel Soybean Intrinsic Protein Gene, GmTIP2;3, Involved in Responding to Osmotic Stress. *Front. Plant Sci.* **2016**, *6*, 1237. [CrossRef]

13. dos Reis, S.P.; Marques, D.N.; Barros, N.L.F.; Costa, C.d.N.M.; de Souza, C.R.B. Genetically engineered food crops to abiotic stress tolerance. In *Genetically Engineered Foods*; Elsevier: Amsterdam, The Netherlands, 2018; pp. 247–279.
14. Lawlor, D.W. Genetic engineering to improve plant performance under drought: Physiological evaluation of achievements, limitations, and possibilities. *J. Exp. Bot.* **2013**, *64*, 83–108. [CrossRef] [PubMed]
15. Rasel, M.; Tahjib-Ul-Arif, M.; Hossain, M.A.; Hassan, L.; Farzana, S.; Brestic, M. Screening of Salt-Tolerant Rice Landraces by Seedling Stage Phenotyping and Dissecting Biochemical Determinants of Tolerance Mechanism multidimensional roles in salt-stressed plants. *J. Plant Growth Regul.* **2020**, 1–16. [CrossRef]
16. Plaut, Z.; Edelstein, M.; Ben-Hur, M. Overcoming Salinity Barriers to Crop Production Using Traditional Methods. *Crit. Rev. Plant Sci.* **2013**, *32*, 250–291. [CrossRef]
17. Gepts, P. The contribution of genetic and genomic approaches to plant domestication studies. *Curr. Opin. Plant Biol.* **2014**, *18*, 51–59. [CrossRef]
18. Shi, J.; Gao, H.; Wang, H.; Lafitte, H.R.; Archibald, R.L.; Yang, M.; Hakimi, S.M.; Mo, H.; Habben, J.E. ARGOS8 variants generated by CRISPR-Cas9 improve maize grain yield under field drought stress conditions. *Plant Biotechnol. J.* **2016**, *15*, 207–216. [CrossRef] [PubMed]
19. Ibrahimova, U.; Zivcak, M.; Gasparovic, K.; Rastogi, A.; Allakhverdiev, S.I.; Yang, X.; Brestic, M. Electron and proton transport in wheat exposed to salt stress: Is the increase of the thylakoid membrane proton conductivity responsible for decreasing the photosynthetic activity in sensitive genotypes? *Photosynth. Res.* **2021**, 1–17. [CrossRef]
20. Singh, B.; Salaria, N.; Thakur, K.; Kukreja, S.; Gautam, S.; Goutam, U. Functional genomic approaches to improve crop plant heat stress tolerance. *F1000Research* **2019**, *8*, 1721. [CrossRef] [PubMed]
21. Yan, K.; Shao, H.; Shao, C.; Chen, P.; Zhao, S.; Brestic, M.; Chen, X. Physiological adaptive mechanisms of plants grown in saline soil and implications for sustainable saline agriculture in coastal zone. *Acta Physiol. Plant.* **2013**, *35*, 2867–2878. [CrossRef]
22. Cheng, R.; Doerge, R.W.; Borevitz, J. Novel Resampling Improves Statistical Power for Multiple-Trait QTL Mapping. *G3 Genes Genomes Genet.* **2017**, *7*, 813–822. [CrossRef]
23. Aktar, S.; Hossain, N.; Azam, M.G.; Billah, M.; Biswas, P.L.; Latif, M.A.; Rohman, M.; Bagum, S.A.; Uddin, M.S. Phenotyping of Hybrid Maize (*Zea mays* L.) at Seedling Stage under Drought Condition. *Am. J. Plant Sci.* **2018**, *9*, 2154–2169. [CrossRef]
24. Sanchez, A.; Subudhi, P.; Rosenow, D.; Nguyen, H. Mapping QTLs associated with drought resistance in sorghum (*Sorghum bicolor* L. Moench). *Plant Mol. Biol.* **2002**, *48*, 713–726. [CrossRef] [PubMed]
25. Sandhu, N.; Singh, A.; Dixit, S.; Cruz, M.T.S.; Maturan, P.C.; Jain, R.K.; Kumar, A. Identification and mapping of stable QTL with main and epistasis effect on rice grain yield under upland drought stress. *BMC Genet.* **2014**, *15*, 63. [CrossRef]
26. Prohens, J.; Gramazio, P.; Plazas, M.; Dempewolf, H.; Kilian, B.; Diez, M.J.; Fita, A.; Herraiz, F.J.; Rodríguez-Burruezo, A.; Soler, S.; et al. Introgressiomics: A new approach for using crop wild relatives in breeding for adaptation to climate change. *Euphytica* **2017**, *213*, 158. [CrossRef]
27. Swamy, B.P.M.; Shamsudin, N.A.A.; Rahman, S.N.A.; Mauleon, R.; Ratnam, W.; Cruz, M.T.S.; Kumar, A. Association Mapping of Yield and Yield-related Traits Under Reproductive Stage Drought Stress in Rice (*Oryza sativa* L.). *Rice* **2017**, *10*, 21. [CrossRef]
28. Abdelraheem, A.; Thyssen, G.N.; Fang, D.D.; Jenkins, J.N.; Mccarty, J.C.; Wedegaertner, T.; Zhang, J. GWAS reveals consistent QTL for drought and salt tolerance in a MAGIC population of 550 lines derived from intermating of 11 Upland cotton (*Gossypium hirsutum*) parents. *Mol. Genet. Genom.* **2020**, *296*, 119–129. [CrossRef] [PubMed]
29. Zhao, Y.; Wang, H.; Shao, B.; Chen, W.; Guo, Z.; Gong, H.; Sang, X.; Wang, J.; Ye, W.W. SSR-based association mapping of salt tolerance in cotton (*Gossypium hirsutum* L.). *Genet. Mol. Res.* **2016**, *15*, gmr.15027370. [CrossRef]
30. Bennett, D.; Reynolds, M.P.; Mullan, D.; Izanloo, A.; Kuchel, H.; Langridge, P.; Schnurbusch, T. Detection of two major grain yield QTL in bread wheat (*Triticum aestivum* L.) under heat, drought and high yield potential environments. *Theor. Appl. Genet.* **2012**, *125*, 1473–1485. [CrossRef] [PubMed]
31. Liu, X.; Fan, Y.; Mak, M.; Babla, M.; Holford, P.; Wang, F.; Chen, G.; Scott, G.; Wang, G.; Shabala, S.; et al. QTLs for stomatal and photosynthetic traits related to salinity tolerance in barley. *BMC Genom.* **2017**, *18*, 1–13. [CrossRef]
32. Fan, Y.; Shabala, S.; Ma, Y.; Xu, R.; Zhou, M. Using QTL mapping to investigate the relationships between abiotic stress tolerance (drought and salinity) and agronomic and physiological traits. *BMC Genom.* **2015**, *16*, 43. [CrossRef] [PubMed]
33. Courtois, B.; McLaren, G.; Sinha, P.; Prasad, K.; Yadav, R.; Shen, L. Mapping QTLs associated with drought avoidance in upland rice. *Mol. Breed.* **2000**, *6*, 55–66. [CrossRef]
34. Bahuguna, R.N.; Gupta, P.; Bagri, J.; Singh, D.; Dewi, A.K.; Tao, L.; Islam, M.; Sarsu, F.; Singla-Pareek, S.L.; Pareek, A. Forward and reverse genetics approaches for combined stress tolerance in rice. *Indian J. Plant Physiol.* **2018**, *23*, 630–646. [CrossRef]
35. Gupta, B.K.; Sahoo, K.K.; Ghosh, A.; Tripathi, A.K.; Anwar, K.; Das, P.; Singh, A.K.; Pareek, A.; Sopory, S.K.; Singla-Pareek, S.L. Manipulation of glyoxalase pathway confers tolerance to multiple stresses in rice. *Plant Cell Environ.* **2017**, *41*, 1186–1200. [CrossRef] [PubMed]
36. Joshi, R.; Prashat, R.; Sharma, P.C.; Singla-Pareek, S.L.; Pareek, A. Physiological characterization of gamma-ray induced mutant population of rice to facilitate biomass and yield improvement under salinity stress. *Indian J. Plant Physiol.* **2016**, *21*, 545–555. [CrossRef]
37. Martinez-Atienza, J.; Jiang, X.; Garciadeblas, B.; Mendoza, I.; Zhu, J.-K.; Pardo, J.M.; Quintero, F.J. Conservation of the Salt Overly Sensitive Pathway in Rice. *Plant Physiol.* **2006**, *143*, 1001–1012. [CrossRef] [PubMed]

38. Mickelbart, M.V.; Hasegawa, P.M.; Bailey-Serres, J. Genetic mechanisms of abiotic stress tolerance that translate to crop yield stability. *Nat. Rev. Genet.* **2015**, *16*, 237–251. [CrossRef]

39. Nutan, K.K.; Kumar, G.; Singla-Pareek, S.L.; Pareek, A. A Salt Overly Sensitive Pathway Member from Brassica juncea BjSOS3 Can Functionally Complement ΔAtsos3 in Arabidopsis. *Curr. Genom.* **2017**, *19*, 60–69. [CrossRef]

40. Olías, R.; Eljakaoui, Z.; Pardo, J.M.; Belver, A. The Na$^+$/H$^+$ exchanger SOS1 controls extrusion and distribution of Na$^+$ in tomato plants under salinity conditions. *Plant Signal. Behav.* **2009**, *4*, 973–976. [CrossRef]

41. Wu, S.-J.; Ding, L.; Zhu, J.-K. SOS1, a genetic locus essential for salt tolerance and potassium acquisition. *Plant Cell* **1996**, *8*, 617–627. [CrossRef]

42. Blumenberg, M. *Introductory Chapter: Transcriptome Analysis*; IntechOpen: London, UK, 2019. [CrossRef]

43. Gahlaut, V.; Jaiswal, V.; Kumar, A.; Gupta, P.K. Transcription factors involved in drought tolerance and their possible role in developing drought tolerant cultivars with emphasis on wheat (*Triticum aestivum* L.). *Theor. Appl. Genet.* **2016**, *129*, 2019–2042. [CrossRef]

44. Bowman, M.J.; Park, W.; Bauer, P.; Udall, J.A.; Page, J.T.; Raney, J.; Scheffler, B.; Jones, D.C.; Campbell, B.T. RNA-Seq Transcriptome Profiling of Upland Cotton (*Gossypium hirsutum* L.) Root Tissue under Water-Deficit Stress. *PLoS ONE* **2013**, *8*, e82634. [CrossRef]

45. Rong, W.; Qi, L.; Wang, A.; Ye, X.; Du, L.; Liang, H.; Xin, Z.; Zhang, Z. The ERF transcription factor Ta ERF 3 promotes tolerance to salt and drought stresses in wheat. *Plant Biotechnol. J.* **2014**, *12*, 468–479. [CrossRef]

46. Gao, S.-Q.; Chen, M.; Xu, Z.-S.; Zhao, C.-P.; Li, L.; Xu, H.-J.; Tang, Y.-M.; Zhao, X.; Ma, Y.-Z. The soybean GmbZIP1 transcription factor enhances multiple abiotic stress tolerances in transgenic plants. *Plant Mol. Biol.* **2011**, *75*, 537–553. [CrossRef]

47. Shin, D.; Moon, S.-J.; Han, S.; Kim, B.-G.; Park, S.R.; Lee, S.-K.; Yoon, H.-J.; Lee, H.E.; Kwon, H.-B.; Baek, D.; et al. Expression of StMYB1R-1, a Novel Potato Single MYB-Like Domain Transcription Factor, Increases Drought Tolerance. *Plant Physiol.* **2010**, *155*, 421–432. [CrossRef]

48. Song, Y.; Ji, D.; Li, S.; Wang, P.; Li, Q.; Xiang, F. The Dynamic Changes of DNA Methylation and Histone Modifications of Salt Responsive Transcription Factor Genes in Soybean. *PLoS ONE* **2012**, *7*, e41274. [CrossRef] [PubMed]

49. Hao, Y.-J.; Wei, W.; Song, Q.-X.; Chen, H.-W.; Zhang, Y.-Q.; Wang, F.; Zou, H.-F.; Lei, G.; Tian, A.; Zhang, W.-K.; et al. Soybean NAC transcription factors promote abiotic stress tolerance and lateral root formation in transgenic plants. *Plant J.* **2011**, *68*, 302–313. [CrossRef] [PubMed]

50. Wang, X.; Zeng, J.; Li, Y.; Rong, X.; Sun, J.; Sun, T.; Li, M.; Wang, L.; Feng, Y.; Chai, R.; et al. Expression of TaWRKY44, a wheat WRKY gene, in transgenic tobacco confers multiple abiotic stress tolerances. *Front. Plant Sci.* **2015**, *6*, 615. [CrossRef] [PubMed]

51. Yoshida, T.; Fujita, Y.; Maruyama, K.; Mogami, J.; Todaka, D.; Shinozaki, K.; Yamaguchi-Shinozaki, K. Four A rabidopsis AREB/ABF transcription factors function predominantly in gene expression downstream of SnRK2 kinases in abscisic acid signalling in response to osmotic stress. *Plant Cell Environ.* **2014**, *38*, 35–49. [CrossRef] [PubMed]

52. Singh, B.; Singh, A.K. *Marker-Assisted Plant Breeding: Principles and Practices*; Springer: Berlin/Heidelberg, Germany, 2015; p. 514. [CrossRef]

53. Agrama, H.A.; Eizenga, G.C.; Yan, W. Association mapping of yield and its components in rice cultivars. *Mol. Breed.* **2007**, *19*, 341–356. [CrossRef]

54. Edae, E.A.; Byrne, P.F.; Manmathan, H.; Haley, S.D.; Reynolds, M.P. Association Mapping and Nucleotide Sequence Variation in Five Drought Tolerance Candidate Genes in Spring Wheat. *Plant Genome* **2013**, *6*, 547–562. [CrossRef]

55. Saeed, M.; Wangzhen, G.; Tianzhen, Z. Association mapping for salinity tolerance in cotton (*Gossypium hirsutum* L.) germplasm from US and diverse regions of China. *Aust. J. Crop Sci.* **2014**, *8*, 338–346.

56. Purcărea, C.; Cachiță-Cosma, D. Studies regarding the effects of salicylic acid on maize (*Zea mays* L.) seedling under salt stress. *Studia Univ. Ser. Tiintele Vietii* **2010**, *20*, 63–68.

57. Wójcik-Jagła, M.; Fiust, A.; Kościelniak, J.; Rapacz, M. Association mapping of drought tolerance-related traits in barley to complement a traditional biparental QTL mapping study. *Theor. Appl. Genet.* **2017**, *131*, 167–181. [CrossRef] [PubMed]

58. Song, S.-Y.; Chen, Y.; Chen, J.; Dai, X.-Y.; Zhang, W.-H. Physiological mechanisms underlying OsNAC5-dependent tolerance of rice plants to abiotic stress. *Planta* **2011**, *234*, 331–345. [CrossRef]

59. Zhang, J.; Hao, C.; Ren, Q.; Chang, X.; Liu, G.; Jing, R. Association mapping of dynamic developmental plant height in common wheat. *Planta* **2011**, *234*, 891–902. [CrossRef] [PubMed]

60. Sehgal, D.; Skot, L.; Singh, R.; Srivastava, R.K.; Das, S.P.; Taunk, J.; Sharma, P.C.; Pal, R.; Raj, B.; Hash, C.T. Exploring Potential of Pearl Millet Germplasm Association Panel for Association Mapping of Drought Tolerance Traits. *PLoS ONE* **2015**, *10*, e0122165. [CrossRef] [PubMed]

61. Sukumaran, S.; Dreisigacker, S.; Lopes, M.; Chavez, P.; Reynolds, M.P. Genome-wide association study for grain yield and related traits in an elite spring wheat population grown in temperate irrigated environments. *Theor. Appl. Genet.* **2014**, *128*, 353–363. [CrossRef]

62. Crossa, J.; Burgueño, J.; Dreisigacker, S.; Vargas, M.; Herrera-Foessel, S.A.; Lillemo, M.; Singh, R.P.; Trethowan, R.; Warburton, M.; Franco, J. Association Analysis of Historical Bread Wheat Germplasm Using Additive Genetic Covariance of Relatives and Population Structure. *Genetics* **2007**, *177*, 1889–1913. [CrossRef]

63. Rasheed, A.; Xia, X.; Ogbonnaya, F.; Mahmood, T.; Zhang, Z.; Mujeeb-Kazi, A.; He, Z. Genome-wide association for grain morphology in synthetic hexaploid wheats using digital imaging analysis. *BMC Plant Biol.* **2014**, *14*, 128. [CrossRef]

64. Rolvien, T.; Kornak, U.; Linke, S.J.; Amling, M.; Oheim, R. Whole-Exome Sequencing Identifies Novel Compound Heterozygous ZNF469 Mutations in Two Siblings with Mild Brittle Cornea Syndrome. *Calcif. Tissue Int.* **2020**, *107*, 294–299. [CrossRef]
65. Lekklar, C.; Pongpanich, M.; Suriya-Arunroj, D.; Chinpongpanich, A.; Tsai, H.; Comai, L.; Chadchawan, S.; Buaboocha, T. Genome-wide association study for salinity tolerance at the flowering stage in a panel of rice accessions from Thailand. *BMC Genom.* **2019**, *20*, 76. [CrossRef]
66. Shi, Y.; Gao, L.; Wu, Z.; Zhang, X.; Wang, M.; Zhang, C.; Zhang, F.; Zhou, Y.; Li, Z. Genome-wide association study of salt tolerance at the seed germination stage in rice. *BMC Plant Biol.* **2017**, *17*, 92. [CrossRef]
67. Hasan, M.; Skalicky, M.; Jahan, M.; Hossain, N.; Anwar, Z.; Nie, Z.; Alabdallah, N.; Brestic, M.; Hejnak, V.; Fang, X.-W. Spermine: Its Emerging Role in Regulating Drought Stress Responses in Plants. *Cells* **2021**, *10*, 261. [CrossRef]
68. Voichek, Y.; Weigel, D. Identifying genetic variants underlying phenotypic variation in plants without complete genomes. *Biol. Res.* **2020**, *52*, 534–540. [CrossRef]
69. Ogbonnaya, F.C.; Rasheed, A.; Okechukwu, E.C.; Jighly, A.; Makdis, F.; Wuletaw, T.; Hagras, A.; Uguru, M.I.; Agbo, C.U. Genome-wide association study for agronomic and physiological traits in spring wheat evaluated in a range of heat prone environments. *Theor. Appl. Genet.* **2017**, *130*, 1819–1835. [CrossRef] [PubMed]
70. Hoang, G.T.; Van Dinh, L.; Nguyen, T.T.; Ta, N.K.; Gathignol, F.; Mai, C.D.; Jouannic, S.; Tran, K.D.; Khuat, T.H.; Do, V.N.; et al. Genome-wide Association Study of a Panel of Vietnamese Rice Landraces Reveals New QTLs for Tolerance to Water Deficit During the Vegetative Phase. *Rice* **2019**, *12*, 1–20. [CrossRef] [PubMed]
71. Ding, F.; Cui, P.; Wang, Z.; Zhang, S.; Ali, S.; Xiong, L. Genome-wide analysis of alternative splicing of pre-mRNA under salt stress in Arabidopsis. *BMC Genom.* **2014**, *15*, 431. [CrossRef] [PubMed]
72. Hübner, S.; Korol, A.B.; Schmid, K.J. RNA-Seq analysis identifies genes associated with differential reproductive success under drought-stress in accessions of wild barley Hordeum spontaneum. *BMC Plant Biol.* **2015**, *15*, 1–14. [CrossRef] [PubMed]
73. Pucholt, P.; Sjödin, P.; Weih, M.; Rönnberg-Wästljung, A.C.; Berlin, S. Genome-wide transcriptional and physiological responses to drought stress in leaves and roots of two willow genotypes. *BMC Plant Biol.* **2015**, *15*, 1–16. [CrossRef]
74. Pham, A.-T.; Maurer, A.; Pillen, K.; Brien, C.; Dowling, K.; Berger, B.; Eglinton, J.K.; March, T.J. Genome-wide association of barley plant growth under drought stress using a nested association mapping population. *BMC Plant Biol.* **2019**, *19*, 134. [CrossRef]
75. Ibrahimova, U.; Kumari, P.; Yadav, S.; Rastogi, A.; Antala, M.; Suleymanova, Z.; Zivcak, M.; Arif, T.U.; Hussain, S.; Abdelhamid, M.; et al. Progress in understanding salt stress response in plants using biotechnological tools. *J. Biotechnol.* **2021**, *329*, 180–191. [CrossRef]
76. Yu, L.-X.; Liu, X.; Boge, W.; Liu, X.-P. Genome-Wide Association Study Identifies Loci for Salt Tolerance during Germination in Autotetraploid Alfalfa (*Medicago sativa* L.) Using Genotyping-by-Sequencing. *Front. Plant Sci.* **2016**, *7*, 956. [CrossRef]
77. Wang, W.; Zhao, X.; Pan, Y.; Zhu, L.; Fu, B.; Li, Z. DNA methylation changes detected by methylation-sensitive amplified polymorphism in two contrasting rice genotypes under salt stress. *J. Genet. Genom.* **2011**, *38*, 419–424. [CrossRef]
78. Hossain, A.; Skalicky, M.; Brestic, M.; Maitra, S.; Alam, M.A.; Abu Syed, M.; Hossain, J.; Sarkar, S.; Saha, S.; Bhadra, P.; et al. Consequences and Mitigation Strategies of Abiotic Stresses in Wheat (*Triticum aestivum* L.) under the Changing Climate. *Agronomy* **2021**, *11*, 241. [CrossRef]
79. Metzker, M.L. Sequencing technologies—the next generation. *Nat. Rev. Genet.* **2010**, *11*, 31. [CrossRef] [PubMed]
80. Unamba, C.I.N.; Nag, A.; Sharma, R.K. Next Generation Sequencing Technologies: The Doorway to the Unexplored Genomics of Non-Model Plants. *Front. Plant Sci.* **2015**, *6*, 1074. [CrossRef] [PubMed]
81. Grada, A.; Weinbrecht, K. Next-Generation Sequencing: Methodology and Application. *J. Investig. Dermatol.* **2013**, *133*, 1–4. [CrossRef]
82. Koboldt, D.C.; Steinberg, K.M.; Larson, D.; Wilson, R.K.; Mardis, E.R. The Next-Generation Sequencing Revolution and Its Impact on Genomics. *Cell* **2013**, *155*, 27–38. [CrossRef] [PubMed]
83. Walter, J.; Hümpel, A. Introduction to epigenetics. In *Epigenetics*; Springer: New York, NY, USA, 2017; pp. 11–29.
84. Chan, S.W.-L.; Henderson, I.; Jacobsen, S.E. Gardening the genome: DNA methylation in Arabidopsis thaliana. *Nat. Rev. Genet.* **2005**, *6*, 351–360. [CrossRef]
85. Singroha, G.; Sharma, P. Epigenetic Modifications in Plants under Abiotic Stress. In *Epigenetics*; IntechOpen: London, UK, 2019.
86. Banerjee, A.; Roychoudhury, A. Epigenetic regulation during salinity and drought stress in plants: Histone modifications and DNA methylation. *Plant Gene* **2017**, *11*, 199–204. [CrossRef]
87. Ding, Y.; Fromm, M.E.; Avramova, Z. Multiple exposures to drought 'train' transcriptional responses in Arabidopsis. *Nat. Commun.* **2012**, *3*, 740. [CrossRef] [PubMed]
88. Fang, H.; Liu, X.; Thorn, G.; Duan, J.; Tian, L. Expression analysis of histone acetyltransferases in rice under drought stress. *Biochem. Biophys. Res. Commun.* **2013**, *443*, 400–405. [CrossRef]
89. Kim, J.-M.; Sasaki, T.; Ueda, M.; Sako, K.; Seki, M. Chromatin changes in response to drought, salinity, heat, and cold stresses in plants. *Front. Plant Sci.* **2015**, *6*, 114. [CrossRef] [PubMed]
90. Ferreira, L.J.; Azevedo, V.S.; Maroco, J.; Oliveira, M.M.; Santos, A.P. Salt Tolerant and Sensitive Rice Varieties Display Differential Methylome Flexibility under Salt Stress. *PLoS ONE* **2015**, *10*, e0124060. [CrossRef] [PubMed]
91. Kaldis, A.; Tsementzi, D.; Tanriverdi, O.; Vlachonasios, K.E. Arabidopsis thaliana transcriptional co-activators ADA2b and SGF29a are implicated in salt stress responses. *Planta* **2010**, *233*, 749–762. [CrossRef]
92. Duan, C.-G.; Wang, C.-H.; Guo, H.-S. Application of RNA silencing to plant disease resistance. *Silence* **2012**, *3*, 5. [CrossRef]

93. Ricaño-Rodríguez, J.; Adame-García, J.; Patlas-Martínez, C.I.; Hipólito-Romero, E.; Ramos-Prado, J.M. Plant gene co-suppression, basis of the molecular machinery of interfering RNA. *Plant Omics* **2016**, *9*, 261–269. [CrossRef]
94. Hunter, C.P. Gene silencing: Shrinking the black box of RNAi. *Curr. Biol.* **2000**, *10*, R137–R140. [CrossRef]
95. Bekele, D.; Tesfaye, K.; Fikre, A. Applications of Virus Induced Gene Silencing (VIGS) in Plant Functional Genomics Studies. *J. Plant Biochem. Physiol.* **2019**, *7*, 1–7. [CrossRef]
96. Guo, Q.; Zhao, L.; Fan, X.; Xu, P.; Xu, Z.; Zhang, X.; Meng, S.; Shen, X.J. Transcription Factor GarWRKY5 Is Involved in Salt Stress Response in Diploid Cotton Species (*Gossypium aridum* L.). *Int. J. Mol. Sci.* **2019**, *20*, 5244. [CrossRef] [PubMed]
97. Kirungu, J.N.; Magwanga, R.O.; Pu, L.; Cai, X.; Xu, Y.; Hou, Y.; Zhou, Y.; Cai, Y.; Hao, F.; Zhou, Z.; et al. Knockdown of Gh_A05G1554 (GhDHN_03) and Gh_D05G1729 (GhDHN_04) Dehydrin genes, Reveals their potential role in enhancing osmotic and salt tolerance in cotton. *Genomics* **2019**, *112*, 1902–1915. [CrossRef]
98. Wang, X.; Ren, Y.; Li, J.; Wang, Z.; Xin, Z.; Lin, T. Knock-down the expression of TaH2B-7D using virus-induced gene silencing reduces wheat drought tolerance. *Biol. Res.* **2019**, *52*, 14. [CrossRef]
99. Oh, D.-H.; Leidi, E.; Zhang, Q.; Hwang, S.-M.; Li, Y.; Quintero, F.J.; Jiang, X.; D'Urzo, M.P.; Lee, S.Y.; Zhao, Y. Loss of halophytism by interference with SOS1 expression. *Plant Physiol.* **2009**, *151*, 210–222. [CrossRef]
100. Zhai, Y.; Wang, H.; Liang, M.; Lu, M. Both silencing-and over-expression of pepper CaATG8c gene compromise plant tolerance to heat and salt stress. *Environ. Exp. Bot.* **2017**, *141*, 10–18. [CrossRef]
101. Bai, C.; Wang, P.; Fan, Q.; Fu, W.-D.; Wang, L.; Zhang, Z.-N.; Song, Z.; Zhang, G.-L.; Wu, J.-H. Analysis of the Role of the Drought-Induced Gene DRI15 and Salinity-Induced Gene SI1 in Alternanthera philoxeroides Plasticity Using a Virus-Based Gene Silencing Tool. *Front. Plant Sci.* **2017**, *8*, 1579. [CrossRef]
102. Li, C.; Yan, J.-M.; Li, Y.-Z.; Zhang, Z.-C.; Wang, Q.-L.; Liang, Y. Silencing the SpMPK1, SpMPK2, and SpMPK3 genes in tomato reduces abscisic acid—mediated drought tolerance. *Int. J. Mol. Sci.* **2013**, *14*, 21983–21996. [CrossRef]
103. Manmathan, H.; Shaner, D.; Snelling, J.; Tisserat, N.; Lapitan, N. Virus-induced gene silencing of Arabidopsis thaliana gene homologues in wheat identifies genes conferring improved drought tolerance. *J. Exp. Bot.* **2013**, *64*, 1381–1392. [CrossRef]
104. Kirungu, J.N.; Magwanga, R.O.; Lu, P.; Cai, X.; Zhou, Z.; Wang, X.; Peng, R.; Wang, K.; Liu, F. Functional characterization of Gh_A08G1120 (GH3.5) gene reveal their significant role in enhancing drought and salt stress tolerance in cotton. *BMC Genet.* **2019**, *20*, 62. [CrossRef] [PubMed]
105. Li, Z.; Li, L.; Zhou, K.; Zhang, Y.; Han, X.; Ding, Y.; Ge, X.; Wang, P.; Li, F.; Ma, Z. GhWRKY6 acts as a negative regulator in both transgenic Arabidopsis and cotton during drought and salt stress. *Front. Genet.* **2019**, *10*, 392. [CrossRef] [PubMed]
106. Belhaj, K.; Chaparro-Garcia, A.; Kamoun, S.; Nekrasov, V. Plant genome editing made easy: Targeted mutagenesis in model and crop plants using the CRISPR/Cas system. *Plant Methods* **2013**, *9*, 39. [CrossRef] [PubMed]
107. Liu, X.; Wu, S.; Xu, J.; Sui, C.; Wei, J. Application of CRISPR/Cas9 in plant biology. *Acta Pharm. Sin. B* **2017**, *7*, 292–302. [CrossRef]
108. Jia, H.; Zhang, Y.; Orbović, V.; Xu, J.; White, F.F.; Jones, J.B.; Wang, N. Genome editing of the disease susceptibility gene CsLOB1 in citrus confers resistance to citrus canker. *Plant Biotechnol. J.* **2016**, *15*, 817–823. [CrossRef]
109. Fernandez i Marti, A.; Dodd, R.S. Using CRISPR as a gene editing tool for validating adaptive gene function in tree landscape genomics. *Front. Ecol. Evol. Dev.* **2018**, *6*, 76. [CrossRef]
110. Debbarma, J.; Sarki, Y.N.; Saikia, B.; Boruah, H.P.D.; Singha, D.L.; Chikkaputtaiah, C. Ethylene response factor (ERF) family proteins in abiotic stresses and CRISPR–Cas9 genome editing of ERFs for multiple abiotic stress tolerance in crop plants: A review. *Mol. Biotechnol.* **2019**, *61*, 153–172. [CrossRef]
111. Nieves-Cordones, M.; Mohamed, S.; Tanoi, K.; Kobayashi, N.I.; Takagi, K.; Vernet, A.; Guiderdoni, E.; Périn, C.; Sentenac, H.; Véry, A.A. Production of low-Cs^+ rice plants by inactivation of the K^+ transporter Os HAK 1 with the CRISPR-Cas system. *Plant J.* **2017**, *92*, 43–56. [CrossRef]
112. Shao, G.; Xie, L.; Jiao, G.; Wei, X.; Sheng, Z.; Tang, S.; Hu, P. CRISPR/CAS9-mediated editing of the fragrant gene Badh2 in rice. *Chin. J. Rice Sci.* **2017**, *31*, 216–222.
113. Wang, F.; Wang, C.; Liu, P.; Lei, C.; Hao, W.; Gao, Y.; Liu, Y.-G.; Zhao, K. Enhanced Rice Blast Resistance by CRISPR/Cas9-Targeted Mutagenesis of the ERF Transcription Factor Gene OsERF922. *PLoS ONE* **2016**, *11*, e0154027. [CrossRef] [PubMed]
114. Zhang, A.; Liu, Y.; Wang, F.; Li, T.; Chen, Z.; Kong, D.; Bi, J.; Zhang, F.; Luo, X.; Wang, J.; et al. Enhanced rice salinity tolerance via CRISPR/Cas9-targeted mutagenesis of the OsRR22 gene. *Mol. Breed.* **2019**, *39*, 47. [CrossRef] [PubMed]
115. Zhang, H.; Zhang, J.; Wei, P.; Zhang, B.; Gou, F.; Feng, Z.; Mao, Y.; Yang, L.; Zhang, H.; Xu, N. The CRISPR/C as9 system produces specific and homozygous targeted gene editing in rice in one generation. *Plant Biotechnol. J.* **2014**, *12*, 797–807. [CrossRef] [PubMed]
116. Zhu, J.; Song, N.; Sun, S.; Yang, W.; Zhao, H.; Song, W.; Lai, J. Efficiency and Inheritance of Targeted Mutagenesis in Maize Using CRISPR-Cas9. *J. Genet. Genom.* **2016**, *43*, 25–36. [CrossRef]
117. Hunter, C.T. CRISPR/Cas9 Targeted Mutagenesis for Functional Genetics in Maize. *Plants* **2021**, *10*, 723. [CrossRef] [PubMed]
118. Kim, D.; Alptekin, B.; Budak, H. CRISPR/Cas9 genome editing in wheat. *Funct. Integr. Genom.* **2017**, *18*, 31–41. [CrossRef]
119. Chen, Y.; Ma, J.; Zhang, X.; Yang, Y.; Zhou, D.; Yu, Q.; Que, Y.; Xu, L.; Guo, J. A Novel Non-specific Lipid Transfer Protein Gene from Sugarcane (NsLTPs), Obviously Responded to Abiotic Stresses and Signaling Molecules of SA and MeJA. *Sugar Tech.* **2016**, *19*, 17–25. [CrossRef]

120. Miao, H.; Sun, P.; Liu, Q.; Liu, J.; Xu, B.; Jin, Z. The AGPase Family Proteins in Banana: Genome-Wide Identification, Phylogeny, and Expression Analyses Reveal Their Involvement in the Development, Ripening, and Abiotic/Biotic Stress Responses. *Int. J. Mol. Sci.* **2017**, *18*, 1581. [CrossRef]

121. Ou, W.; Mao, X.; Huang, C.; Tie, W.; Yan, Y.; Ding, Z.; Wu, C.; Xia, Z.; Wang, W.; Zhou, S.; et al. Genome-Wide Identification and Expression Analysis of the KUP Family under Abiotic Stress in Cassava (*Manihot esculenta* Crantz). *Front. Physiol.* **2018**, *9*, 17. [CrossRef]

122. Ye, J.; Yang, H.; Shi, H.; Wei, Y.; Tie, W.; Ding, Z.; Yan, Y.; Luo, Y.; Xia, Z.; Wang, W.; et al. The MAPKKK gene family in cassava: Genome-wide identification and expression analysis against drought stress. *Sci. Rep.* **2017**, *7*, 1–12. [CrossRef] [PubMed]

123. He, P.; Zhao, P.; Wang, L.; Zhang, Y.; Wang, X.; Xiao, H.; Yu, J.; Xiao, G. The PIN gene family in cotton (*Gossypium hirsutum*): Genome-wide identification and gene expression analyses during root development and abiotic stress responses. *BMC Genom.* **2017**, *18*, 507. [CrossRef] [PubMed]

124. Dass, A.; Abdin, M.Z.; Reddy, V.S.; Leelavathi, S. Isolation and characterization of the dehydration stress-inducible GhRDL1 promoter from the cultivated upland cotton (*Gossypium hirsutum*). *J. Plant Biochem. Biotechnol.* **2016**, *26*, 113–119. [CrossRef]

125. Arroyo-Herrera, A.; Yáñez, F.; Castano, E.; Santamaria, J.; Pereira-Santana, A.; Espadas-Alcocer, J.; Teyer, L.F.S.; Espadas-Gil, F.; Alcaraz, L.D.; López-Gómez, R.; et al. A novel Dreb2-type gene from Carica papaya confers tolerance under abiotic stress. *Plant Cell Tissue Organ Cult.* **2015**, *125*, 119–133. [CrossRef]

126. Kumar, V.V.S.; Verma, R.K.; Yadav, S.K.; Yadav, P.; Watts, A.; Rao, M.V.; Chinnusamy, V. CRISPR-Cas9 mediated genome editing of drought and salt tolerance (OsDST) gene in indica mega rice cultivar MTU1010. *Physiol. Mol. Biol. Plants* **2020**, *26*, 1099–1110. [CrossRef] [PubMed]

127. Li, R.; Liu, C.; Zhao, R.; Wang, L.; Chen, L.; Yu, W.; Zhang, S.; Sheng, J.; Shen, L. CRISPR/Cas9-Mediated SlNPR1 mutagenesis reduces tomato plant drought tolerance. *BMC Plant Biol.* **2019**, *19*, 38. [CrossRef]

128. Liao, S.; Qin, X.; Luo, L.; Han, Y.; Wang, X.; Usman, B.; Nawaz, G.; Zhao, N.; Liu, Y.; Li, R. CRISPR/Cas9-Induced Mutagenesis of Semi-Rolled Leaf1,2 Confers Curled Leaf Phenotype and Drought Tolerance by Influencing Protein Expression Patterns and ROS Scavenging in Rice (*Oryza sativa* L.). *Agronomy* **2019**, *9*, 728. [CrossRef]

129. Anzalone, A.V.; Randolph, P.B.; Davis, J.R.; Sousa, A.A.; Koblan, L.; Levy, J.M.; Chen, P.; Wilson, C.; Newby, G.A.; Raguram, A.; et al. Search-and-replace genome editing without double-strand breaks or donor DNA. *Nature* **2019**, *576*, 149–157. [CrossRef] [PubMed]

130. Wiens, C.P.; Shabala, L.; Zhang, J.; Pottosin, I.; Bose, J.; Zhu, M.; Fuglsang, A.T.; Velarde, A.; Massart, A.; Hill, C. Cell-type specific H$^+$-ATPase activity enables root K$^+$ retention and mediates acclimatation to salinity. *Plant Physiol.* **2016**, *172*, 2445–2458.

131. Gill, M.B.; Zeng, F.; Shabala, L.; Zhang, G.; Fan, Y.; Shabala, S.; Zhou, M. Cell-Based Phenotyping Reveals QTL for Membrane Potential Maintenance Associated with Hypoxia and Salinity Stress Tolerance in Barley. *Front. Plant Sci.* **2017**, *8*, 1941. [CrossRef] [PubMed]

132. Monneveux, P.; Reynolds, M.P.; Aguilar, J.G.; Singh, R.P.; Weber, W.E. Effects of the 7DL.7Ag translocation from Lophopyrum elongatum on wheat yield and related morphophysiological traits under different environments. *Plant Breed.* **2003**, *122*, 379–384. [CrossRef]

133. Ashraf, M.; Foolad, M.R. Crop breeding for salt tolerance in the era of molecular markers and marker-assisted selection. *Plant Breed.* **2012**, *132*, 10–20. [CrossRef]

134. thi Lang, N.; Buu, B.; Ismail, A. Molecular mapping and marker-assisted selection for salt tolerance in rice (*Oryza sativa* L.). *OmonRice* **2008**, *16*, 50–56.

135. Prince, S.J.; Beena, R.; Gomez, S.M.; Senthivel, S.; Babu, R.C. Mapping Consistent Rice (*Oryza sativa* L.) Yield QTLs under Drought Stress in Target Rainfed Environments. *Rice* **2015**, *8*, 25. [CrossRef] [PubMed]

136. Zhai, L.; Zheng, T.; Wang, X.; Wang, Y.; Chen, K.; Wang, S.; Xu, J.; Li, Z. QTL mapping and candidate gene analysis of peduncle vascular bundle related traits in rice by genome-wide association study. *Rice* **2018**, *11*, 13. [CrossRef]

137. Shamsudin, N.A.A.; Swamy, B.P.M.; Ratnam, W.; Cruz, M.T.S.; Raman, A.; Kumar, A. Marker assisted pyramiding of drought yield QTLs into a popular Malaysian rice cultivar, MR219. *BMC Genet.* **2016**, *17*, 1–14. [CrossRef]

138. Badu-Apraku, B.; Talabi, A.O.; Fakorede, M.A.B.; Fasanmade, Y.; Gedil, M.; Magorokosho, C.; Asiedu, R. Yield gains and associated changes in an early yellow bi-parental maize population following genomic selection for Striga resistance and drought tolerance. *BMC Plant Biol.* **2019**, *19*, 1–17. [CrossRef]

139. Wang, J.; Li, R.; Mao, X.; Jing, R. Functional Analysis and Marker Development of TaCRT-D Gene in Common Wheat (*Triticum aestivum* L.). *Front. Plant Sci.* **2017**, *8*, 1557. [CrossRef]

140. Bimpong, I.K.; Manneh, B.; Sock, M.; Diaw, F.; Amoah, N.K.A.; Ismail, A.M.; Gregorio, G.; Singh, R.K.; Wopereis, M. Improving salt tolerance of lowland rice cultivar 'Rassi' through marker-aided backcross breeding in West Africa. *Plant Sci.* **2016**, *242*, 288–299. [CrossRef]

141. Gu, J.; Yin, X.; Zhang, C.; Wang, H.; Struik, P.C. Linking ecophysiological modelling with quantitative genetics to support marker-assisted crop design for improved yields of rice (*Oryza sativa*) under drought stress. *Ann. Bot.* **2014**, *114*, 499–511. [CrossRef] [PubMed]

142. Gill, S.; Tuteja, N. Reactive oxygen species and antioxidant machinery in abiotic stress tolerance in crop plants. *Plant Physiol. Biochem.* **2010**, *48*, 909–930. [CrossRef] [PubMed]

143. Mhamdi, A.; Van Breusegem, F. Reactive oxygen species in plant development. *Development* **2018**, *145*, dev164376. [CrossRef]

144. You, J.; Chan, Z. ROS Regulation During Abiotic Stress Responses in Crop Plants. *Front. Plant Sci.* **2015**, *6*, 1092. [CrossRef]
145. Keunen, E.; Peshev, D.; Vangronsveld, J.; Ende, W.V.D.; Cuypers, A. Plant sugars are crucial players in the oxidative challenge during abiotic stress: Extending the traditional concept. *Plant Cell Environ.* **2013**, *36*, 1242–1255. [CrossRef]
146. Foyer, C.H.; Noctor, G. Redox Homeostasis and Antioxidant Signaling: A Metabolic Interface between Stress Perception and Physiological Responses. *Plant Cell* **2005**, *17*, 1866–1875. [CrossRef]
147. Lu, W.; Chu, X.; Li, Y.; Wang, C.; Guo, X. Cotton GhMKK1 Induces the Tolerance of Salt and Drought Stress, and Mediates Defence Responses to Pathogen Infection in Transgenic Nicotiana benthamiana. *PLoS ONE* **2013**, *8*, e68503. [CrossRef] [PubMed]
148. Ning, J.; Li, X.; Hicks, L.; Xiong, L. A Raf-Like MAPKKK Gene DSM1 Mediates Drought Resistance through Reactive Oxygen Species Scavenging in Rice. *Plant Physiol.* **2009**, *152*, 876–890. [CrossRef]
149. Du, H.; Wang, N.; Cui, F.; Li, X.; Xiao, J.; Xiong, L. Characterization of the β-Carotene Hydroxylase Gene DSM2 Conferring Drought and Oxidative Stress Resistance by Increasing Xanthophylls and Abscisic Acid Synthesis in Rice. *Plant Physiol.* **2010**, *154*, 1304–1318. [CrossRef]
150. Nadarajah, K.K. ROS Homeostasis in Abiotic Stress Tolerance in Plants. *Int. J. Mol. Sci.* **2020**, *21*, 5208. [CrossRef]
151. Dongdong, L.; Ming, Z.; Lili, H.; Xiaobo, C.; Yang, G.; Xingqi, G.; Han, L. GhMAPKKK49, a novel cotton (*Gossypium hirsutum* L.) MAPKKK gene, is involved in diverse stress responses. *Acta Physiol. Plant.* **2015**, *38*, 13. [CrossRef]
152. Gao, M.; Liu, J.; Bi, D.; Zhang, Z.; Cheng, F.; Chen, S.; Zhang, Y. MEKK1, MKK1/MKK2 and MPK4 function together in a mitogen-activated protein kinase cascade to regulate innate immunity in plants. *Cell Res.* **2008**, *18*, 1190–1198. [CrossRef]
153. Bundó, M.; Coca, M. Enhancing blast disease resistance by overexpression of the calcium-dependent protein kinaseOsCPK4in rice. *Plant Biotechnol. J.* **2015**, *14*, 1357–1367. [CrossRef] [PubMed]
154. Asano, T.; Hayashi, N.; Kobayashi, M.; Aoki, N.; Miyao, A.; Mitsuhara, I.; Ichikawa, H.; Komatsu, S.; Hirochika, H.; Kikuchi, S.; et al. A rice calcium-dependent protein kinase OsCPK12 oppositely modulates salt-stress tolerance and blast disease resistance. *Plant J.* **2011**, *69*, 26–36. [CrossRef] [PubMed]
155. Yu, T.-F.; Zhao, W.-Y.; Fu, J.-D.; Liu, Y.-W.; Chen, M.; Zhou, Y.-B.; Ma, Y.-Z.; Xu, Z.-S.; Xi, Y.-J. Genome-Wide Analysis of CDPK Family in Foxtail Millet and Determination of SiCDPK24 Functions in Drought Stress. *Front. Plant Sci.* **2018**, *9*, 651. [CrossRef]
156. Deng, X.; Hu, W.; Wei, S.; Zhou, S.; Zhang, F.; Han, J.; Chen, L.; Li, Y.; Feng, J.; Fang, B.; et al. TaCIPK29, a CBL-Interacting Protein Kinase Gene from Wheat, Confers Salt Stress Tolerance in Transgenic Tobacco. *PLoS ONE* **2013**, *8*, e69881. [CrossRef] [PubMed]
157. Wang, R.-K.; Li, L.-L.; Cao, Z.-H.; Zhao, Q.; Li, M.; Zhang, L.-Y.; Hao, Y.-J. Molecular cloning and functional characterization of a novel apple MdCIPK6L gene reveals its involvement in multiple abiotic stress tolerance in transgenic plants. *Plant Mol. Biol.* **2012**, *79*, 123–135. [CrossRef]
158. Hu, D.-G.; Ma, Q.-J.; Sun, C.-H.; Sun, M.-H.; You, C.-X.; Hao, Y.-J. Overexpression of MdSOS2L1, a CIPK protein kinase, increases the antioxidant metabolites to enhance salt tolerance in apple and tomato. *Physiol. Plant.* **2015**, *156*, 201–214. [CrossRef]
159. Singh, N.K.; Shukla, P.; Kirti, P.B. A CBL-interacting protein kinase AdCIPK5 confers salt and osmotic stress tolerance in transgenic tobacco. *Sci. Rep.* **2020**, *10*, 1–14. [CrossRef]
160. Li, C.-H.; Wang, G.; Zhao, J.-L.; Zhang, L.-Q.; Ai, L.-F.; Han, Y.-F.; Sun, D.-Y.; Zhang, S.-W.; Sun, Y. The receptor-like kinase SIT1 mediates salt sensitivity by activating MAPK3/6 and regulating ethylene homeostasis in rice. *Plant Cell* **2014**, *26*, 2538–2553. [CrossRef]
161. Ueno, Y.; Yoshida, R.; Kishi-Kaboshi, M.; Matsushita, A.; Jiang, C.-J.; Goto, S.; Takahashi, A.; Hirochika, H.; Takatsuji, H. Abiotic Stresses Antagonize the Rice Defence Pathway through the Tyrosine-Dephosphorylation of OsMPK6. *PLoS Pathog.* **2015**, *11*, e1005231. [CrossRef]
162. Zhu, D.; Chang, Y.; Pei, T.; Zhang, X.; Liu, L.; Li, Y.; Zhuang, J.; Yang, H.; Qin, F.; Song, C.; et al. MAPK-like protein 1 positively regulates maize seedling drought sensitivity by suppressing ABA biosynthesis. *Plant J.* **2019**, *102*, 747–760. [CrossRef] [PubMed]
163. Yan, Y.; Wang, L.; Ding, Z.; Tie, W.; Ding, X.; Zeng, C.; Wei, Y.; Zhao, H.; Peng, M.; Hu, W. Genome-Wide Identification and Expression Analysis of the Mitogen-Activated Protein Kinase Gene Family in Cassava. *Front. Plant Sci.* **2016**, *7*, 1294. [CrossRef] [PubMed]
164. Zhang, M.; Pan, J.; Kong, X.; Zhou, Y.; Liu, Y.; Sun, L.; Li, D. ZmMKK3, a novel maize group B mitogen-activated protein kinase kinase gene, mediates osmotic stress and ABA signal responses. *J. Plant Physiol.* **2012**, *169*, 1501–1510. [CrossRef]
165. You, J.; Zong, W.; Hu, H.; Li, X.; Xiao, J.; Xiong, L. A SNAC1-regulated protein phosphatase gene OsPP18 modulates drought and oxidative stress tolerance through ABA-independent reactive oxygen species scavenging in rice. *Plant Physiol.* **2014**, *166*, 2100–2114. [CrossRef] [PubMed]
166. Deveshwar, P.; Prusty, A.; Sharma, S.; Tyagi, A.K. Phytohormone-Mediated Molecular Mechanisms Involving Multiple Genes and QTL Govern Grain Number in Rice. *Front. Genet.* **2020**, *11*, 586462. [CrossRef] [PubMed]
167. Huang, X.-Y.; Chao, D.-Y.; Gao, J.-P.; Zhu, M.-Z.; Shi, M.; Lin, H.-X. A previously unknown zinc finger protein, DST, regulates drought and salt tolerance in rice via stomatal aperture control. *Genes Dev. Chang.* **2009**, *23*, 1805–1817. [CrossRef]
168. Zhang, H.; Liu, Y.; Wen, F.; Yao, D.; Wang, L.; Guo, J.; Ni, L.; Zhang, A.; Tan, M.; Jiang, M. A novel rice C2H2-type zinc finger protein, ZFP36, is a key player involved in abscisic acid-induced antioxidant defence and oxidative stress tolerance in rice. *J. Exp. Bot.* **2014**, *65*, 5795–5809. [CrossRef]
169. Jan, A.; Maruyama, K.; Todaka, D.; Kidokoro, S.; Abo, M.; Yoshimura, E.; Shinozaki, K.; Nakashima, K.; Yamaguchi-Shinozaki, K. OsTZF1, a CCCH-Tandem Zinc Finger Protein, Confers Delayed Senescence and Stress Tolerance in Rice by Regulating Stress-Related Genes. *Plant Physiol.* **2013**, *161*, 1202–1216. [CrossRef] [PubMed]

170. Shen, H.; Liu, C.; Zhang, Y.; Meng, X.; Zhou, X.; Chu, C.; Wang, X. OsWRKY30 is activated by MAP kinases to confer drought tolerance in rice. *Plant Mol. Biol.* **2012**, *80*, 241–253. [CrossRef] [PubMed]
171. Manjunath, K.; Mahadeva, A.; Sreevathsa, R.; Ramachandra Swamy, N.; Swamy, N. In Vitro Screening and Identification of Putative Sunflower (*Helianthus annus* L.) Transformants Expressing ECNAC1 Gene by Salt Stress Method. *Trends Biosci.* **2013**, *6*, 108–111.
172. Lee, S.; Seo, P.J.; Lee, H.-J.; Park, C.-M. A NAC transcription factor NTL4 promotes reactive oxygen species production during drought-induced leaf senescence in Arabidopsis. *Plant J.* **2012**, *70*, 831–844. [CrossRef] [PubMed]
173. Babitha, K.; Vemanna, R.S.; Nataraja, K.N.; Udayakumar, M. Overexpression of EcbHLH57 transcription factor from *Eleusine coracana* L. in tobacco confers tolerance to salt, oxidative and drought stress. *PLoS ONE* **2015**, *10*, e0137098. [CrossRef]
174. Zhangwu, H.; Liu, W.; Wan, L.; Li, F.; Dai, L.; Li, D.; Zhang, Z.; Huang, R. Functional analyses of ethylene response factor JERF3 with the aim of improving tolerance to drought and osmotic stress in transgenic rice. *Transgenic Res.* **2010**, *19*, 809–818. [CrossRef]
175. McKersie, B.D.; Bowley, S.R.; Harjanto, E.; Leprince, O. Water-Deficit Tolerance and Field Performance of Transgenic Alfalfa Overexpressing Superoxide Dismutase. *Plant Physiol.* **1996**, *111*, 1177–1181. [CrossRef]
176. Zhang, Q.; Cui, M.; Xin, X.; Ming, X.; Jing, L.; WU, J.-X. Overexpression of a cytosolic ascorbate peroxidase gene, OsAPX2, increases salt tolerance in transgenic alfalfa. *J. Integr. Agric.* **2014**, *13*, 2500–2507. [CrossRef]
177. Islam, T.; Manna, M.; Reddy, M.K. Glutathione Peroxidase of Pennisetum glaucum (PgGPx) Is a Functional Cd^{2+} Dependent Peroxiredoxin that Enhances Tolerance against Salinity and Drought Stress. *PLoS ONE* **2015**, *10*, e0143344. [CrossRef]
178. Oberschall, A.; Deák, M.; Török, K.; Sass, L.; Vass, I.; Kovács, I.; Fehér, A.; Dudits, D.; Horváth, G.V. A novel aldose/aldehyde reductase protects transgenic plants against lipid peroxidation under chemical and drought stresses. *Plant J.* **2000**, *24*, 437–446. [CrossRef]
179. Nepal, N.; Yactayo-Chang, J.P.; Medina-Jiménez, K.; Acosta-Gamboa, L.M.; González-Romero, M.E.; Arteaga-Vázquez, M.A.; Lorence, A. Mechanisms underlying the enhanced biomass and abiotic stress tolerance phenotype of an Arabidopsis MIOX over-expresser. *Plant Direct* **2019**, *3*, e00165. [CrossRef]
180. Yang, Q.; Liu, K.; Niu, X.; Wang, Q.; Wan, Y.; Yang, F.; Li, G.; Wang, Y.; Wang, R. Genome-wide Identification of PP2C Genes and Their Expression Profiling in Response to Drought and Cold Stresses in Medicago truncatula. *Sci. Rep.* **2018**, *8*, 1–14. [CrossRef]
181. Zhou, L.; Liu, Z.; Liu, Y.; Kong, D.; Li, T.; Yu, S.; Mei, H.; Xu, X.; Liu, H.; Chen, L.; et al. A novel gene OsAHL1 improves both drought avoidance and drought tolerance in rice. *Sci. Rep.* **2016**, *6*, 30264. [CrossRef]
182. Wen, F.; Qin, T.; Wang, Y.; Dong, W.; Zhang, A.; Tan, M.; Jiang, M. OsHK3 is a crucial regulator of abscisic acid signaling involved in antioxidant defense in rice. *J. Integr. Plant Biol.* **2014**, *57*, 213–228. [CrossRef] [PubMed]
183. Yuan, B.; Chen, M.; Li, S. Isolation and Identification of *Ipomoea cairica* (L.) Sweet Gene IcSRO1 Encoding a SIMILAR TO RCD-ONE Protein, Which Improves Salt and Drought Tolerance in Transgenic Arabidopsis. *Int. J. Mol. Sci.* **2020**, *21*, 1017. [CrossRef]
184. Ye, N.; Zhu, G.; Liu, Y.; Li, Y.; Zhang, J. ABA Controls H2O2 Accumulation Through the Induction of OsCATB in Rice Leaves Under Water Stress. *Plant Cell Physiol.* **2011**, *52*, 689–698. [CrossRef]
185. Yamauchi, T.; Yoshioka, M.; Fukazawa, A.; Mori, H.; Nishizawa, N.K.; Tsutsumi, N.; Yoshioka, H.; Nakazono, M. An NADPH Oxidase RBOH Functions in Rice Roots during Lysigenous Aerenchyma Formation under Oxygen-Deficient Conditions. *Plant Cell* **2017**, *29*, 775–790. [CrossRef] [PubMed]
186. Chen, X.; Wang, J.; Zhu, M.; Jia, H.; Liu, D.; Hao, L.; Guo, X. A cotton Raf-like MAP3K gene, GhMAP3K40, mediates reduced tolerance to biotic and abiotic stress in Nicotiana benthamiana by negatively regulating growth and development. *Plant Sci.* **2015**, *240*, 10–24. [CrossRef] [PubMed]
187. Na, Y.-J.; Choi, H.-K.; Park, M.Y.; Choi, S.-W.; Vo, K.T.X.; Jeon, J.-S.; Kim, S.Y. OsMAPKKK63 is involved in salt stress response and seed dormancy control. *Plant Signal. Behav.* **2019**, *14*, e1578633. [CrossRef]
188. Wang, G.; Liang, Y.-H.; Zhang, J.-Y.; Cheng, Z.-M. Cloning, molecular and functional characterization by overexpression in Arabidopsis of MAPKK genes from grapevine (*Vitis vinifera*). *BMC Plant Biol.* **2020**, *20*, 194. [CrossRef]
189. Jagodzik, P.; Tajdel-Zielinska, M.; Cieśla, A.; Marczak, M.; Ludwikow, A. Mitogen-Activated Protein Kinase Cascades in Plant Hormone Signaling. *Front. Plant Sci.* **2018**, *9*, 1387. [CrossRef] [PubMed]
190. Farooq, S.; Azam, F. Co-Existence of Salt and Drought Tolerance in Triticeae. *Hereditas* **2004**, *135*, 205–210. [CrossRef] [PubMed]
191. Agarwal, P.; Baranwal, V.K.; Khurana, P. Genome-wide analysis of bZIP transcription factors in wheat and functional characterization of a TabZIP under abiotic stress. *Sci. Rep.* **2019**, *9*, 1–18. [CrossRef] [PubMed]
192. Huang, X.-S.; Liu, J.-H.; Chen, X.-J. Overexpression of PtrABF gene, a bZIP transcription factor isolated from Poncirus trifoliata, enhances dehydration and drought tolerance in tobacco via scavenging ROS and modulating expression of stress-responsive genes. *BMC Plant Biol.* **2010**, *10*, 230. [CrossRef]
193. Joo, J.; Lee, Y.H.; Song, S.I. Overexpression of the rice basic leucine zipper transcription factor OsbZIP12 confers drought tolerance to rice and makes seedlings hypersensitive to ABA. *Plant Biotechnol. Rep.* **2014**, *8*, 431–441. [CrossRef]
194. Kang, J.-Y.; Choi, H.-I.; Im, M.-Y.; Kim, S.Y. Arabidopsis Basic Leucine Zipper Proteins That Mediate Stress-Responsive Abscisic Acid Signaling. *Plant Cell* **2002**, *14*, 343–357. [CrossRef]
195. Li, Q.; Zhao, H.; Wang, X.; Kang, J.; Lv, B.; Dong, Q.; Li, C.; Chen, H.; Wu, Q. Tartary Buckwheat Transcription Factor FtbZIP5, Regulated by FtSnRK2.6, Can Improve Salt/Drought Resistance in Transgenic Arabidopsis. *Int. J. Mol. Sci.* **2020**, *21*, 1123. [CrossRef]

196. Liu, C.; Mao, B.; Ou, S.; Wang, W.; Liu, L.; Wu, Y.; Chu, C.; Wang, X. OsbZIP71, a bZIP transcription factor, confers salinity and drought tolerance in rice. *Plant Mol. Biol.* **2013**, *84*, 19–36. [CrossRef]
197. Lu, G.; Gao, C.; Zheng, X.; Han, B. Identification of OsbZIP72 as a positive regulator of ABA response and drought tolerance in rice. *Planta* **2008**, *229*, 605–615. [CrossRef]
198. Ma, H.; Liu, C.; Li, Z.; Ran, Q.; Xie, G.; Wang, B.; Fang, S.; Chu, J.; Zhang, J. ZmbZIP4 Contributes to Stress Resistance in Maize by Regulating ABA Synthesis and Root Development. *Plant Physiol.* **2018**, *178*, 753–770. [CrossRef]
199. Tang, N.; Zhang, H.; Li, X.; Xiao, J.; Xiong, L. Constitutive activation of transcription factor OsbZIP46 improves drought tolerance in rice. *Plant Physiol.* **2012**, *158*, 1755–1768. [CrossRef]
200. Xiang, Y.; Tang, N.; Du, H.; Ye, H.; Xiong, L. Characterization of OsbZIP23 as a key player of the basic leucine zipper transcription factor family for conferring abscisic acid sensitivity and salinity and drought tolerance in rice. *Plant Physiol.* **2008**, *148*, 1938–1952 [CrossRef]
201. Babitha, K.; Ramu, S.; Pruthvi, V.; Mahesh, P.; Nataraja, K.N.; Udayakumar, M. Co-expression of at bHLH17 and at WRKY 28 confers resistance to abiotic stress in Arabidopsis. *PLoS ONE* **2013**, *22*, 327–341.
202. Dombrecht, B.; Xue, G.P.; Sprague, S.J.; Kirkegaard, J.A.; Ross, J.J.; Reid, J.B.; Fitt, G.P.; Sewelam, N.; Schenk, P.M.; Manners, J.M.; et al. MYC2 Differentially Modulates Diverse Jasmonate-Dependent Functions inArabidopsis. *Plant Cell* **2007**, *19*, 2225–2245. [CrossRef] [PubMed]
203. Dong, Y.; Wang, C.; Han, X.; Tang, S.; Liu, S.; Xia, X.; Yin, W. A novel bHLH transcription factor PebHLH35 from Populus euphratica confers drought tolerance through regulating stomatal development, photosynthesis and growth in Arabidopsis. *Biochem. Biophys. Res. Commun.* **2014**, *450*, 453–458. [CrossRef] [PubMed]
204. Krishnamurthy, P.; Vishal, B.; Khoo, K.; Rajappa, S.; Loh, C.-S.; Kumar, P.P. Expression of AoNHX1 increases salt tolerance of rice and Arabidopsis, and bHLH transcription factors regulate AtNHX1 and AtNHX6 in Arabidopsis. *Plant Cell Rep.* **2019**, *38*, 1299–1315. [CrossRef] [PubMed]
205. Le Hir, R.; Castelain, M.; Chakraborti, D.; Moritz, T.; Dinant, S.; Bellini, C. At bHLH68 transcription factor contributes to the regulation of ABA homeostasis and drought stress tolerance in Arabidopsis thaliana. *Physiol. Plant.* **2017**, *160*, 312–327. [CrossRef]
206. Seo, J.-S.; Joo, J.; Kim, M.-J.; Kim, Y.-K.; Nahm, B.H.; Song, S.I.; Cheong, J.-J.; Lee, J.S.; Kim, J.-K.; Choi, Y.D. OsbHLH148, a basic helix-loop-helix protein, interacts with OsJAZ proteins in a jasmonate signaling pathway leading to drought tolerance in rice. *Plant J.* **2011**, *65*, 907–921. [CrossRef]
207. Yao, P.; Sun, Z.; Li, C.; Zhao, X.; Li, M.; Deng, R.; Huang, Y.; Zhao, H.; Chen, H.; Wu, Q. Overexpression of Fagopyrum ta-taricum FtbHLH2 enhances tolerance to cold stress in transgenic Arabidopsis. *Front. Plant Sci.* **2018**, *125*, 85–94.
208. Yao, P.-F.; Li, C.-L.; Zhao, X.-R.; Li, M.-F.; Zhao, H.-X.; Guo, J.-Y.; Cai, Y.; Chen, H.; Wu, Q. Overexpression of a Tartary Buckwheat Gene, FtbHLH3, Enhances Drought/Oxidative Stress Tolerance in Transgenic Arabidopsis. *Front. Plant Sci.* **2017**, *8*, 625. [CrossRef] [PubMed]
209. Shoucheng, C.; Chai, S.; McIntyre, C.; Xue, G.-P. Overexpression of a predominantly root-expressed NAC transcription factor in wheat roots enhances root length, biomass and drought tolerance. *Plant Cell Rep.* **2017**, *37*, 225–237. [CrossRef]
210. Jeong, J.S.; Kim, Y.S.; Baek, K.H.; Jung, H.; Ha, S.-H.; Choi, Y.D.; Kim, M.; Reuzeau, C.; Kim, J.-K. Root-Specific Expression of OsNAC10 Improves Drought Tolerance and Grain Yield in Rice under Field Drought Conditions. *Plant Physiol.* **2010**, *153*, 185–197. [CrossRef]
211. Jeong, J.S.; Kim, Y.S.; Redillas, M.C.F.R.; Jang, G.; Jung, H.; Bang, S.W.; Choi, Y.D.; Ha, S.-H.; Reuzeau, C.; Kim, J.-K. OsNAC5overexpression enlarges root diameter in rice plants leading to enhanced drought tolerance and increased grain yield in the field. *Plant Biotechnol. J.* **2012**, *11*, 101–114. [CrossRef]
212. Lee, D.K.; Chung, P.J.; Jeong, J.S.; Jang, G.; Bang, S.W.; Jung, H.; Kim, Y.S.; Ha, S.H.; Choi, Y.D.; Kim, J.K. The rice Os NAC 6 transcription factor orchestrates multiple molecular mechanisms involving root structural adaptions and nicotianamine biosynthesis for drought tolerance. *Plant Biotechnol. J.* **2017**, *15*, 754–764. [CrossRef]
213. Ohnishi, T.; Sugahara, S.; Yamada, T.; Kikuchi, K.; Yoshiba, Y.; Hirano, H.-Y.; Tsutsumi, N. OsNAC6, a member of the NAC gene family, is induced by various stresses in rice. *Genes Genet. Syst.* **2005**, *80*, 135–139. [CrossRef]
214. Redillas, M.C.; Jeong, J.S.; Kim, Y.S.; Jung, H.; Bang, S.W.; Choi, Y.D.; Ha, S.H.; Reuzeau, C.; Kim, J.K. The overexpression of OsNAC9 alters the root architecture of rice plants enhancing drought resistance and grain yield under field conditions. *Plant Biotechnol. J.* **2012**, *10*, 792–805. [CrossRef] [PubMed]
215. Shahnejat-Bushehri, S.; Mueller-Roeber, B.; Balazadeh, S. Arabidopsis NAC transcription factor JUNGBRUNNEN1 affects thermomemory-associated genes and enhances heat stress tolerance in primed and unprimed conditions. *Plant Signal. Behav.* **2012**, *7*, 1518–1521. [CrossRef] [PubMed]
216. Tran, L.-S.; Nakashima, K.; Sakuma, Y.; Simpson, S.D.; Fujita, Y.; Maruyama, K.; Fujita, M.; Seki, M.; Shinozaki, K.; Yamaguchi-Shinozaki, K. Isolation and Functional Analysis of Arabidopsis Stress-Inducible NAC Transcription Factors That Bind to a Drought-Responsive cis-Element in the early responsive to dehydration stress 1 Promoter. *Plant Cell* **2004**, *16*, 2481–2498. [CrossRef]
217. Xu, Z.; Gong, Z.; Wang, C.; Xue, F.; Zhang, H.; Ji, W. Wheat NAC transcription factor TaNAC29 is involved in response to salt stress. *Plant Physiol. Biochem.* **2015**, *96*, 356–363. [CrossRef]
218. Yang, X.; Kim, M.Y.; Ha, J.; Lee, S.-H. Overexpression of the Soybean NAC Gene GmNAC109 Increases Lateral Root Formation and Abiotic Stress Tolerance in Transgenic Arabidopsis Plants. *Front. Plant Sci.* **2019**, *10*, 1036. [CrossRef]

219. Zhang, H.; Ma, F.; Wang, X.; Liu, S.; Saeed, U.H.; Hou, X.; Zhang, Y.; Luo, D.; Meng, Y.; Zhang, W.; et al. Molecular and Functional Characterization of CaNAC035, an NAC Transcription Factor From Pepper (*Capsicum annuum* L.). *Front. Plant Sci.* **2020**, *11*, 14. [CrossRef]
220. Dubouzet, J.G.; Sakuma, Y.; Ito, Y.; Kasuga, M.; Dubouzet, E.G.; Miura, S.; Seki, M.; Shinozaki, K.; Yamaguchi-Shinozaki, K. OsDREBgenes in rice, *Oryza sativa* L., encode transcription activators that function in drought-, high-salt- and cold-responsive gene expression. *Plant J.* **2003**, *33*, 751–763. [CrossRef] [PubMed]
221. Hsieh, E.-J.; Cheng, M.-C.; Lin, T.-P. Functional characterization of an abiotic stress-inducible transcription factor AtERF53 in Arabidopsis thaliana. *Plant Mol. Biol.* **2013**, *82*, 223–237. [CrossRef]
222. Jisha, V.; Dampanaboina, L.; Vadassery, J.; Mithöfer, A.; Kappara, S.; Ramanan, R. Overexpression of an AP2/ERF Type Transcription Factor OsEREBP1 Confers Biotic and Abiotic Stress Tolerance in Rice. *PLoS ONE* **2015**, *10*, e0127831. [CrossRef] [PubMed]
223. Li, J.; Guo, X.; Zhang, M.; Wang, X.; Zhao, Y.; Yin, Z.; Zhang, Z.; Wang, Y.; Xiong, H.; Zhang, H.; et al. OsERF71 confers drought tolerance via modulating ABA signaling and proline biosynthesis. *Plant Sci.* **2018**, *270*, 131–139. [CrossRef] [PubMed]
224. Pan, Y.; Seymour, G.B.; Lu, C.; Hu, Z.; Chen, X.; Chen, G. An ethylene response factor (ERF5) promoting adaptation to drought and salt tolerance in tomato. *Plant Cell Rep.* **2011**, *31*, 349–360. [CrossRef] [PubMed]
225. Park, S.; Lee, C.; Doherty, C.J.; Gilmour, S.J.; Kim, Y.; Thomashow, M.F. Regulation of the Arabidopsis CBF regulon by a complex low-temperature regulatory network. *Plant J.* **2015**, *82*, 193–207. [CrossRef] [PubMed]
226. Qin, F.; Kakimoto, M.; Sakuma, Y.; Maruyama, K.; Osakabe, Y.; Tran, L.S.P.; Shinozaki, K.; Yamaguchi-Shinozaki, K. Regu-lation and functional analysis of ZmDREB2A in response to drought and heat stresses in *Zea mays* L. *Plant J.* **2007**, *50*, 54–69. [CrossRef]
227. Yao, Y.; He, R.J.; Xie, Q.L.; Zhao, X.H.; Deng, X.M.; He, J.B.; Song, L.; He, J.; Marchant, A.; Chen, X.Y. Ethylene Response Factor 74 (ERF74) plays an essential role in controlling a respiratory burst oxidase homolog D (RbohD)-dependent mechanism in response to different stresses in Arabidopsis. *New Phytol.* **2017**, *213*, 1667–1681. [CrossRef]
228. Zhang, B.; Su, L.; Hu, B.; Li, L.J. Expression of AhDREB1, an AP2/ERF transcription factor gene from peanut, is af-fected by histone acetylation and increases abscisic acid sensitivity and tolerance to osmotic stress in Arabidopsis. *Int. J. Mol. Sci.* **2018**, *19*, 1441. [CrossRef]
229. Zhang, G.; Chen, M.; Li, L.; Xu, Z.; Chen, X.; Guo, J.; Ma, Y. Overexpression of the soybean GmERF3 gene, an AP2/ERF type transcription factor for increased tolerances to salt, drought, and diseases in transgenic tobacco. *J. Exp. Bot.* **2009**, *60*, 3781–3796. [CrossRef]
230. Zhao, C.; Zhang, Z.; Xie, S.; Si, T.; Li, Y.; Zhu, J.-K. Mutational Evidence for the Critical Role of CBF Transcription Factors in Cold Acclimation in Arabidopsis. *Plant Physiol.* **2016**, *171*, 2744–2759. [CrossRef]
231. Cui, M.H.; Yoo, K.S.; Hyoung, S.; Nguyen, H.T.K.; Kim, Y.Y.; Kim, H.J.; Ok, S.H.; Yoo, S.D.; Shin, J.S. An Arabidopsis R2R3-MYB transcription factor, AtMYB20, negatively regulates type 2C serine/threonine protein phosphatases to enhance salt tolerance. *FEBS Lett.* **2013**, *587*, 1773–1778. [CrossRef]
232. Jung, C.; Seo, J.S.; Han, S.W.; Koo, Y.J.; Kim, C.H.; Song, S.I.; Nahm, B.H.; Choi, Y.D.; Cheong, J.-J. Overexpression of AtMYB44 Enhances Stomatal Closure to Confer Abiotic Stress Tolerance in Transgenic Arabidopsis. *Plant Physiol.* **2007**, *146*, 323–324. [CrossRef]
233. Lee, S.B.; Kim, H.; Kim, R.J.; Suh, M.C. Overexpression of Arabidopsis MYB96 confers drought resistance in Camelina sativa via cuticular wax accumulation. *Plant Cell Rep.* **2014**, *33*, 1535–1546. [CrossRef] [PubMed]
234. Liao, Y.; Zou, H.-F.; Wang, H.-W.; Zhang, W.-K.; Ma, B.; Zhang, J.-S.; Chen, S.-Y. Soybean GmMYB76, GmMYB92, and GmMYB177 genes confer stress tolerance in transgenic Arabidopsis plants. *Cell Res.* **2008**, *18*, 1047–1060. [CrossRef] [PubMed]
235. Song, S.; Qi, T.; Huang, H.; Ren, Q.; Wu, D.; Chang, C.; Peng, W.; Liu, Y.; Peng, J.; Xie, D. The Jasmonate-ZIM domain proteins interact with the R2R3-MYB transcription factors MYB21 and MYB24 to affect Jasmonate-regulated stamen development in Arabidopsis. *Plant Cell* **2011**, *23*, 1000–1013. [CrossRef] [PubMed]
236. Tang, Y.; Bao, X.; Zhi, Y.; Wu, Q.; Guo, Y.; Yin, X.; Zeng, L.; Li, J.; Zhang, J.; He, W.; et al. Overexpression of a MYB Family Gene, OsMYB6, Increases Drought and Salinity Stress Tolerance in Transgenic Rice. *Front. Plant Sci.* **2019**, *10*, 168. [CrossRef]
237. Vannini, C.; Locatelli, F.; Bracale, M.; Magnani, E.; Marsoni, M.; Osnato, M.; Mattana, M.; Baldoni, E.; Coraggio, I. Overex-pression of the rice Osmyb4 gene increases chilling and freezing tolerance of Arabidopsis thaliana plants. *Plant J.* **2004**, *37*, 115–127. [CrossRef]
238. Xiong, H.; Li, J.; Liu, P.; Duan, J.; Zhao, Y.; Guo, X.; Li, Y.; Zhang, H.; Ali, J.; Li, Z. Overexpression of OsMYB48-1, a Novel MYB-Related Transcription Factor, Enhances Drought and Salinity Tolerance in Rice. *PLoS ONE* **2014**, *9*, e92913. [CrossRef]
239. Zhu, N.; Cheng, S.; Liu, X.; Du, H.; Dai, M.; Zhou, D.-X.; Yang, W.; Zhao, Y. The R2R3-type MYB gene OsMYB91 has a function in coordinating plant growth and salt stress tolerance in rice. *Plant Sci.* **2015**, *236*, 146–156. [CrossRef] [PubMed]
240. He, G.-H.; Xu, J.-Y.; Wang, Y.-X.; Liu, J.-M.; Li, P.-S.; Chen, M.; Ma, Y.-Z.; Xu, Z.-S. Drought-responsive WRKY transcription factor genes TaWRKY1 and TaWRKY33 from wheat confer drought and/or heat resistance in Arabidopsis. *BMC Plant Biol.* **2016**, *16*, 1–16. [CrossRef]
241. Li, H.; Gao, Y.; Xu, H.; Dai, Y.; Deng, D.; Chen, J. ZmWRKY33, a WRKY maize transcription factor conferring enhanced salt stress tolerances in Arabidopsis. *Plant Growth Regul.* **2013**, *70*, 207–216. [CrossRef]
242. Mzid, R.; Zorrig, W.; ben Ayed, R.; Ben Hamed, K.; Ayadi, M.; Damak, Y.; Lauvergeat, V.; Hanana, M. The grapevine VvWRKY2 gene enhances salt and osmotic stress tolerance in transgenic Nicotiana tabacum. *3 Biotech* **2018**, *8*, 277. [CrossRef]

243. Qiu, Y.; Yu, D. Over-expression of the stress-induced OsWRKY45 enhances disease resistance and drought tolerance in Arabidopsis. *Environ. Exp. Exp. Bot.* **2009**, *65*, 35–47. [CrossRef]

244. Villano, C.; Esposito, S.; D'Amelia, V.; Garramone, R.; Alioto, D.; Zoina, A.; Aversano, R.; Carputo, D. WRKY genes family study reveals tissue-specific and stress-responsive TFs in wild potato species. *Sci. Rep.* **2020**, *10*, 1–12. [CrossRef] [PubMed]

245. Wu, X.; Shiroto, Y.; Kishitani, S.; Ito, Y.; Toriyama, K. Enhanced heat and drought tolerance in transgenic rice seedlings overexpressing OsWRKY11 under the control of HSP101 promoter. *Plant Cell Rep.* **2009**, *28*, 21–30. [CrossRef]

246. Mittler, R.; Vanderauwera, S.; Gollery, M.; Van Breusegem, F. Reactive oxygen gene network of plants. *Trends Plant Sci.* **2004**, *9*, 490–498. [CrossRef]

247. Lin, K.-H.; Sei, S.-C.; Su, Y.-H.; Chiang, C.-M. Overexpression of the Arabidopsis and winter squash superoxide dismutase genes enhances chilling tolerance via ABA-sensitive transcriptional regulation in transgenic Arabidopsis. *Plant Signal. Behav.* **2019**, *14*, 1685728. [CrossRef]

248. Zhou, C.; Zhu, C.; Fu, H.; Li, X.; Chen, L.; Lin, Y.; Lai, Z.; Guo, Y. Genome-wide investigation of superoxide dismutase (SOD) gene family and their regulatory miRNAs reveal the involvement in abiotic stress and hormone response in tea plant (Camellia sinensis). *PLoS ONE* **2019**, *14*, e0223609. [CrossRef]

249. Caverzan, A.; Passaia, G.; Rosa, S.B.; Ribeiro, C.W.; Lazzarotto, F.; Margis-Pinheiro, M. Plant responses to stresses: Role of ascorbate peroxidase in the antioxidant protection. *Genet. Mol. Biol.* **2012**, *35*, 1011–1019. [CrossRef]

250. Pandey, S.; Fartyal, D.; Agarwal, A.; Shukla, T.; James, D.; Kaul, T.; Negi, Y.K.; Arora, S.; Reddy, M.K. Abiotic stress toler-ance in plants: Myriad roles of ascorbate peroxidase. *Front. Plant Sci.* **2017**, *8*, 581. [CrossRef]

251. Liu, J.-X.; Feng, K.; Duan, A.-Q.; Li, H.; Yang, Q.-Q.; Xu, Z.-S.; Xiong, A.-S. Isolation, purification and characterization of an ascorbate peroxidase from celery and overexpression of the AgAPX1 gene enhanced ascorbate content and drought tolerance in Arabidopsis. *BMC Plant Biol.* **2019**, *19*, 1–13. [CrossRef]

252. Liu, Z.; Bao, H.; Cai, J.; Han, J.; Zhou, L. A Novel Thylakoid Ascorbate Peroxidase from Jatrophacurcas Enhances Salt Tolerance in Transgenic Tobacco. *Int. J. Mol. Sci.* **2013**, *15*, 171–185. [CrossRef] [PubMed]

253. Chen, Y.; Cai, J.; Yang, F.; Zhou, B.; Zhou, L. Ascorbate peroxidase from Jatropha curcas enhances salt tolerance in transgenic Arabidopsis. *Genet. Mol. Res.* **2015**, *14*, 4879–4889. [CrossRef] [PubMed]

254. Filiz, E.; Ozyigit, I.I.; Saracoglu, I.A.; Uras, M.E.; Sen, U.; Yalcin, B. Abiotic stress-induced regulation of antioxidant genes in different Arabidopsis ecotypes: Microarray data evaluation. *Biotechnol. Biotechnol. Equip.* **2019**, *33*, 128–143. [CrossRef]

255. Nie, Q.; Gao, G.-L.; Fan, Q.-J.; Qiao, G.; Wen, X.-P.; Liu, T.; Peng, Z.-J.; Cai, Y.-Q. Isolation and characterization of a catalase gene "HuCAT3" from pitaya (Hylocereus undatus) and its expression under abiotic stress. *Gene* **2015**, *563*, 63–71. [CrossRef] [PubMed]

256. Zhou, Y.; Liu, S.; Yang, Z.; Yang, Y.; Jiang, L.; Hu, L. CsCAT3, a catalase gene from Cucumis sativus, confers resistance to a variety of stresses to Escherichia coli. *Biotechnol. Biotechnol. Equip.* **2017**, *31*, 886–896. [CrossRef]

257. Milla, M.A.R.; Maurer, A.; Huete, A.R.; Gustafson, J.P. Glutathione peroxidase genes in Arabidopsis are ubiquitous and reg-ulated by abiotic stresses through diverse signaling pathways. *Plant J.* **2003**, *36*, 602–615. [CrossRef]

258. Zhou, Y.; Li, J.; Wang, J.; Yang, W.; Yang, Y. Identification and Characterization of the Glutathione Peroxidase (GPX) Gene Family in Watermelon and Its Expression under Various Abiotic Stresses. *Agronomy* **2018**, *8*, 206. [CrossRef]

259. Diao, Y.; Xu, H.; Li, G.; Yu, A.; Yu, X.; Hu, W.; Zheng, X.; Li, S.; Wang, Y.; Hu, Z. Cloning a glutathione peroxidase gene from Nelumbo nucifera and enhanced salt tolerance by overexpressing in rice. *Mol. Biol. Rep.* **2014**, *41*, 4919–4927. [CrossRef] [PubMed]

260. Wang, X.; Fang, G.; Yang, J.; Li, Y. A Thioredoxin-Dependent Glutathione Peroxidase (OsGPX5) Is Required for Rice Normal Development and Salt Stress Tolerance. *Plant Mol. Biol. Rep.* **2017**, *35*, 333–342. [CrossRef]

261. Sultana, S.; Khew, C.Y.; Morshed, M.; Namasivayam, P.; Napis, S.; Ho, C.-L. Overexpression of monodehydroascorbate reductase from a mangrove plant (AeMDHAR) confers salt tolerance on rice. *J. Plant Physiol.* **2011**, *169*, 311–318. [CrossRef] [PubMed]

262. Eltayeb, A.E.; Kawano, N.; Badawi, G.H.; Kaminaka, H.; Sanekata, T.; Shibahara, T.; Inanaga, S.; Tanaka, K. Overexpression of monodehydroascorbate reductase in transgenic tobacco confers enhanced tolerance to ozone, salt and polyethylene glycol stresses. *Planta* **2006**, *225*, 1255–1264. [CrossRef] [PubMed]

263. Kabir, M.H.; Wang, M.-H. Expression Pattern of Two Dehydroascorbate Reductase Genes from Tomato (*Solanum lycopersicum* L.) in Response to Stress. *J. Korean Soc. Appl. Biol. Chem.* **2010**, *53*, 668–676. [CrossRef]

264. Hao, Z.; Wang, X.; Zong, Y.; Wen, S.; Cheng, Y.; Li, H. Enzymatic activity and functional analysis under multiple abiotic stress conditions of a dehydroascorbate reductase gene derived from Liriodendron Chinense. *Environ. Exp. Bot.* **2019**, *167*, 103850. [CrossRef]

265. Zhang, Y.; Li, Z.; Peng, Y.; Wang, X.; Peng, D.; Li, Y.; McPhee, D.J.; Zhang, X.; Ma, X.; Huang, L.; et al. Clones of FeSOD, MDHAR, DHAR Genes from White Clover and Gene Expression Analysis of ROS-Scavenging Enzymes during Abiotic Stress and Hormone Treatments. *Molecules* **2015**, *20*, 20939–20954. [CrossRef]

 *plants*

Article

Sequential Application of Antioxidants Rectifies Ion Imbalance and Strengthens Antioxidant Systems in Salt-Stressed Cucumber

Mahmoud F. Seleiman [1,2,*], Wael M. Semida [3], Mostafa M. Rady [4], Gamal F. Mohamed [4], Khaulood A. Hemida [5], Bushra Ahmed Alhammad [6], Mohamed M. Hassan [7,8] and Ashwag Shami [9]

[1] Plant Production Department, College of Food and Agriculture Sciences, King Saud University, P.O. Box 2460, Riyadh 11451, Saudi Arabia

[2] Department of Crop Sciences, Faculty of Agriculture, Menoufia University, Shibin El-Kom 32514, Egypt

[3] Horticulture Department, Faculty of Agriculture, Fayoum University, Fayoum 63514, Egypt; wms00@fayoum.edu.eg

[4] Botany Department, Faculty of Agriculture, Fayoum University, Fayoum 63514, Egypt; mmr02@fayoum.edu.eg (M.M.R.); gfm00@fayoum.edu.eg (G.F.M.)

[5] Botany Department, Faculty of Science, Fayoum University, Fayoum 63514, Egypt; kah00@fayoum.edu.eg

[6] Biology Department, College of Science and Humanity Studies, Prince Sattam Bin Abdulaziz University, Al Kharj Box 292, Riyadh 11942, Saudi Arabia; b.alhamad@psau.edu.sa

[7] Department of Biology, College of Science, Taif University, P.O. Box 11099, Taif 21944, Saudi Arabia; m.khyate@tu.edu.sa

[8] Department of Genetics, Faculty of Agriculture, Menoufia University, Sheben El-Kom, Menoufia 32514, Egypt

[9] Biology Department, College of Sciences, Princess Nourah Bint Abdulrahman University, Riyadh 11617, Saudi Arabia; AYShami@pnu.edu.sa

* Correspondence: mseleiman@ksu.edu.sa or mahmoud.seleiman@agr.menofia.edu.eg; Tel.: +96-655-315-3351

Received: 28 November 2020; Accepted: 14 December 2020; Published: 16 December 2020

Abstract: Exogenous antioxidant applications enable salt-stressed plants to successfully cope with different environmental stresses. The objectives of this investigation were to study the effects of sequential treatments of proline (Pro), ascorbic acid (AsA), and/or glutathione (GSH) on 100 mM NaCl-stressed cucumber transplant's physio-biochemical and growth traits as well as systems of antioxidant defense. Under salinity stress, different treatment of AsA, Pro, or/and GSH improved growth characteristics, stomatal conductance (gs), enhanced the activities of glutathione reductase (GR), superoxide dismutase (SOD), ascorbate peroxidase (APX), and catalase (CAT) as well as increased contents of AsA, Pro, and GSH. However, sequential application of antioxidants (GSH-Pro-AsA) significantly exceeded all individual applications, reducing leaf and root Cd^{2+} and Na^+ contents in comparison to the control. In plants grown under NaCl-salt stress, growth characteristics, photosynthetic efficiency, membrane stability index (MSI), relative water content (RWC), contents of root and leaf K^+ and Ca^{2+}, and ratios of K^+/Na^+ and Ca^{2+}/Na^+ were notably reduced, while leaf contents of non-enzymatic and enzymatic antioxidants, as well as root and leaf Cd^{2+} and Na^+ concentrations were remarkably increased. However, AsA, Pro, or/and GSH treatments significantly improved all investigated growth characteristics, photosynthetic efficiency, RWC and MSI, as well as AsA, Pro, and GSH, and enzymatic activity, leaf and root K^+ and Ca^{2+} contents and their ratios to Na^+, while significantly reduced leaf and root Cd^{2+} and Na^+ contents.

Keywords: sequential application of antioxidants; salinity; *Cucumis sativus*; photosynthetic efficiency; antioxidant defense systems

1. Introduction

Cucumber (*Cucumis sativus* L.) is one of the most economically important and widely distributed vegetable crops, as well as being useful for human health and nutrition because it has several important mineral nutrients and vitamins and has reasonable amounts of proteins and carbohydrates [1]. In addition, it has many natural antioxidants and vitamins used as a nutritive source and for medical purposes such as headaches and anti-acne lotions due to its analgesic activities [2]. Also, cucumber can have antioxidants phenolic, especially flavonoids can be involved in reactive oxygen species (ROS) detoxification.

Cucumber has been found to be moderately sensitive to salinity [3]. In particular, salt stress causes inhibition of cucumber plant growth and productivity through affecting the key physiological and biochemical processes [4–7]. Due to the climatic changes occurring presently, salinization of plant-growing media is gradually aggravated, which will lead to further inhibition and loss in plant growth and yield, respectively. Therefore, considerable attention has been paid to boost salinity tolerance in cucumber plants by using many exogenous supports [6,8,9].

Generally, salinity negatively affects growth and restricts the productivity and quality of crops, especially in arid and semi-arid agricultural regions [10–15]. Plant growth and productivity are declined at varying degrees subjecting to the salt stress levels [1,16,17]. Salt stress causes plant performance reduction through the dysfunction occurred in the biochemical and physiological and processes, and the antioxidant defenses as a result of ROS excessive production such as $^1O_2$, $O_2^{\bullet-}$, H_2O_2, and OH^-, as well as the instabilities of cell membranes and lipid peroxidation occurred due to stress of increased Na+ ions along with increased ROS [18–22].

Plants have different antioxidative mechanisms for ROS scavenging through inspiring the antioxidative enzymes for instance CAT, GR, SOD, and APX, which are functioned along with the non-enzymatic antioxidants to mitigate the adverse impacts of different stresses in plant species [23–25]. The main non-enzymatic plant antioxidants are ascorbate, glutathione, carotenoids, tocopherols, and phenolic compounds. Non-enzymatic antioxidants such as AsA, GSH, and Pro can represent a fundamental portion of the endogenous defense mechanisms and adaptive strategies to fight the stress. They are often insufficient to cope with stress; therefore, plants are exogenously provided by these antioxidants singly or in combinations to promote and enhance their antioxidant defense systems in contrast to different environmental stresses, including salinity [26–28]. As reported in these works, the exogenous treatments (i.e., AsA, Pro, and GSH) applied singly or in sequences had boosted plant performance (i.e., growth and yield) through a positive stimulation of the key physio-biochemical and molecular attributes and increasing the capacity of antioxidant defenses. In addition, the above works have reported ROS detoxification, cell membrane stability, ion balance, and Na^+ ions decline.

To our knowledge, only little previous works [24,28,29] showed that the sequential application of Pro, AsA, and GSH is proved to be more effective strategy for plants to efficiently cope with various abiotic stresses (i.e., heavy metals, drought, and salinity) than their individual or combined applications. Therefore, the aims of the current investigation were to assess the effects of exogenously applied AsA, Pro and GSH in a sequence method in comparison with individual treatments on the growth, efficiency of photosynthesis, tissue water content, and membrane health in terms of MSI and the contents of Cd^{2+} and Na^+ as well as non-enzymatic and enzymatic defenses of cucumber transplants.

2. Materials and Methods

2.1. Experimental Setup

Seeds of cucumber (hybrid Bahi®) were separated to five groups, each consisted of 80 seeds. Then, seeds were soaked in distilled water (group 1), 0.5 mM AsA solution (group 2), 0.5 mM Pro solution (group 3), or 0.5 mM GSH solution (group 4). The duration of soaking was 4 h. Furthermore, seeds of group number 5 were soaked in sequential application of different antioxidants as follows: 0.5 mM AsA for 90 min, 0.5 mM Pro for 80 min, and then 0.5 mM GSH for 70 min in a sequence, respectively. Concentrations of antioxidants applied singly or in AsA-Pro-GSH and soaking duration were selected according to our data (not shown) of a preliminary study. In addition, AsA-Pro-GSH was generated greatest response among different tested sequences of our preliminary study (data not shown).

In each group, the seeds were divided into two sub-groups (i.e., n = 40 seeds) to represent 10 treatments. After germinating the seeds of all groups, irrigation was applied using a distilled water for all first sub-groups in the same time, while all second sub-groups were irrigated with 100 mM NaCl solution. Foam trays (209 cells) were used for the current work conducted on 28 May 2017 for 28 d. Trays were arranged in an open greenhouse under which the conditions were 62–68% humidity, $29 \pm 3/19 \pm 1$ °C day/night temperatures, and average 13/11 h day/night length. The light intensity was the intensity of natural sunlight throughout the season (28th of May–24th of June). Peat moss and vermiculite mixture (1:2 *v/v*) was a germinating and growing medium of transplants. The completely randomized designs was used for the experiment with three foam trays/replicates for each treatment. Twenty-eight days after sowing, transplants of cucumber were collected for different measurements of morphology and physiological traits and biochemical attributes.

2.2. Measurements

After 28 days, seedlings (n = 20) were randomly selected from each treatment to record growth attributes i.e., leaf area, shoot length, stem diameter, and shoot FW (fresh weight) and DW (dry weight). Leaf area was determined as detailed by [30].

Stomatal conductance (gs) $mmol^{-2}$ S^{-1} was measured by leaf porometer (Decagon Devices Inc., Pullman, WA, USA). The measurements were conducted 10, 14, 18, and 22 days after onset of NaCl applications from 7:00 am: 5:00 pm with 2 h intervals. SPAD chlorophyll meter (SPAD-502; Minolta, Osaka, Japan) was used to measure relative contents of chlorophyll. Chlorophyll fluorescence (performance index, PI; = maximum quantum yield of PSII = efficiency of the photosystem 2, Fv/Fm; Fv/F0), as a convenient tool to assess photosynthetic efficiency, was determined according to [31] using Handy PEA (Hansatech Instruments Ltd., Kings Lynn, UK). Leaves relative water content (RWC) and cell membrane stability index (MSI) were measured according to [32].

The method detailed by [33] was followed to determine leaf content of free proline after extraction of 500 mg fresh leaf using 3%, *v/v*, sulphosalicylic acid. Following the [34] method, leaf AsA was extracted and its content was determined. Leaf GSH content was determined as detailed in the [35] method. Protein content and activity of SOD (EC 1.15.1.1) was analyzed as described by Bradford [36] and Kono [37] respectively. CAT (EC 1.11.1.6) activity was analyzed as described by Aebi [38]. Potassium phosphate (KH_2PO_4; a buffer) and hydrogen peroxide (H_2O_2; a substrate) were used, and the absorbance was read using spectrophotometer at 240 nm. The activity of APX (EC 1.11.1.11) was analyzed using the method of [39] and the absorbance was read using spectrophotometer at 290 nm. GR (EC 1.6.4.1) activity was analyzed and the NADPH oxidation was monitored for three absorbance readings recorded at 340 nm [39].

A weight of 100 mg dried powdered leaf or root samples was digested for 12 h by using 80% HClO4 (2 mL) + concentrated H2SO4 (10 mL). The digested samples were then diluted each to 100 mL. The digested leaf samples were used to analyze the concentrations of K^+ and Ca^{2+} via flame photometry [40]. In addition, the digested leaf and root samples were used to analyze Cd^{2+} contents via Atomic Absorption Spectrophotometer (a Perkin-Elmer, Model 3300) as detailed in [41] method.

2.3. Statistical Analysis

The data obtained from the effects of exogenously applied AsA, Pro and GSH in a sequence method in comparison with individual treatments on growth, physiological and biochemical traits as well as Cd^{2+} and Na^+ contents of cucumber transplants gown under 100 mM NaCl-salt stress were statistically analyzed using the GLM procedure of Gen STAT (version 11; VSN International Ltd., Oxford, UK). Least significant differences (LSD) was calculated to compare the differences among means at 5% probability ($p \leq 0.05$).

3. Results

Concerning the antioxidant applications under normal conditions, the three antioxidants; AsA, Pro, and GSH that applied individually for cucumber seed improved all of the investigated growth characteristics (Table 1), gs (Table 2) and endogenous contents of AsA, Pro, and GSH (Table 3), as well as the enzymatic activities of SOD, CAT, GR, and APX (Table 4). On the other side, all of these individual antioxidant applications did not affect Fv/Fm, and PI and chlorophyll content (Table 2), RWC and MSI (Table 3), leaf and root contents of the poisonous elements Cd^{2+} and Na^+, and the beneficial elements K^+ and Ca^{2+} (Figures 1 and 2). However, AsA-Pro-GSH treatment leads to better results for most of investigated parameters than those obtained with the individual applications, along with significant reductions of Cd^{2+} contents of leaves and roots upon comparing to the controls (Figure 1).

Regarding the stress impacts of salinity, treatment with 100 mM NaCl considerably suppressed cucumber transplant growth parameters compared to those grown under control. For instance, it reduced the length of plant shoot by 24.4%, leaf area by 26.3%, stem diameter by 16.7%, shoot FW by 17.5%, and shoot DW by 23.1% (Table 1), photosynthetic efficiency (i.e., gs by 26.4%, SPAD chlorophyll by 19.7%, Fv/Fm by 8.4%, and PI by 16.8%; Table 2), leaf RWC and MSI (by 15.0 and 16.6%, respectively; Table 3), respectively compared to the controls. However, treatment with 100 mM NaCl significantly increased leaf contents of antioxidants, which are non-enzymatic and have low-molecular-weights such as Pro by 22.2%, AsA by 95.3%, and GSH by 28.6% (Table 3), leaf enzymatic activities such as SOD by 8.7%, CAT by 26.1%, GR by 18.5%, and APX by 20.0%, respectively in comparison to the plants grown under control treatment (Table 4).

Table 1. Effects of seed single or sequential application of antioxidants on growth traits of 100 mM NaCl-salt-stressed cucumber transplants.

Treatments	Shoot Length (cm)	Leaf Area (cm²)	Stem Diameter (mm)	Shoot FW (g)	Shoot DW (g)
Control	15.6 ± 0.4 cd	64.0 ± 2.2 cd	3.6 ± 0.1 b	1.71 ± 0.07 b	0.13 ± 0.02 d
AsA	17.0 ± 0.6 abc	70.6 ± 5.8 ab	3.8 ± 0.1 ab	2.00 ± 0.16 a	0.15 ± 0.03 bc
Pro	16.6 ± 0.7 bcd	69.3 ± 3.6 bc	3.7 ± 0.1 b	2.03 ± 0.08 a	0.15 ± 0.04 bc
GSH	17.9 ± 0.7 ab	70.8 ± 3.3 ab	3.8 ± 0.1 ab	2.02 ± 0.08 a	0.15 ± 0.02 bc
AsA-Pro-GSH	18.3 ± 0.6 a	75.5 ± 5.0 a	4.0 ± 0.1 a	2.03 ± 0.14 a	0.17 ± 0.04 a
NaCl	11.8 ± 0.6 e	47.2 ± 2.0 e	3.0 ± 0.1 c	1.41 ± 0.05 c	0.10 ± 0.02 e
NaCl + AsA	16.8 ± 0.3 abc	63.4 ± 4.2 d	3.6 ± 0.1 b	1.95 ± 0.12 a	0.14 ± 0.02 cd
NaCl + Pro	15.4 ± 0.6 cd	61.1 ± 2.3 d	3.6 ± 0.1 b	1.98 ± 0.06 a	0.13 ± 0.02 d
NaCl + GSH	15.1 ± 0.5 d	64.1 ± 3.6 cd	3.6 ± 0.1 b	1.98 ± 0.09 a	0.14 ± 0.03 cd
NaCl + AsA-Pro-GSH	17.9 ± 0.4 ab	70.7 ± 2.5 ab	3.7 ± 0.1 b	2.00 ± 0.07 a	0.16 ± 0.04 ab

Means with the same letter within each trait are not significantly differed at $p \leq 0.05$. AsA = ascorbic acid, Pro = proline, GSH = glutathione, and AsA-Pro-GSH = sequential application of different antioxidants.

Table 2. Effects of seed single or sequential application of antioxidants on photosynthetic efficiency of 100 mM NaCl-salt-stressed cucumber transplants.

Treatments	Stomatal Conductance (mmol⁻² S⁻¹)	SPAD Chlorophyll	F$_v$/F$_m$	Performance Index (%)
Control	193 ± 5 d	42.1 ± 2.2 ab	0.83 ± 0.06 a	3.64 ± 0.24 ab
AsA	241 ± 7 abc	42.4 ± 1.0 ab	0.83 ± 0.05 a	3.75 ± 0.10 ab
Pro	247 ± 5 abc	42.8 ± 0.6 ab	0.83 ± 0.05 a	3.73 ± 0.12 ab
GSH	254 ± 6 a	42.9 ± 0.9 ab	0.83 ± 0.04 a	3.78 ± 0.14 a
AsA-Pro-GSH	257 ± 8 a	44.7 ± 1.4 a	0.84 ± 0.05 a	3.78 ± 0.12 a
NaCl	142 ± 4 e	33.8 ± 2.0 e	0.76 ± 0.04 b	3.03 ± 0.35 c
NaCl + AsA	232 ± 8 c	38.9 ± 1.8 cd	0.82 ± 0.05 a	3.46 ± 0.30 b
NaCl + Pro	234 ± 5 bc	38.6 ± 1.9 d	0.82 ± 0.04 a	3.46 ± 0.29 b
NaCl + GSH	234 ± 7 bc	39.0 ± 0.8 cd	0.82 ± 0.04 a	3.48 ± 0.31 b
NaCl + AsA-Pro-GSH	252 ± 6 ab	41.4 ± 1.5 bc	0.84 ± 0.05 a	3.72 ± 0.19 ab

Means with the same letter within each trait are not significantly differed at $p \leq 0.05$. AsA = ascorbic acid, Pro = proline, GSH = glutathione, and AsA-Pro-GSH = sequential application of different antioxidants. F$_v$/F$_m$ = maximum quantum yield of PSII = efficiency of the photosystem 2.

Table 3. Effects of seed single or sequential application of antioxidants on leaf relative water content (RWC), membrane stability index (MSI), and non-enzymatic antioxidant system of cucumber transplants grown under 100 mM NaCl-salt stress.

Treatments	RWC (%)	MSI (%)	Free Proline (mg g^{-1} DW)	AsA (µmol Ascorbate g^{-1} DW)	GSH (nmol GSH g^{-1} DW)
Control	87.6 ± 0.7 [abc]	76.3 ± 1.1 [a]	0.27 ± 0.00 [f]	1.48 ± 0.04 [g]	0.14 ± 0.01 [g]
AsA	88.3 ± 1.0 [abc]	78.5 ± 0.4 [a]	0.31 ± 0.01 [ef]	3.54 ± 0.08 [d]	0.19 ± 0.01 [f]
Pro	89.2 ± 0.8 [a]	78.3 ± 0.5 [a]	0.65 ± 0.04 [c]	1.85 ± 0.02 [f]	0.20 ± 0.00 [f]
GSH	88.7 ± 2.1 [ab]	79.0 ± 1.1 [a]	0.32 ± 0.02 [e]	1.97 ± 0.04 [f]	0.35 ± 0.01 [d]
AsA-Pro-GSH	91.3 ± 2.6 [a]	79.7 ± 0.8 [a]	0.76 ± 0.00 [b]	4.27 ± 0.08 [c]	0.40 ± 0.00 [c]
NaCl	74.5 ± 0.4 [d]	63.6 ± 2.1 [c]	0.33 ± 0.00 [e]	2.89 ± 0.08 [e]	0.18 ± 0.01 [fg]
NaCl + AsA	82.7 ± 0.9 [c]	69.8 ± 0.7 [b]	0.40 ± 0.02 [d]	6.32 ± 0.05 [b]	0.25 ± 0.00 [e]
NaCl + Pro	83.2 ± 2.7 [bc]	69.9 ± 1.5 [b]	0.80 ± 0.01 [b]	3.34 ± 0.03 [d]	0.28 ± 0.01 [e]
NaCl + GSH	82.8 ± 1.2 [bc]	70.4 ± 0.8 [b]	0.41 ± 0.01 [d]	3.40 ± 0.04 [d]	0.49 ± 0.01 [b]
NaCl + AsA-Pro-GSH	87.8 ± 1.4 [abc]	76.8 ± 1.0 [a]	0.87 ± 0.01 [a]	7.58 ± 0.13 [a]	0.59 ± 0.01 [a]

Means with the same letter within each trait are not significantly differed at $p \leq 0.05$. AsA = ascorbic acid, Pro = proline, GSH = glutathione, and AsA-Pro-GSH = sequential application of different antioxidants.

Table 4. Effects of seed single or sequential application of antioxidants on enzymatic antioxidant system of cucumber transplants grown under 100 mM NaCl-salt stress.

Treatments	Superoxide Dismutase	Catalase	Glutathione Reductase	Ascorbate Peroxidase
	µmol mg^{-1} Protein min^{-1}			
Control	0.23 ± 0.01 [e]	0.23 ± 0.00 [d]	0.27 ± 0.01 [d]	0.30 ± 0.01 [e]
AsA	0.29 ± 0.01 [cd]	0.28 ± 0.01 [c]	0.32 ± 0.01 [c]	0.43 ± 0.02 [c]
Pro	0.28 ± 0.01 [d]	0.30 ± 0.02 [c]	0.34 ± 0.02 [c]	0.42 ± 0.01 [c]
GSH	0.28 ± 0.01 [d]	0.28 ± 0.01 [c]	0.34 ± 0.02 [c]	0.42 ± 0.02 [c]
AsA-Pro-GSH	0.31 ± 0.02 [bcd]	0.35 ± 0.00 [b]	0.41 ± 0.02 [b]	0.52 ± 0.02 [b]
NaCl	0.25 ± 0.01 [e]	0.29 ± 0.00 [c]	0.32 ± 0.01 [c]	0.36 ± 0.01 [d]
NaCl + AsA	0.31 ± 0.01 [bcd]	0.35 ± 0.01 [b]	0.39 ± 0.02 [b]	0.49 ± 0.02 [b]
NaCl + Pro	0.31 ± 0.02 [bcd]	0.37 ± 0.00 [b]	0.39 ± 0.01 [b]	0.50 ± 0.02 [b]
NaCl + GSH	0.33 ± 0.02 [b]	0.37 ± 0.01 [b]	0.40 ± 0.02 [b]	0.50 ± 0.02 [b]
NaCl + AsA-Pro-GSH	0.38 ± 0.02 [a]	0.41 ± 0.02 [a]	0.47 ± 0.02 [a]	0.55 ± 0.02 [a]

Means with the same letter within each trait are not significantly differed at $p \leq 0.05$. AsA = ascorbic acid, Pro = proline, GSH = glutathione, and AsA-Pro-GSH = sequential application of different antioxidants.

Treatment with 100 mM NaCl resulted in a reduction for leaf and root contents of Ca^{2+} and K^+ by 26.8 and 36.3%, and 52.0 and 57.2% in comparison to controls, respectively (Figure 1). Moreover, it reduced the ratios of Ca^{2+}/Na^+ and K^+/Na^+ of leaves and roots by 87.2 and 90.6%, and 91.6 and 93.8% compared to the controls, respectively (Figure 2). Nevertheless, salt stress treatment with 100 mM NaCl resulted in an increments in the contents of root and leaf Cd^{2+} and Na^+ by 340.1 and 759.3%, and 585.1 and 472.0%, respectively compared to the controls (Figures 1 and 2).

However, the application of non-enzymatic antioxidants for cucumber seed attenuated the stress impacts of 100 mM NaCl salinity by modifying, positively, the above salt stress effects. Where compared to NaCl treatment, the individual use of each of AsA, Pro, and GSH for cucumber seed markedly elevated all of the investigated growth characteristics, photosynthetic efficiency, RWC, MSI, endogenous AsA, Pro, and GSH levels (further increase), and the activity of antioxidant enzymes; CAT, APX, GR, and SOD (further increase), the leaf and root Ca^{2+} and K^+ contents and their ratios to Na^+, while considerably reduced Cd^{2+} and Na^+ contents in leaves and roots. However, AsA-Pro-GSH treatment significantly exceeded all of the individual applications for all of the investigated parameters. This AsA-Pro-GSH applied sequentially elevated the length of the shoot by 51.7%, total area of leaves by 49.8%, stem diameter by 23.3%, shoot FW by 41.8%, shoot DW by 60.0%, gs by 77.5%, SPAD chlorophyll by 22.5%, Fv/Fm by 10.5%, PI by 22.8%, RWC by 17.9%, and MSI by 20.8%. This best treatment also increased the activity of SOD by 52.0%, CAT by 41.4%, GR by 46.9%, and APX by 52.8%, the content of Pro by 163.6%, AsA by 162.3%, GSH by 227.8%, root K^+ by 128.3%, leaf K^+ by 105.9%, root Ca^{2+} by 51.0%, and leaf Ca^{2+} by 34.1%, and the ratio of root K^+/Na^+ by 510.5%, leaf K^+/Na^+ by 484.1%, root Ca^{2+}/Na^+ by 300.0%, and leaf Ca^{2+}/Na^+ by 280.4% compared to NaCl treatment without antioxidants. AsA, Pro, and GSH applied individually or sequentially were more noticeable under salt stress than their applications under the control condition, especially for the AsA-Pro-GSH applied sequentially.

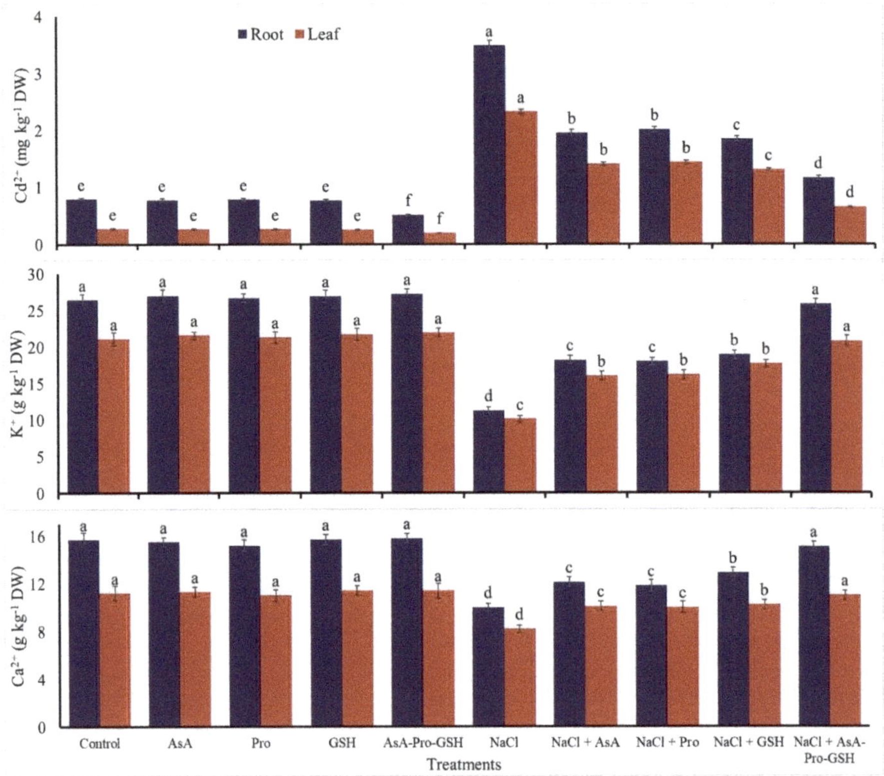

Figure 1. Effects of seed single or sequential application of antioxidants on Cd^{2+}, K^+ and Ca^{2+} contents in roots and leaves of 100 mM NaCl-salt-stressed cucumber transplants. Means with the same letter within each trait are not significantly differed at $p \leq 0.05$. AsA = ascorbic acid, Pro = proline, GSH = glutathione, and AsA-Pro-GSH = sequential application of different antioxidants.

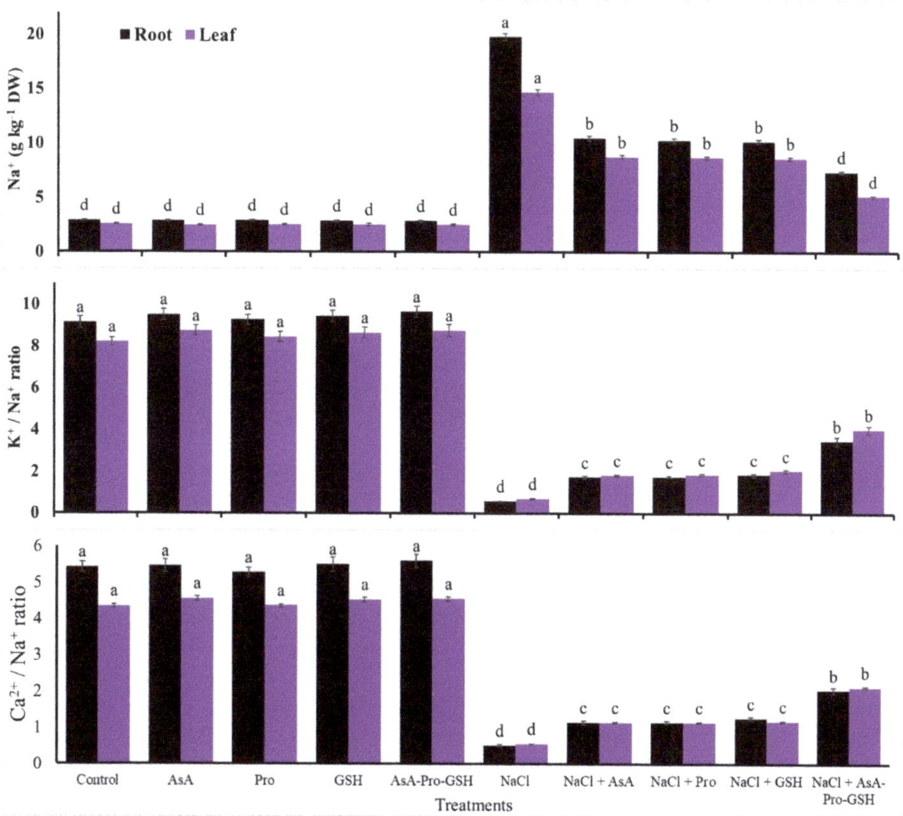

Figure 2. Effects of seed single or sequential application of antioxidants on the content of Na^+ and its relation to K_+ and Ca^{2+} in roots and leaves of 100 mM NaCl-salt-stressed cucumber transplants. Means with the same letter within each trait are not significantly differed at $p \leq 0.05$. AsA = ascorbic acid, Pro = proline, GSH = glutathione, and AsA-Pro-GSH = sequential application of different antioxidants.

4. Discussion

Salinity significantly limited cucumber transplant growth characteristics (Table 1). This growth limitation might be due to alterations in osmotic potential emerging from the limited availability of water [42]. An increased concentration of salts restricts the ability of the plant to absorb enough water and forces it to absorb Na^+ and Cl^- in excessive amounts, early emerging osmotic stress and accumulating ionic Na^+ and Cl^- stresses. This mainly impairs the metabolic processes, negatively affecting photosynthetic efficiency and limiting the growth [43]. However, sequence use of AsA-Pro-GSH conferred significant results, mitigating the devastating impacts of salinity stress. AsA, Pro, or GSH applied exogenously as individual treatments have been reported to overcome the disastrous impacts of salinity on metabolic processes related to plant growth [21,44–46]. However, the exogenous use of AsA, Pro, and GSH as AsA-Pro-GSH treatment has rarely been shown to attenuate the salt stress influences [28]. AsA, proline, and GSH as antioxidants have great power to counter and rectify the damages constructed by the salt-induced ROS. This enables stressed plants to qualify a complex antioxidative system to maximize defensive strategies in plant cells to tolerate the NaCl-induced oxidative stress [18,28,47]. Soaked seeds using AsA-Pro-GSH generated significantly better growth results than seeds soaked in individual AsA, Pro, or GSH solution. This may be attributed to that application of the three antioxidants in an integrative sequential method confers a balance for the so-called "AsA-GSH cycle" to effectively control the ROS radicals along with the osmoprotectant Pro to save water for cellular processes under salt stress in favor of plant growth. As proved for the AsA-Pro-GSH treatment in our earlier study [28], AsA was the most helpful when applied first because it has been involved in several biological activity types associated with oxidative stress resistance in plant cells, where it functions as a donor/acceptor in electron transport in the chloroplasts or at the cellular plasma membranes, in addition to acting as an antioxidant and enzymatic cofactor, etc. [48]. In chloroplasts, AsA is oxidized to monodehydroascorbate (MDA) in the presence of APX in the "Halliwell-Asada" pathway. MDA might trigger dehydroascorbate (DHA), which is reduced with MDA to cause regeneration of the ascorbate pool. This scavenging type can occur close to the PSII, leading to minimizing the ROS escape risk and/or their reacting with each other [49]. Also, the AsA acts as the "terminal antioxidant" due to the redox potential of the AsA/MDA pair (280 mv), which is lower than the redox potential of most bio-radicals [50]. These into the elucidation of [51] that the biosynthesis of AsA from hexose phosphate and its inclusion in protecting against the stress effects of photo-oxidation suggests that there could be links between the size of AsA pool and the photosynthesis process.

Under salt (100 mM NaCl) stress, photosynthesis efficiency attributes (i.e., gs, SPAD chlorophyll, Fv/Fm, and PI; Table 2) of cucumber transplants were significantly decreased. This may be due to the inhibited or insufficient nutrient uptake [52], which accompanied by the increased uptake of undesired elements (i.e., Na^+ and Cd^{2+}) as shown in our results (Tables 4 and 5). It has also been explained that the diminished contents of chlorophyll could be attributed to the inhibited biosynthesis of chlorophyll and/or the increased chlorophyll-degrading enzyme chlorophyllase [53]. Since chlorophyll is reported as an indicator of oxidative damage, ROSs induced by salt also contribute to chlorophyll degradation and loss of pigment. Fv/Fm ratio reflects the master photochemistry of PSII capacity that is very susceptible to different stress impacts induced by environmental stressors [54]. The ratio of Fv/Fm shows, firmly, a noteworthy susceptibility to stress influences of salinity and it is reported as an indicator of photo-inhibition and/or other PS2 complexes injuries [55]. In the current work, our findings displayed that in the same time of which various antioxidant applications did not affect the photosynthetic efficiency attributes under normal conditions, they significantly improved these attributes in salt-stressed cucumber transplants, with a preference for the AsA-Pro-GSH treatment. The improvements in the values of SPAD chlorophyll, gs, PI, and Fv/Fm were positively reflected in transplants growth improvement. This may be elucidated according to the integrative roles giving integrative mode of actions of AsA, Pro, and GSH as a major mechanism in mitigating the impacts of salt stress in plants [28].

Our findings reported that RWC and MSI were markedly decreased under NaCl-salt stress (Table 3), indicating the destruction of the membrane stability and increasing the membrane lipid peroxidation [22]. The undesirable alterations in the membranes that underwent stress were identified as inorganic leakage, in which salt stress was demonstrated repeatedly to induce peroxidative damage to cellular plasma membranes [56]. However, pretreatment with antioxidants, especially sequenced AsA-Pro-GSH showed significant RWC and MSI improvements under NaCl-salt stress, preventing partially or the peroxidative damage to plasma membranes especially due to the fact that MSI value of cucumber tissue was reached equally to the MSI unstressed control value that was also true for the RWC value (Table 3). [57] showed that AsA treatment prevented the peroxidation of membrane lipids and reduced malondialdehyde production as a final product of membrane lipid peroxidation, positively modifying the membrane properties and functions to minimize inorganic leakage and consequently improve the stability of cell membranes [28]. The same results were explained both with Pro [58] and GSH [21]. RWC is a proper measurement of tissue status of water in the term physiological result of cellular water scarcity, while water potential is a measurement of plant water transport in the soil-plant-atmosphere continuum [59]. It is a key marker for various studies of salinity stress. It is also a general measurement used to assess the balance of water in plant tissues over periods of water deficit and measures the amount of leafy water in a plant as a portion of the total volumetric water, which can be held in the leaf at its full aqueous capacity. In plant cells, RWC maintaining allows the metabolic activities to continue by osmotic adjustments and other attributes of salinity and/or drought adaptation [60]. The use of AsA-Pro-GSH resulted in a significant increase of RWC, helping positive modifying of the plasma membrane that is reported as a target of unacceptable environmental stressors. As found with AsA-Pro-GSH treatment, it is often proved that maintaining cell membrane stability and integrity is a key component to achieving satisfactory plant performance [28,61]. The favorable effects of the sequential AsA-Pro-GSH treatment on MSI and RWC can be explained based on the integrative positive modification of osmotic adjustment by Pro addition and improving the efficiency of AsA-GSH cycle by AsA and GSH addition for scavenging the ROS effectively. Additionally, the advantageous effects, in this regard, of AsA, Pro, and GSH, which form a pivotal component of the abiotic stress response in plant cells [21,60].

Table 5. Changes (%) in seedling physiological, biochemical and growth traits, and antioxidant system components relative to the control in cucumber under normal (N) and saline (100 mM NaCl) conditions. Three color scale heatmap as follow: yellow = midpoint of control and different traits with insignificant values compared to control, green = changes over control, and red = changes below control.

Parameters	Control (N)	Treatments								
		N+ AsA	N+ Pro	N+ GSH	N+ A.P.G	NaCl (S)	S+ AsA	S+ Pro	S+ GSH	S+ A.P.G
Shoot length	15.6	9.0	6.4	15	17	−24	7.7	−1.3	−3.2	15
Leaf area	64.0	10	8.3	11	18	−26	−0.9	−4.5	0.2	10
Stem diameter	3.6	5.6	2.8	5.6	11	−17	0.0	0.0	0.0	2.8
Shoot FW	1.71	17	19	18	19	−18	14	16	16	19
Shoot DW	0.13	15	15	15	31	−23	7.7	0.0	7.7	23
gs	193	25	28	32	33	−26	20	21	21	31
SPAD value	42.1	0.7	1.7	1.9	6.2	−20	−7.6	−8.3	-7.4	-1.7
Fv/Fm	0.83	0.0	0.0	0.0	1.2	−8.4	−1.2	−1.2	−1.2	1.2
PI	3.64	3.0	2.5	3.8	3.8	−17	−4.9	−4.9	−4.4	2.2
RWC	87.6	0.8	1.8	1.3	4.2	−15	−5.6	−5.0	−5.5	0.2
MSI	76.3	2.9	2.6	3.5	4.5	−17	−8.5	−8.4	−7.7	0.7
Free proline	0.27	15	141	19	181	22	48	196	52	222
Ascorbate	1.48	139	25	33	189	95	327	126	130	412
Glutathione	0.14	36	43	150	186	29	79	100	250	321
SOD activity	0.23	26	22	22	35	8.7	35	35	43	65
CAT activity	0.23	22	30	22	52	26	52	61	61	78
GR activity	0.27	19	26	26	52	19	44	44	48	74
APX activity	0.30	43	40	40	73	20	63	67	67	83
Cd^{2+} content	0.53	−31	−2.5	−5.6	-33	550	283	292	257	94
K^+ content	23.8	2.3	1.0	2.4	3.4	−55	−28	−28	−22	−1.9
Ca^{2+} content	13.5	−0.4	-2.5	0.9	1.2	−31	-16	−18	−13	−2.8
Na^+ content	2.73	−2.8	−1.1	−1.9	-2.5	529	253	248	245	129
K^+/Na^+ ratio	8.67	5.4	2.2	4.4	6.2	−93	−80	−79	−77	−57
Ca^{2+}/Na^+ ratio	4.90	2.7	−2.6	2.8	3.9	−89	−77	−76	−75	−56

AsA = ascorbate, Pro = proline, GSH = glutathione, A.P.G. = ascrobate-proline-glutathione, gs = stomatal conductance, PI = performance index, RWC = relative water content, MSI = membrane stability index, CAT = catalase, GR = glutathione reductase, SOD = superoxide dismutase, APX = ascorbate peroxidase, Cd^{2+} = cadmium, Ca^{2+} = calcium, K^+ = potassium, and Na^+ = sodium.

Conferring several mechanisms to overcome salt stress effects, Pro, AsA, and GSH, as low-molecular-weight antioxidants with others, comprise a major part of the plant defense system, rendering safeguarding roles to withstand oxidative stress effects [62]. Due to the fact that Pro acts as osmo- and/or enzyme protectant and might make itself as a reserve of N, as well as it considers as a free radical scavenger, there is a potent connection between its level in plant cells and the ability to withstand the effects of stress. [63]. Our findings showed a noticeable proline level in cucumber transplants (Table 3), positively reflecting in transplant growth, tissue water content, and photosynthetic efficiency (Tables 1 and 2) by AsA, Pro, and GSH pretreatment, especially when applied in a sequencing method. This result can be obtained on account of the expression of key genes-encoding enzymes for Pro biosynthesis (the biosynthetic P5CS) and oxidizing enzyme low activities [28,64]. In addition, Pro acts to modify toxic ions (Cd^{2+}; Figure 1 and Na^+; Figure 2) and organic solute contents [64]. It was also effective in districting the oxidative damages of NaCl by the plant's antioxidant system, which includes various enzymes and low-molecular-weight antioxidants (Tables 3 and 4). Another mechanism that may be acted in transplants under NaCl stress, the Pro/P5C cycle transfers the NAD(P)H equivalents, which are reduced to the mitochondria to support the maintenance of the pool of $NAD(P)^+$ [65]. In the salt-stressed leaves of the plant, there is marked ProDH activity and decreased levels of Pro and P5C. These compounds (Pro and P5C) can act as sources of N due to the increase in soluble protein and N compound contents that contribute to the osmotic adjustment [64].

As shown in Table 3, AsA and GSH levels in AsA-, Pro-, and/or GSH-pretreated salt-stressed cucumber transplants were found to be markedly greater in comparison with the control levels. The AsA-Pro-GSH as an integrative treatment was more effective than individuals. These results are in a parallel line with those of [28]. Minutely, the pool of AsA and GSH must be balanced along with appropriate APX activity to improve the plant's antioxidant ability to avoid damage of oxidative stress [66]. AsA is a highly potent ROS scavenger due to its electron donation ability in various reactions that occurred enzymatically and non-enzymatically. It keeps safe the cell membrane integrity by direct ROS ($O_2^{\bullet-}$ and OH^-) scavenging [67]. Since the AsA-GSH cycle contains both AsA and GSH as major components, they can control the level of H_2O_2 in plant cells. By forming AsA and GSH, GR together with MDHAR and DHAR the all are mostly responsible for providing substrates for APX [68]. Under the stress of salinity, Desoky et al. [69] have reported increases in the state of redox activity of AsA and GSH in conjunction with an increase in their contents to reduce the level of H_2O_2. The reactions occurring in the transformation of oxidized glutathione; catalyzes the reduction of oxidized glutathione to reduced glutathione; GSH are catalyzed by GR.

In addition to non-enzyme-based antioxidants that play important roles in counteracting the effects of salinity, plants make use of antioxidant enzymes. This has become evident that comparatively raised activities of enzymes that scavenge different ROS have been reached in wheat seedlings [69]. Results of the current study (Table 4) have supported this finding. Also, the antioxidative enzymes assayed in the present work such as CAT, APX, GR, and SOD have a special role in mitigating the effects of oxidative stress stimulated by NaCl salinity. The activities of these enzymes were measured in cucumber transplants under normal or 100 mM NaCl stress in response to AsA, Pro, or GSH (singly) or AsA-Pro-GSH (sequential method as antioxidative integration). Our results showed that the activities of all enzymes raised under the stress of NaCl salinity and further increased with antioxidants pretreatment, especially with AsA-Pro-GSH pretreatment. Because SOD is an effective $O_2^{\bullet-}$ scavenger, it is the plant's first defense employee against ROS [28], demonstrating the SOD defensive role for biosystems. Besides, CAT is considered to be the main scavenger of H_2O_2, producing H_2O and O_2. It may also be a protective agent against the formation of OH^- radicals, which peroxidize cell membrane lipids and severely affect plant growth [70]. As CAT does, APX eliminates H_2O_2, and its elevated activity has also been noticed under salinity in different plant species [25,69]. Salt stress stimulates excess ROS accumulation in plant cells such as H_2O_2, which its metabolism depends on several antioxidant enzymes that functionally interconnected to eliminate H_2O_2 from cells of stressed plants [71,72].

AsA, Pro, and GSH contribute to reduce the salt-induced K^+ efflux and increase Ca^{2+}/Na^+ and K^+/Na^+ ratios in the transplants stressed with NaCl salinity (Figures 1 and 2), conferring proper ionic homeostasis in salt-stressed transplants. This suggests an effective mechanism in the roots of transplants stressed with NaCl salinity to avert Na^+ xylem tonnage. Another efficient mechanism that may occur by applied antioxidants is the compartmentalization of excess Na^+ authorizing a maximal K^+ influx to transplant leaves [24,73,74], leading to an increased cytosolic ratio of K^+/Na^+ (along with Ca^{2+}/Na^+ ratio) that represents a crucial indicator for tolerance to salinity stress in plants [69]. An adaptive mechanism such as elevated K^+ re-uptake lets plant cell to avert starvation of K^+ under increased salts. Also, the three applied antioxidants may have integrated crucial roles as the main mechanism in reducing Cd^{2+} content in transplants (Figure 1). This may be due to the partitioning of Cd^{2+} to different organs to overcome the toxicity impacts of Cd^{2+} in the transplants of cucumber. Exogenous AsA-Pro-GSH improved ROS removal and metal ions chelation activities, establishing an important portion of the plant cell response to abiotic stress [24].

In general, sequential pretreatment with AsA-Pro-GSH resulted in the best findings as compared to their individuals, attenuating the stress harmful impacts of NaCl salinity. Exogenously applied AsA-Pro-GSH as a sequential pretreatment has been proved to attenuate the stress adverse impacts of 100 mM NaCl salinity on metabolic processes related to plant growth [28]. These antioxidants also reduced the Na^+ and Cd^{2+} ion contents within plants (Figures 1 and 2) due to the decreased uptake of these injurious ions and/or compartmentalization of them into transplant organs.

If a look is taken at the potential plant salt tolerance mechanisms proved in several works and the supporting role of exogenous AsA-Pro-GSH pretreatment performed in the current study, it has been found that: (1) Accumulation of osmotic adjustment substances is an important salt tolerance mechanism. This mechanism is supported by the exogenous addition of Pro in the AsA-Pro-GSH application, wherein plant cells Pro is considered to be a primary substance for osmotic adjustment and is also acted as a scavenger of ROS, a buffer of redox reactions, and/or molecular chaperone. These potential functions of Pro contribute to stabilizing the structures of plasma membranes and proteins under the conditions of stress, reflecting positive results in our study. (2) Selective absorption of ions and their compartmentalization is another pivotal mechanism of tolerance to salinity stress in plants. A high cytosolic K+/Na+ ratio contributes to the cells emptying of Na+ ions or conveying them to the region of inactive metabolism, the translocation of Na^+ ions to the extracellular zone by the Na^+/H^+ antiporter at plasma membranes, and/or the partitioning of Na^+ by the Na^+/H^+ antiporter in the vacuoles [75]. Besides, the gene *A. thaliana* AtNHX1 encodes the Na1yH1 antiporter at the tonoplast and is functioned by compartmentalization of Na^+ into the cell vacuoles [76]. These results are in a parallel line with our results concerning the reduction of Na^+ ion and the increase of K^+ ion and K^+/Na^+ ratio by AsA-Pro-GSH application. (3) Enzymatically or non-enzymatically scavenging of the ROS by various antioxidants is a very crucial salt tolerance mechanism. The exogenous application of AsA, Pro, and GSH (non-enzymatic, low-molecular-weight antioxidants) in the sequential AsA-Pro-GSH treatment significantly supported their endogenous concentrations to effectively scavenge the ROS, reflecting in positive results in our study. These important mechanisms may be supported by another crucial salt tolerance mechanism; (4) Salt tolerance genes. Tolerance to the effects of salinity stress is a polygenic genetic trait. The plasma membrane Na^+/H^+ antiporter gene SOS1 and vacuolar Na^+/H^+ antiporter gene AtNHX1 in *A. thaliana*, in addition to rice OsbZIP71 gene and wheat genes Ta-UPnP, TaZNF, TaSST, TaDUF1, and TaSP are proved to promote the tolerance to the impacts of salinity stress in transgenic plants [75,77,78].

Improvements occurred by exogenous antioxidants in the seedling photosynthetic efficiency (Tables 2 and 5), the relative content of water and stability index of cell membranes (Tables 3 and 5), endogenous levels of AsA, free proline, and glutathione (Tables 3 and 5), and the activities of antioxidative enzymes (Tables 4 and 5) positively affected the growth traits of salt-stressed cucumber seedlings. Besides, the increased K^+ and Ca^{2+} contents contributed to higher ratios of Ca^{2+}/Na^+ and K^+/Na^+, which were associated with a markedly lower Na^+ content with the exogenous use of AsA-Pro-GSH (Table 5; Figures 1 and 2) and all finally contributed to the increases in growth traits of salt-stressed cucumber seedlings (Tables 1 and 5).

5. Conclusions

When applied exogenously as individuals or sequentially to soaking the seeds of cucumber, AsA, Pro, and GSH significantly enhanced the transplant water content and the cell membranes stabilities. Besides, photosynthetic activity, nutrients contents, and antioxidant defense systems were also improved along with the decrease of Na^+ and Cd^{2+} contents under 100 mM NaCl-salt stress. These positive findings have led to the healthy growth of cucumber transplants. Pretreatment with AsA-Pro-GSH applied sequentially was more effective than any of the three individuals; Pretreatment with AsA, Pro, or GSH. Therefore, the elevation of the tolerance to salinity stress impacts in the transplants of cucumber, and the improvement of transplant growth and health would occur with pretreatment with AsA-Pro-GSH applied sequentially upon growth under 100 mM NaCl-salt stress.

Author Contributions: Conceptualization, W.M.S., M.F.S., M.M.R., G.F.M., K.A.H., B.A.A., A.S. and M.M.H.; Data curation, W.M.S., M.M.R., G.F.M., K.A.H., and M.M.H.; Formal analysis, W.M.S., M.F.S., M.M.R., G.F.M., K.A.H. and B.A.A., A.S.; Investigation, W.M.S., M.F.S., M.M.R., G.F.M., and K.A.H., B.A.A., A.S.; Methodology, W.M.S., M.M.R., G.F.M., K.A.H.; Resources, W.M.S., M.F.S., M.M.R., G.F.M., and K.A.H.; Software, M.F.S., M.M.R., B.A.A. and M.M.H., A.S.; Writing—original draft, W.M.S., M.F.S., M.M.R., G.F.M., K.A.H., B.A.A., A.S. and M.M.H.; Writing—review and editing, W.M.S., M.F.S., M.M.R., G.F.M., K.A.H., B.A.A., A.S. and M.M.H. All authors have read and agree to the published version of the manuscript.

Funding: The current work was funded by Taif University Researchers Supporting Project number (TURSP-2020/59), Taif university, Taif, Saudi Arabia. This research was funded by the Deanship of Scientific Research at Princess Nourah bint Abdulrahman University through the Fast-track Research Funding Program.

Acknowledgments: The authors extend their appreciation to Taif University for funding current work by Taif University Researchers Supporting Project number (TURSP-2020/59), Taif University, Taif, Saudi Arabia. This research was funded by the Deanship of Scientific Research at Princess Nourah bint Abdulrahman University through the Fast-track Research Funding Program.

Conflicts of Interest: The authors declare no conflict of interest.

References

1. Rady, M.O.A.; Semida, W.M.; El-Mageed, T.A.A.; Hemida, K.A.; Rady, M.M. Up-regulation of antioxidative defense systems by glycine betaine foliar application in onion plants confer tolerance to salinity stress. *Sci. Hortic.* **2018**, *240*, 614–622. [CrossRef]
2. Kumar, N.; Pal, M.; Singh, A.; SaiRam, R.K.; Srivastava, G.C. Exogenous proline alleviates oxidative stress and increase vase life in rose (Rosa hybrida L. 'Grand Gala'). *Sci. Hortic.* **2010**, *127*, 79–85. [CrossRef]
3. Jones, R.W., Jr.; Pike, L.M.; Yourman, L.F. Salinty influences cucumber growth and yield. *J. Am. Soc. Hortic. Sci.* **1989**, *114*, 547–551.
4. Abu-zinada, I.A. Effect of salinity levels and application stage on cucumber and soil under greenhouse condition. *Int. J. Agric. Crop Sci.* **2015**, *8*, 73–80.
5. Furtana, G.B.; Tipirdamaz, R. Physiological and antioxidant response of three cultivars of cucumber (Cucumis sativus L.) to salinity. *Turk. J. Biol.* **2010**, *34*, 287–296. [CrossRef]
6. Peykanpour, E.; AM, G.; Fallahzade, J.; Najarian, M. Interactive effects of salinity and ozonated water on yield components of cucumber. *Plant Soil Environ.* **2016**, *62*, 361–366. [CrossRef]
7. Wan, S.; Kang, Y.; Wang, D.; Liu, S. Effect of saline water on cucumber (Cucumis sativus L.) yield and water use under drip irrigation in North China. *Agric. Water Manag.* **2010**, *98*, 105–113. [CrossRef]
8. Wu, Y.; Jin, X.; Liao, W.; Hu, L.; Dawuda, M.M.; Zhao, X. 5-Aminolevulinic acid (ALA) alleviated salinity stress in cucumber seedlings by enhancing chlorophyll synthesis pathway. *Front. Plant Sci.* **2018**, *9*, 1–16. [CrossRef]
9. Yildirim, B.; Yasar, F.; Özpay, T.; Türközü, D.; Terziodlu, Ö.; Tamkoç, A. Variations in response to salt stress among field pea genotypes (Pisum sativum sp.arvense L.). *J. Anim. Vet. Adv.* **2008**, *7*, 907–910.
10. Seleiman, M.F.; Kheir, A.S. Saline soil properties, quality and productivity of wheat grown with bagasse ash and thiourea in different climatic zones. *Chemosphere* **2018**, *193*, 538–546. [CrossRef]
11. Seleiman, M.F. Use of plant nutrients in improving abiotic stress tolerance in wheat. In *Wheat Production in Changing Environments: Management, Adaptation and Tolerance*; Hasanuzzaman, M., Nahar, K., Hossain, A., Eds.; Springer Nature: Singapore, 2019; pp. 481–495.
12. Seleiman, M.F.; Kheir, A.M.S.; Al-Dhumri, S.; Alghamdi, A.G.; Omar, E.H.; Aboelsoud, H.M.; Abdella, K.A.; AbouElHassan, W.H. Exploring optimal tillage improved soil characteristics and productivity of wheat irrigated with different water qualities. *Agronomy* **2019**, *9*, 233. [CrossRef]
13. Al-Ashkar, I.; Alderfasi, A.; Ben Romdhane, W.; Seleiman, M.F.; El-Said, R.A.; Al-Doss, A. Morphological and genetic diversity within salt tolerance detection in eighteen wheat genotypes. *Plants* **2020**, *9*, 287. [CrossRef]
14. Taha, R.S.; Seleiman, M.F.; Alotaibi, M.; Alhammad, B.A.; Rady, M.M.; Mahdi, A.H.A. Exogenous potassium treatments elevate salt tolerance and performances of glycine max L. by boosting antioxidant defense system under actual saline field conditions. *Agronomy* **2020**, *10*, 1741. [CrossRef]

15. Ding, Z.; Kheir, A.S.; Ali, O.A.; Hafez, E.; Elshamey, E.A.; Zhou, Z.; Wang, B.; Lin, X.; Ge, Y.; Fahmy, A.F.; et al. Vermicompost and deep tillage system to improve saline-sodic soil quality and wheat productivity. *J. Environ. Manag.* **2020**, *277*, 111388. [CrossRef]
16. Bargaz, A.; Nassar, R.M.A.; Rady, M.M.; Gaballah, M.S.; Thompson, S.M.; Brestic, M.; Schmidhalter, U.; Abdelhamid, M.T. Improved salinity tolerance by phosphorus fertilizer in two phaseolus vulgaris recombinant inbred lines contrasting in their P-efficiency. *J. Agron. Crop Sci.* **2016**. [CrossRef]
17. Semida, W.M.; Abd El-Mageed, T.A.; Hemida, K.; Rady, M.M. Natural bee-honey based biostimulants confer salt tolerance in onion via modulation of the antioxidant defence system. *J. Hortic. Sci. Biotechnol.* **2019**, *94*, 632–642. [CrossRef]
18. Abd El-Mageed, T.A.; Semida, W.M.; Mohamed, G.F.; Rady, M.M. Combined effect of foliar-applied salicylic acid and deficit irrigation on physiological—Anatomical responses, and yield of squash plants under saline soil. *S. Afr. J. Bot.* **2016**, *106*, 8–16. [CrossRef]
19. Apel, K.; Hirt, H. Reactive oxygen species: Metabolism, oxidative stress, and signal transduction. *Annu. Rev. Plant Biol.* **2004**, *55*, 373–399. [CrossRef]
20. Baker, N.R. Chlorophyll fluorescence: A probe of photosynthesis in vivo. *Annu. Rev. Plant Biol.* **2008**, *113*, 59–89. [CrossRef]
21. Rady, M.M.; Taha, S.S.; Kusvuran, S. Integrative application of cyanobacteria and antioxidants improves common bean performance under saline conditions. *Sci. Hortic.* **2018**, *62*, 361–366. [CrossRef]
22. Zhu, J.; Bie, Z.; Li, Y. Physiological and growth responses of two different salt-sensitive cucumber cultivars to NaCl stress. *Soil Sci. Plant Nutr.* **2008**, *54*, 400–407. [CrossRef]
23. Parida, A.K.; Das, A.B. Salt tolerance and salinity effects on plants: A review. *Ecotoxicol. Environ. Saf.* **2005**, *60*, 324–349. [CrossRef] [PubMed]
24. Semida, W.M.; Hemida, K.A.; Rady, M.M. Sequenced ascorbate-proline-glutathione seed treatment elevates cadmium tolerance in cucumber transplants. *Ecotoxicol. Environ. Saf.* **2018**, *154*, 171–179. [CrossRef] [PubMed]
25. Desoky, E.M.; El-maghraby, L.M.M.; Awad, A.E.; Abdo, A.I.; Rady, M.M.; Semida, W.M. Fennel and ammi seed extracts modulate antioxidant defence system and alleviate salinity stress in cowpea (Vigna unguiculata). *Sci. Hortic.* **2020**, *272*, 109576. [CrossRef]
26. Ejaz, B.; Sajid, Z.A.; Aftab, F. Effect of exogenous application of ascorbic acid on antioxidant enzyme activities, proline contents, and growth parameters of Saccharum spp. hybrid cv. HSF-240 under salt stress. *Turk. J. Biol.* **2012**, *36*, 630–640. [CrossRef]
27. Rady, M.M.; Taha, R.S.; Semida, W.M.; Alharby, H.F. Modulation of salt stress effects on vicia faba l. plants grown on a reclaimed-saline soil by salicylic acid application. *Rom. Agric. Res.* **2017**, *34*, 175–185.
28. Rady, M.M.; Hemida, K.A. Sequenced application of ascorbate-proline-glutathione improves salt tolerance in maize seedlings. *Ecotoxicol. Environ. Saf.* **2016**, *133*, 252–259. [CrossRef]
29. Rady, M.M.; Hemida, K.A. Modulation of cadmium toxicity and enhancing cadmium-tolerance in wheat seedlings by exogenous application of polyamines. *Ecotoxicol. Environ. Saf.* **2015**, *119*, 178–185. [CrossRef]
30. Semida, W.M.; El-Mageed, T.A.A.; Mohamed, S.E.; El-Sawah, N.A. Combined effect of deficit irrigation and foliar-applied salicylic acid on physiological responses, yield, and water-use efficiency of onion plants in saline calcareous soil. *Arch. Agron. Soil Sci.* **2017**, *63*, 1227–1239. [CrossRef]
31. Maxwell, K.; Johnson, G.N. Chlorophyll fluorescence—A practical guide. *J. Exp. Bot.* **2000**, *51*, 659–668. [CrossRef]
32. Rady, M.M. Effect of 24-epibrassinolide on growth, yield, antioxidant system and cadmium content of bean (Phaseolus vulgaris L.) plants under salinity and cadmium stress. *Sci. Hortic.* **2011**, *129*, 232–237. [CrossRef]
33. Bates, L.S.; Waldeen, R.P.; Teare, I.D. Rapid determination of free proline for water—Stress studies. *Plant Soil* **1973**, *207*, 205–207. [CrossRef]
34. Kampfenkel, K.; Vanmontagu, M.; Inze, D. Extraction and determination of ascorbate and dehydroascorbate from plant tissue. *Anal. Biochem.* **1995**, *225*, 165–167. [CrossRef] [PubMed]
35. Griffith, O.W. Determination of glutathione and glutathione disulfide using glutathione reductase and 2-vinylpyridine. *Anal. Biochem.* **1980**, *106*, 207–212. [CrossRef]
36. Bradford, M.M. A rapid and sensitive method for the quantitation of microgram quantities of protein utilizing the principle of protein-dye binding. *Anal. Biochem.* **1976**, *72*, 248–254. [CrossRef]

37. Kono, Y. Generation of superoxide radical during autoxidation of hydroxylamine and an assay for superoxide dismutase. *Arch. Biochem. Biophys.* **1978**, *186*, 189–195. [CrossRef]
38. Aebi, H. Catalase in vitro. *Methods Enzymol.* **1984**, *105*, 121–126.
39. Rao, M.V.; Paliyath, G.; Ormrod, P. Ultraviolet-9- and ozone-lnduced biochemical changes in antioxidant enzymes of Arabidopsis thaliana. *Plant Physiol.* **1996**, *110*, 125–136. [CrossRef]
40. Williams, V.; Twine, S. Flame photometric method for sodium, potassium and calcium. In *Modern Methods of Plant Analysis*; Springer: Berlin/Heidelberg, Germany, 1960; pp. 3–5.
41. Chapman, H.D.; Pratt, P.F. *Methods of Analysis for Soil, Plants and Water*; University of California, Division of Agricultural Science: Berkeley, CA, USA, 1961.
42. Mousavi, S.A.R.; Chauvin, A.; Pascaud, F.; Kellenberger, S.; Farmer, E.E. GLUTAMATE RECEPTOR-LIKE genes mediate leaf-to-leaf wound signalling. *Nature* **2013**, *500*, 422. [CrossRef]
43. Munns, R.; Tester, M. Mechanisms of salinity tolerance. *Annu. Rev. Plant Biol.* **2008**, *59*, 651–681. [CrossRef]
44. Aliniaeifard, S.; Hajilou, J.; Tabatabaei, S.J.; Sifi-Kalhor, M. Effects of ascorbic acid and reduced glutathione on the alleviation of salinity stress in olive plants. *Int. J. Fruit Sci.* **2016**, *16*, 395–409. [CrossRef]
45. Rady, M.M.; Semida, W.M.; Hemida, K.A.; Abdelhamid, M.T. The effect of compost on growth and yield of Phaseolus vulgaris plants grown under saline soil. *Int. J. Recycl. Org. Waste Agric.* **2016**, *5*, 311–321. [CrossRef]
46. Shalata, A.; Neumann, P.M. Exogenous ascorbic acid (vitamin C) increases resistance to salt stress and reduces lipid peroxidation. *J. Exp. Bot.* **2001**, *52*, 2207–2211. [CrossRef] [PubMed]
47. Wutipraditkul, N.; Wongwean, P.; Buaboocha, T. Alleviation of salt-induced oxidative stress in rice seedlings by proline and/or glycinebetaine. *Biol. Plant.* **2015**, *59*, 547–553. [CrossRef]
48. Conklin, P.L. Recent advances in the role and biosynthesis of ascorbic acid in plants. *Plant Cell Environ.* **2001**, *24*, 383–394. [CrossRef]
49. Foyer, C.H.; Noctor, G. Redox homeostasis and antioxidant signaling: A metabolic interface between Stress perception and physiological responses. *Plant Cell Online* **2005**, *17*, 1866–1875. [CrossRef]
50. Scandalios, J.G.; Guan, L.; Polidoros, A.N. Catalases in plants: Gene structure, properties, regulation, and expression. *Cold Spring Harb. Monogr. Ser.* **1997**, *34*, 343–406.
51. Reddy, K.R.; Kakani, V.G.; Zhao, D.; Koti, S.; Gao, W. Interactive effects of ultraviolet-B radiation and temperature on cotton physiology, growth, development and hyperspectral reflectance. *Photochem. Photobiol.* **2004**, *79*, 416–427. [CrossRef]
52. Kanokwan, S.; Tanatorn, S.; Aphichart, K. Effect of salinity stress on antioxidative enzyme activities in tomato cultured in vitro. *Pak. J. Bot.* **2015**, *47*, 1–10.
53. Rezende, R.A.L.S.; Rodrigues, F.A.; Soares, J.D.R.; Silveira, H.R.O.; Pasqual, M.; Dias, G.M.G. Salt stress and exogenous silicon influence physiological and anatomical features of in vitro-grown cape gooseberry. *Cienc. Rural* **2018**, *48*. [CrossRef]
54. Sheng, M.; Tang, M.; Chen, H.; Yang, B.; Zhang, F.; Huang, Y. Influence of arbuscular mycorrhizae on photosynthesis and water status of maize plants under salt stress. *Mycorrhiza* **2008**, *18*, 287–296. [CrossRef]
55. Ranjbarfordoei, A.; Samson, R.; Van Damme, P. Chlorophyll fluorescence performance of sweet almond [Prunus dulcis (Miller) D. Webb] in response to salinity stress induced by NaCl. *Photosynthetica* **2006**, *44*, 513–522. [CrossRef]
56. Younis, M.E.; Mohamed, S.; Tourky, N. Influence of salinity and adaptive compounds on growth parameters, carbohydrates, amino EISSN: Acids and nucleic acids in two cultivars of Vicia faba contrasting in salt tolerance. *Plant Knowl. J.* **2014**, *3*, 47–56.
57. Dolatabadian, A.; Jouneghani, R.S. Impact of exogenous ascorbic acid on antioxidant activity and some physiological traits of common bean subjected to salinity stress. *Not. Bot. Horti Agrobot. Cluj-Napoca* **2009**, *37*, 165–172. [CrossRef]
58. Valizadeh, M.; Asgari, A.; Shiri, M. Proline, glycine betaine, total phenolics and pigment contents in response to osmotic stress in maize seedlings. *J. BioSci. Biotechnol.* **2015**, *4*, 313–319.
59. Darvishan, M.; Tohidi-moghadam, H.R.; Zahedi, H. The effects of foliar application of ascorbic acid (vitamin C) on physiological and biochemical changes of corn (Zea mays L) under irrigation withholding in different growth stages. *Maydica* **2013**, *58*, 195–200.
60. Slabbert, M.M.; Krüger, G.H.J. Antioxidant enzyme activity, proline accumulation, leaf area and cell membrane stability in water stressed Amaranthus leaves. *S. Afr. J. Bot.* **2014**, *95*, 123–128. [CrossRef]

61. Rady, M.M.; Abd El-Mageed, T.A.; Abdelhamid, M.T.; Abd El-Azeem, M.M.M. Integrative potassium humate and biochar application reduces salinity effects and contaminants and improves growth and yield of eggplant grown under saline conditions. *Int. J. Empir. Educ. Res.* **2018**, *1*, 37–56. [CrossRef]

62. Valero, E.; Macià, H.; De la Fuente, I.M.; Hernández, J.A.; González-Sánchez, M.I.; García-Carmona, F. Modeling the ascorbate-glutathione cycle in chloroplasts under light/dark conditions. *BMC Syst. Biol.* **2016**, *10*, 11. [CrossRef]

63. Sairam, R.K.; Tyagi, A. Physiology and molecular biology of salinity stress tolerance in plants. *Curr. Sci.* **2004**, *86*, 407–421.

64. Rady, M.M.; Kuşvuran, A.; Alharby, H.F.; Alzahrani, Y.; Kuşvuran, S. Pretreatment with proline or an organic bio-stimulant induces salt tolerance in wheat plants by improving antioxidant redox state and enzymatic activities and reducing the oxidative stress. *J. Plant Growth Regul.* **2019**, *38*, 449–462. [CrossRef]

65. Miller, G.; Honig, A.; Stein, H.; Suzuki, N.; Mittler, R.; Zilberstein, A. Unraveling Δ1-pyrroline-5-carboxylate-proline cycle in plants by uncoupled expression of proline oxidation enzymes. *J. Biol. Chem.* **2009**, *84*, 19–36. [CrossRef] [PubMed]

66. Foyer, C.H.; Noctor, G. Ascorbate and glutathione: The heart of the redox hub. *Plant Physiol.* **2011**, *155*, 2–18. [CrossRef] [PubMed]

67. Semida, W.M.; Rady, M.M. Pre-soaking in 24-epibrassinolide or salicylic acid improves seed germination, seedling growth, and anti-oxidant capacity in Phaseolus vulgaris L. grown under NaCl stress. *J. Hortic. Sci. Biotechnol.* **2014**, *89*, 338–344. [CrossRef]

68. Zhou, Y.; Wen, Z.; Zhang, J.; Chen, X.; Cui, J.; Xu, W.; Liu, H.-y. Exogenous glutathione alleviates salt-induced oxidative stress in tomato seedlings by regulating glutathione metabolism, redox status, and the antioxidant system. *Sci. Hortic.* **2017**, *220*, 90–101. [CrossRef]

69. Desoky, E.S.M.; Elrys, A.S.; Rady, M.M. Integrative moringa and licorice extracts application improves Capsicum annuum fruit yield and declines its contaminant contents on a heavy metals-contaminated saline soil. *Ecotoxicol. Environ. Saf.* **2019**, *169*, 50–60. [CrossRef]

70. Abogadallah, G.M. Antioxidative defense under salt stress. *Plant Signal. Behav.* **2010**, *5*, 369–374. [CrossRef]

71. Kim, S.-Y.; Lim, J.-H.; Park, M.-R.; Kim, Y.-J.; Park, T.-I.; Seo, Y.-W.; Choi, K.-G.; Yun, S.-J. Enhanced antioxidant enzymes are associated with reduced hydrogen peroxide in barley roots under saline stress. *BMB Rep.* **2005**, *38*, 218–224. [CrossRef]

72. Desoky, E.M.; Merwad, A.M.; Semida, W.M.; Ibrahim, S.A.; El-saadony, M.T.; Rady, M.M. Heavy metals-resistant bacteria (HM-RB): Potential bioremediators of heavy metals-stressed Spinacia oleracea plant. *Ecotoxicol. Environ. Saf.* **2020**, *198*, 110685. [CrossRef]

73. Assaha, D.V.M.; Ueda, A.; Saneoka, H.; Al-Yahyai, R.; Yaish, M.W. The Role of Na+ and K+ Transporters in Salt Stress Adaptation in Glycophytes. *Front. Physiol.* **2017**, *8*, 509. [CrossRef]

74. Abd El-mageed, T.A.; Abd El-mageed, S.A.; Semida, W.M.; Rady, M.O.A. Silicon Defensive Role in Maize (Zea mays L.) against Drought Stress and Metals-Contaminated Irrigation Water. *Silicon* **2020**. [CrossRef]

75. Liang, W.; Ma, X.; Wan, P.; Liu, L. Plant salt-tolerance mechanism: A review. *Biochem. Biophys. Res. Commun.* **2018**, *495*, 286–291. [CrossRef] [PubMed]

76. Gaxiola, R.A.; Rao, R.; Sherman, A.; Grisafi, P.; Alper, S.L.; Fink, G.R. The Arabidopsis thaliana proton transporters, AtNhx1 and Avp1, can function in cation detoxification in yeast. *Proc. Natl. Acad. Sci. USA* **1999**, *96*, 1480–1485. [CrossRef] [PubMed]

77. He, C.; Yan, J.; Shen, G.; Fu, L.; Holaday, A.S.; Auld, D.; Blumwald, E.; Zhang, H. Expression of an Arabidopsis vacuolar sodium/proton antiporter gene in cotton improves photosynthetic performance under salt conditions and increases fiber yield in the field. *Plant Cell Physiol.* **2005**, *46*, 1848–1854. [CrossRef]

78. Liu, C.; Mao, B.; Ou, S.; Wang, W.; Liu, L.; Wu, Y.; Chu, C.; Wang, X. OsbZIP71, a bZIP transcription factor, confers salinity and drought tolerance in rice. *Plant Mol. Biol.* **2014**, *84*, 19–36. [CrossRef]

Publisher's Note: MDPI stays neutral with regard to jurisdictional claims in published maps and institutional affiliations.

 *plants*

Article

Oxidative Stress Responses of Some Endemic Plants to High Altitudes by Intensifying Antioxidants and Secondary Metabolites Content

Ahmed M. Hashim [1], Basmah M. Alharbi [2], Awatif M. Abdulmajeed [3], Amr Elkelish [4], Wael N. Hozzein [5,6,*] and Heba M. Hassan [1]

[1]   Botany Department, Faculty of Science, Ain Shams University, Cairo 11865, Egypt; hashim-a-m@sci.asu.edu.eg (A.M.H.); hebamito@yahoo.com (H.M.H.)

[2]   Biology Department, Faculty of Science, Tabuk University, Tabuk 71421, Saudi Arabia; b.alharbi@ut.edu.sa

[3]   Biology Department, Faculty of Science, Tabuk University, Umluj 41912, Saudi Arabia; awabdulmajeed@ut.edu.sa

[4]   Botany Department, Faculty of Science, Suez Canal University, Ismailia 41522, Egypt; amr.elkelish@science.suez.edu.eg

[5]   Bioproducts Research Chair, Zoology Department, College of Science, King Saud University, Riyadh 11451, Saudi Arabia

[6]   Botany and Microbiology Department, Faculty of Science, Beni-Suef University, Beni-Suef 62511, Egypt

*   Correspondence: whozzein@ksu.edu.sa; Tel.: +20-1024824643

Received: 21 May 2020; Accepted: 29 June 2020; Published: 9 July 2020

Abstract: Most endemic plant species have limited altitudinal ranges. At higher altitudes, they are subjected to various environmental stresses. However, these plants use unique defense mechanisms at high altitudes as a convenient survival strategy. The changes in antioxidant defense system and accumulation of different secondary metabolites (SMs) were investigated as depending on altitude in five endemic endangered species (*Nepeta septemcrenata, Origanum syriacum* subsp. *Sinaicum, Phlomis aurea, Rosa arabica*, and *Silene schimperiana*) naturally growing in Saint Katherine protectorate (SKP). Leaves were collected from different sites between 1600 and 2200 m above sea level to assess the biochemical and physiological variations in response to high altitudes. At higher altitudes, the soil pH and micronutrient soil content decreased, which can be attributed to lower mineralization processes at lower pH. Total phenols, ascorbic acid, proline, flavonoids, and tannins increased in response to different altitudes. SMs progressively increased in the studied species, associated with a significant decrease in the levels of antioxidant enzyme activity. *R. arabica*, as the most threatened plant, showed the maximum response compared with other species. There was an increase in photosynthetic pigments, which was attained via the increase in chlorophyll a, chlorophyll b, and carotenoid contents. There was a significant increase in total soluble sugars and total soluble protein content in response to different altitudes. SDS-PAGE of leaf proteins showed alteration in the protein profile between different species and the same species grown at a different altitude. These five species can adapt to high-altitude habitats by various physiological mechanisms, which can provide a theoretical basis for the future conservation of these endangered endemic species in SKP.

Keywords: altitudinal variation; antioxidant activity; bioactive compounds; endemic species; oxidative damage

---

1. Introduction

Mountains are characterized by unique biodiversity. Most plant species at high altitudes are isolated and have a small number of niche habitats in comparison to lowland plants [1]. Consequently, the plant populations of the mountains have a higher percentage of endemism than lowlands plants [2,3]. Endemic species are most frequently endangered at high altitudes because of their limited altitudinal distribution ranges [4,5]. Climate limitations primarily affect the high-altitude ecosystems and most plants flourish only near their climatic limits [6,7].

The region above 2800 m above sea level (a. s. l.) in Saint Katherine protectorate (SKP), Sinai Peninsula, is a notable area, not only for its natural landscapes, but also its diversity of medicinal plants of national and global interest [8,9]. SKP contains a wide range of habitats that are a consequence of varying climatic conditions, altitudes, and topography [10]. Being one of the most floristically diverse locations in the Middle East, SKP accounts for about 44% of Egypt's endemic plant species [11,12]. Sinai is home to approximately 1285 species, of which approximately 800 species are recorded in its southern region [13]. Previous studies have identified several endemic species among Egyptian flora [14]. Approximately 60 endemic plant species have been tabulated in Egypt, including 31 species on the Sinai Peninsula (i.e., 51.6% of Egyptian endemism) [15]. Twenty-four of these are endemic to South Sinai, and are known for their medicinal properties and use in traditional therapy and remedies [16].

Due to the high altitude of SKP, harsh and complex climatic conditions reduce the distribution and diversity of plants [8,17]. These hard conditions include reduced O_2 and CO_2, strong winds, high solar irradiance, shallow rocky soils, low temperatures, and low water and nutrient contents [18]. Although these stress conditions could retard plant growth, many species have survived and developed different adaptive mechanisms under these circumstances [19].

Accumulation of secondary metabolites (SMs) has been reported to play a vital role in tolerance against different environmental biotic and abiotic stresses [20–24]. Actually, these SMs help in adaptation of stressed plants to different environmental conditions; for example, accumulation of flavonoids and phenols as antioxidants; increase in proline and soluble protein contents against drought; and increased carotenoids compared to chlorophyll to reduce high light intensity damage, through upregulation of the xanthophyll-cycle pigment pool [25,26]. Several factors, such as age, season, and nutrition status, are determinants affecting the quantity of secondary metabolites in plants [27,28]. Similarly, environmental factors significantly cause quantitative and qualitative variations in the production of secondary compounds among plants and population groups [29]; these factors include altitude, high/low temperature, drought, and light conditions [30,31].

Combinations of high-altitude environmental stress lead to increased reactive oxygen species (ROS) production, which increases the risk of oxidative damage [32–34]. Many enzymes of antioxidative activity, such as ascorbate peroxidase, catalase, superoxide dismutase, and glutathione reductase, have the capacity to scavenge different ROS, including superoxide, hydroxyl radicals, and singlet oxygen [35–37]. It was previously confirmed that high mountain plants species are acclimated to high irradiance and chilling stress [38]. High antioxidant and carotenoid levels have been positively associated with the altitude of alpine plants. However, in high mountain plants, the ability of the antioxidant system and the xanthophyll cycle differed considerably [39].

In spite of the ecological significance of these endemic species and many of their promising characteristics, their ecophysiology has not been closely studied, particularly at the SKP high altitudes [40,41]. Consequently, knowing how such adaptations influence plant growth is relevant for researchers who wish to predict the response of endemic species in a future scenario of climate change. In the present work, five endemic and endangered species samples were collected from three different sites with varying altitudes in SKP. This study investigated the SMs and antioxidant defense capacity in plant samples from different altitudes, to test the hypothesis that plants at high altitudes have greater defense capacity than those at lower altitudes. In addition, the present work intended to investigate the eco-physiological responses of some endemic species as a result of their exposure to the stresses prevailing at high altitudes of the SKP mountains.

2. Results

2.1. Soil Analysis

Analysis of soil samples at different heights of SKP mountains showed that soil samples were slightly alkaline, and the pH value decreased with an increase in altitude (Table 1), showing a significant decrease at the highest altitude (2000–2200 m a.s.l.). Elemental analysis of the soil samples (Table 1) showed a reduction in some macronutrients (Na^+, SO_4^-, and Cl^-), while HCO_3^-, Mg^{+2}, and Ca^{++} were higher at the middle altitude. Finally, K^+ increased with the increase in altitude.

Table 1. Chemical analysis of soil samples collected from Saint Katherine protectorate (SKP) mountains at different altitudes (1600–1800, 1800–2000, and 2000–2200 m a.s.l.). Data are means of three replications ± SE.

Altitude (m a.s.l.)	pH Value	Anions (ppm)				Cations (ppm)		
		HCO_3^-	SO_4^-	Cl^-	Ca^{++}	Mg^{++}	Na^+	K^+
1600–1800	8.2 ± 0.12 a	165.6 ± 7.5 b	95.6 ± 2 a	34 ± 4.9 a	79.5 ± 1.5 a	6.6 ± 0.4 c	28.0 ± 1.26 a	7 ± 1.15 b
1800–2000	7.8 ± 0.03 ab	228.7 ± 11.4a	59.1 ± 1.7 b	31.2 ± 5.1 a	42.2 ± 3.5 b	10.8 ± 0.37 a	8.6 ± 1.33 b	14.3 ± 0.88 b
2000–2200	7.6 ± 0.22 b	190.6 ± 5.6 b	18.6 ± 4.5 c	27.5 ± 1.3 a	74.0 ± 0.85 a	8.9 ± 0.29 b	8.0 ± 1.15 b	33.3 ± 3.84 a

Columns with different letters are significantly different at $p < 0.05$.

2.2. Phytochemical Assay

2.2.1. Change in the Photosynthetic Pigments

The photosynthetic pigment is shown in Figure 1. Chlorophyll a was considerably increased by moving to a higher altitude (Figure 1A). In addition, chlorophyll b was significantly raised with higher altitude, with the exception of *N. septemcrenata*, which exhibited lower content of chlorophyll b at 1800–2000 m a.s.l. Finally, as altitude increased, carotenoids showed a gradual rise in content, similar to that of chlorophyll a, in the five plant species, except for *S. schimperiana* (Figure 1B).

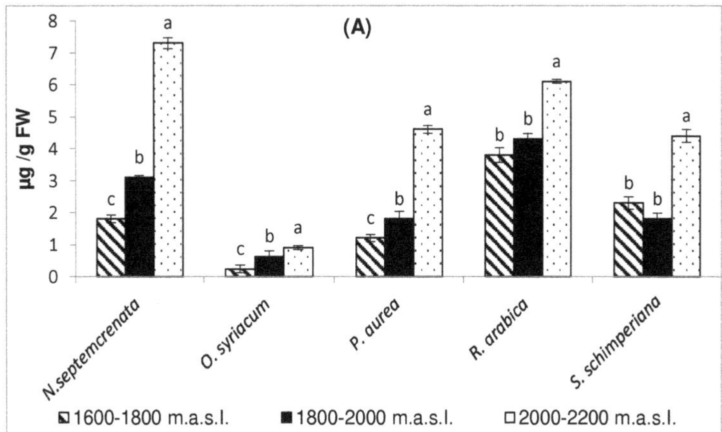

Figure 1. *Cont.*

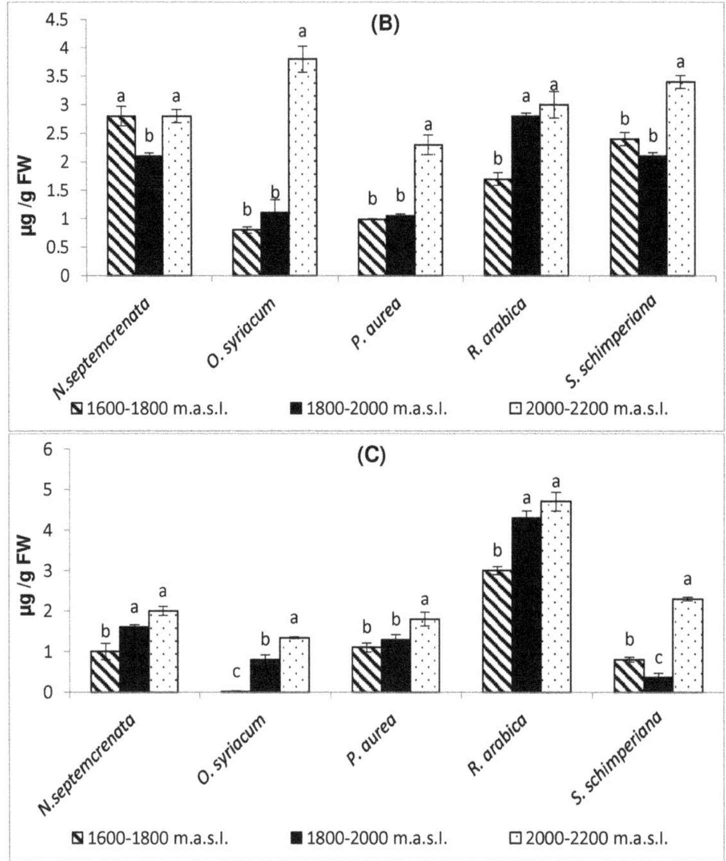

Figure 1. Effect of altitude on the photosynthetic pigments chlorophyll a (**A**), chlorophyll b (**B**), and carotenoids (**C**). Each value is the mean of three replicates ± SE. Bars with different letters are significantly different at $p \leq 0.05$.

2.2.2. Changes in the Content of Major Secondary Metabolites

It is clear from Figure 2 that, with few exceptions, as the altitude in the SKP mountains increased from 1600 to 2200 m a.s.l. there was a significant increase in the content of total phenols, flavonoids, and tannins of the five plant species under investigation. Within the five plant species, *Rosa arabica* recorded the highest values of total phenols, flavonoids, and tannins at the highest altitude (2000–2200 m a.s.l.).

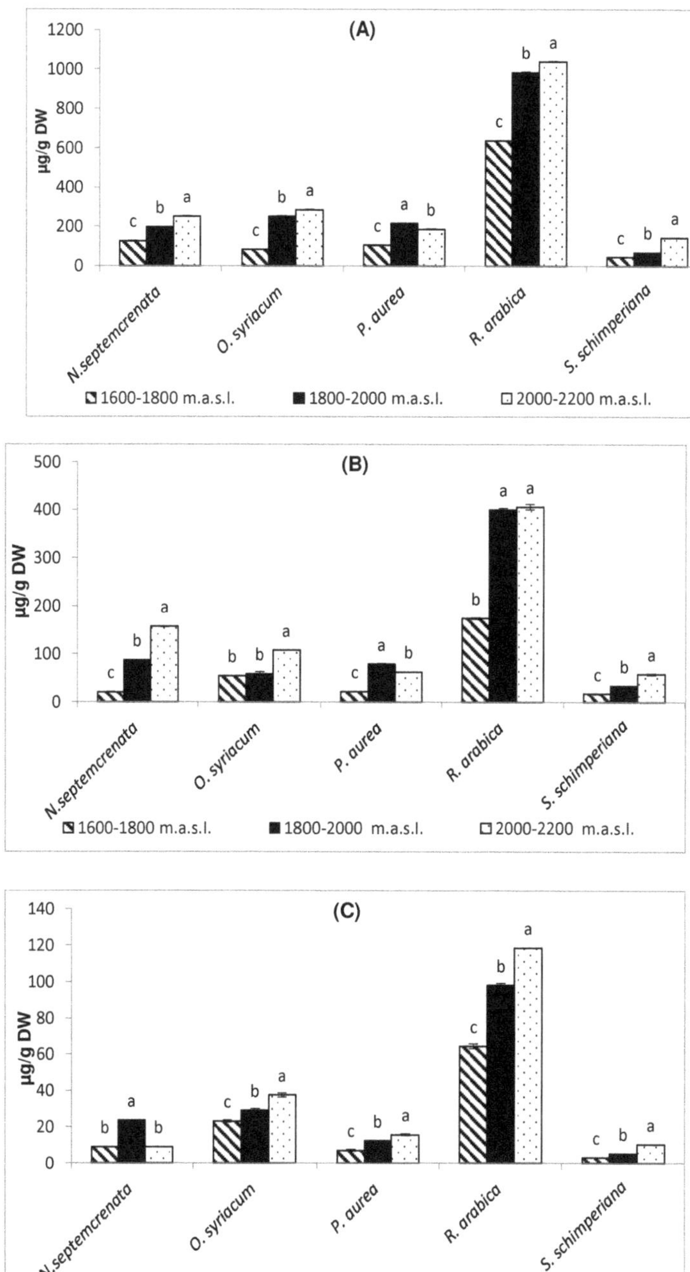

Figure 2. Effect of altitude on the contents of total phenols (**A**), flavonoids (**B**), and tannins (**C**). Each value is the mean of three replicates ± SE. Bars with different letters are significantly different at $p \leq 0.05$.

2.2.3. Changes in the Content of Total Soluble Sugars

In all studied species, total soluble sugar content increased in all samples from higher altitudes compared to those from lower altitudes. These increases reached 60 and 76.85 µg/g DW in *N. septemcrenata* and *R. arabica*, respectively, grown at 2000–2200 m a.s.l. (Figure 3).

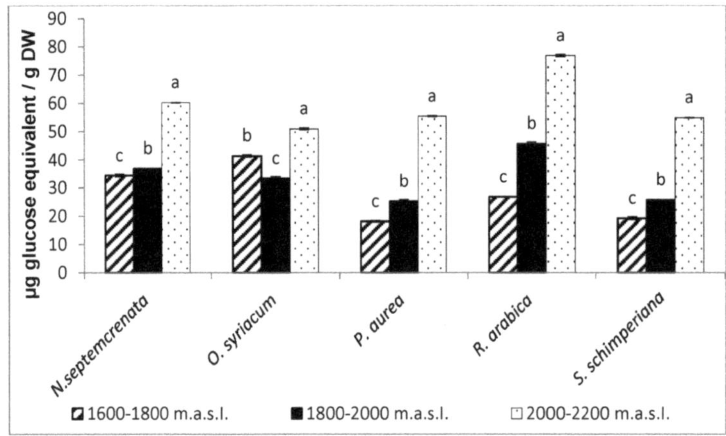

Figure 3. Effect of altitude on the total soluble sugars. Each value is the mean of three replicates ± SE. Bars with different letters are significantly different at $p \leq 0.05$.

2.3. Biochemical Analysis

2.3.1. Changes in the Content of Total Soluble Proteins

Total soluble proteins in response to high altitudes showed high concentrations in all sampled plants at the three studied sites. *N. septtemcrenata* and *R. arabica* attained the highest values of total soluble proteins (363.6 and 365.93 µg/g FW), respectively (Figure 4).

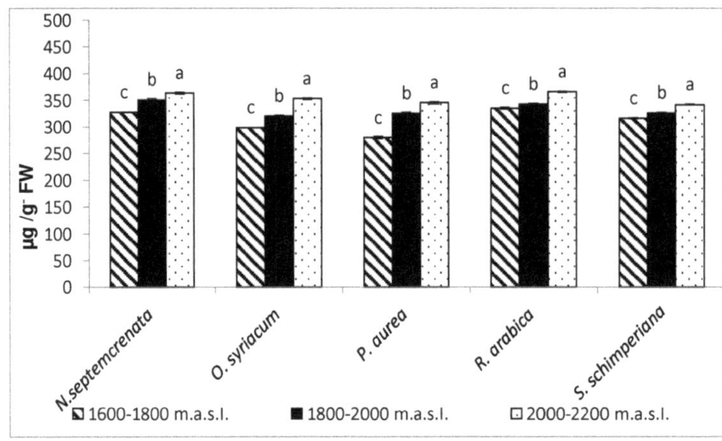

Figure 4. Effect of altitude on the total soluble proteins. Each value is the mean of three replicates ± SE. Bars with different letters are significantly different at $p \leq 0.05$.

2.3.2. Determination of Protein Banding Pattern

Plate 1 shows the change in protein banding patterns of the five species under investigation in response to a change in altitude of the SKP mountains from 1600 to 2200 m a.s.l. SDS-PAGE (Figure 5) revealed the presence of common bands in the leaves of different plant species grown at the same altitude of the SKP mountains. Common bands detected in plants grown at the relatively low altitude (1600–1800 m a.s.l.) were 70.63, 48, 40, 35, 17, 15, and 11 KDa. Nine protein bands (135, 100, 75, 48, 35, 25, 20, 15, and 11 KDa) were detected in protein extracts of plants grown at 1800–2000 m a.s.l. and nine protein bands (180, 135, 100, 48, 35, 24, 18, 15, and 11 KDa) in plant species grown at the highest altitude (2000–2200 m a.s.l.).

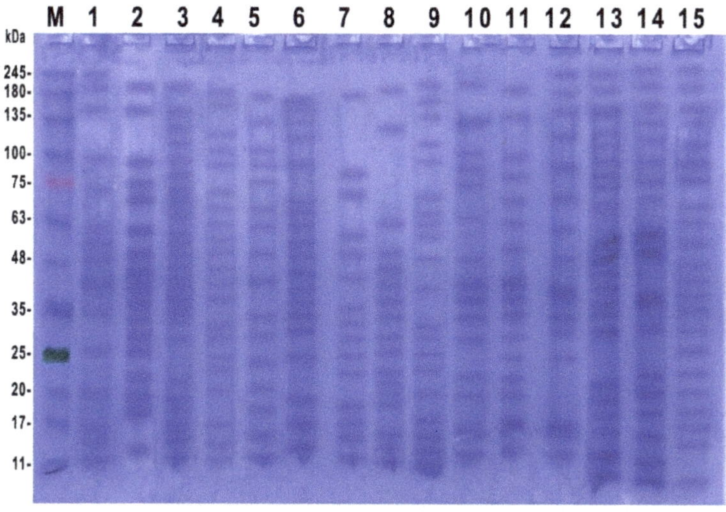

| | | Altitude | | |
Plant species	1600–1800 m a.s.l.	1800–2000 m a.s.l.	2000–2200 m a.s.l.
N. septemcrenata	Lane 1	Lane 2	Lane 3
O. syriacum	Lane 4	Lane 5	Lane 6
P. aurea	Lane 7	Lane 8	Lane 9
R. arabica	Lane 10	Lane 11	Lane 12
S. schimperiana	Lane 13	Lane 14	Lane 15

Figure 5. Change in protein banding pattern of different five plant species grown at different altitudes of SKP mountains.

2.3.3. Changes in the Proline Content

Data shown in Figure 6 indicate that, among all species and at different altitude levels, *Rosa arabica* recorded the highest value of proline content (1244.7 µg/100 g FW), especially at the highest altitude (2000–2200 m a.s.l.).

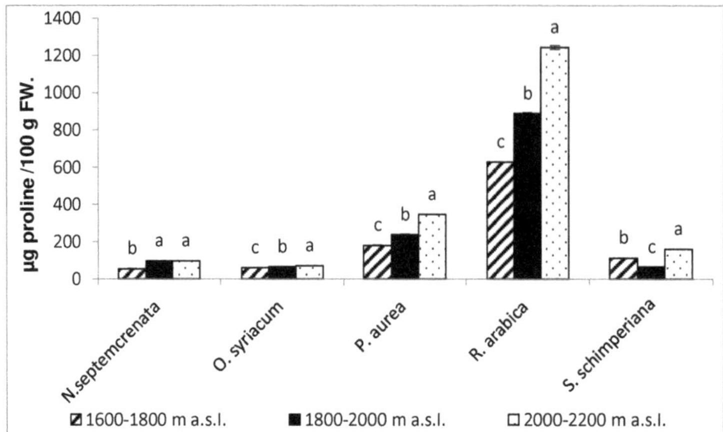

Figure 6. Change in proline contents for five plant species grown at different altitudes of SKP mountains. Each value is the mean of three replicates ± SE. Bars with different letters are significantly different at $p \leq 0.05$.

2.3.4. Changes in Total Antioxidant Capacity and Malondialdehyde (MDA) Contents

The increase in altitude induced a significant increase in total antioxidant capacity of the five investigated species, with the highest value in *R. arabica* (111.7 µg/g FW) at the highest altitude (2000–2200 m a.s.l.); Figure 7A. Concomitant with the increase in total antioxidant capacity, the content of MDA was mostly significantly decreased. MDA was used as an indicator for lipid peroxidation and hence oxidative stress; the increase was more pronounced in *S. schimperiana* (located at altitude 1800–2000 m a.s.l.) and *R. arabica* (located at first altitude 1600–1800 m a.s.l.) with values of 2.05 and 1.5 µmoles/g FW, respectively, as shown in Figure 7B.

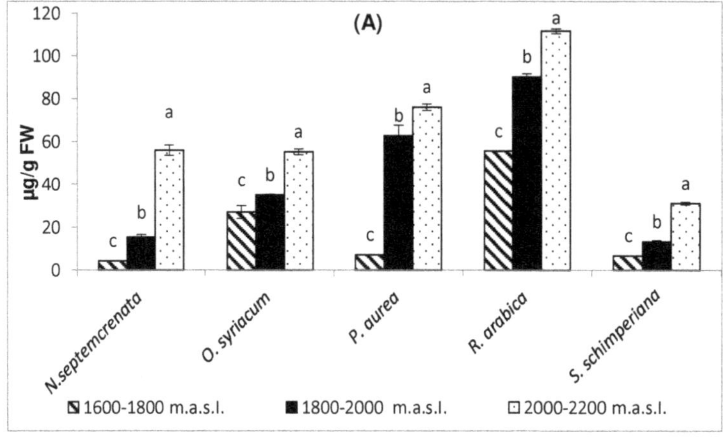

Figure 7. *Cont.*

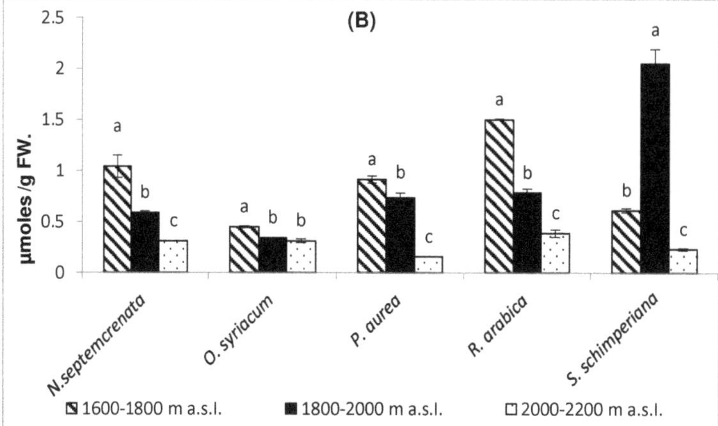

Figure 7. Effect of altitude on the total antioxidant capacity (**A**) and the content of malondialdehyde (**B**). Each value is the mean of three replicates ± SE. Bars with different letters are significantly different at $p \leq 0.05$.

2.3.5. Change in Activity Level of Some Antioxidant Enzymes

The present findings reveal that catalase (CAT), superoxide dismutase (SOD), polyphenol peroxidase (POD), and ascorbate peroxidase (APX) activity level decreased with altitude increase in all species, with the exceptions of *O. syriacum* (Figure 8A,C) and *R. arabica* (Figure 8A–D). CAT, POD, and APX activities of *R. arabica* increased significantly with increase in altitude and recorded their highest values (1.75 mM/g FW/min, 16.4 amount of quinon/g FW/min, and 16.5 mM ascorbate oxidized/g FW/min, respectively), at the highest altitude (2000–2200 m a.s.l.), whereas SOD activity increased to the maximum value (9.89 unit/mg protein) in *P. aurea* at the lowest altitude in our study (1600–1800 m a.s.l.).

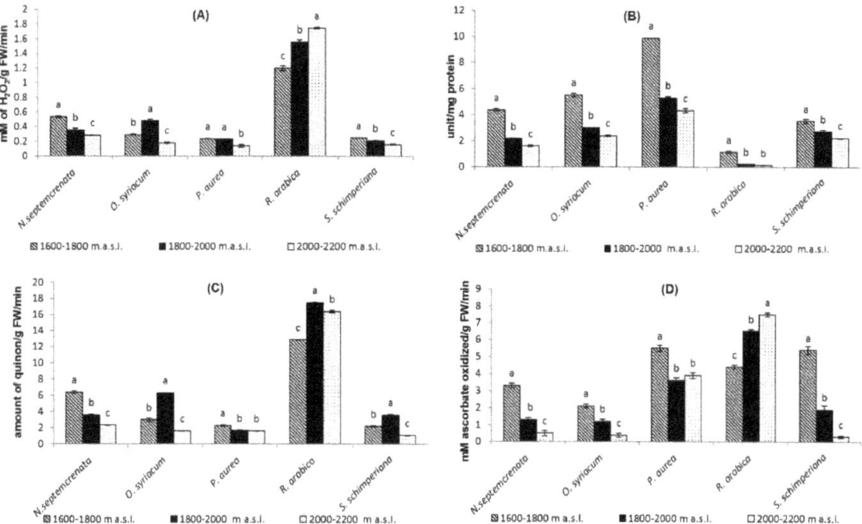

Figure 8. Effect of altitude on the activity of some antioxidant enzymes, catalase (**A**), superoxide dismutase (**B**), ascorbic acid peroxidase (**C**), and polyphenol peroxidase (**D**). Each value is the mean of three replicates ± SE. Bars with different letters are significantly different at $p \leq 0.05$.

2.3.6. Change in Ascorbic Acid Contents

Data represented in Figure 9 indicate that there was a significant increase in the content of ascorbic acid in most plant species with increase in altitude, with the exceptions of *O. syraicum* and *P. aurea*. The highest values (1.8 and 2.1 µmole/g FW) were recorded in *N. septemcrenata* and *Rosa arabica*, respectively, at the highest altitude (2000–2200 m a.s.l.).

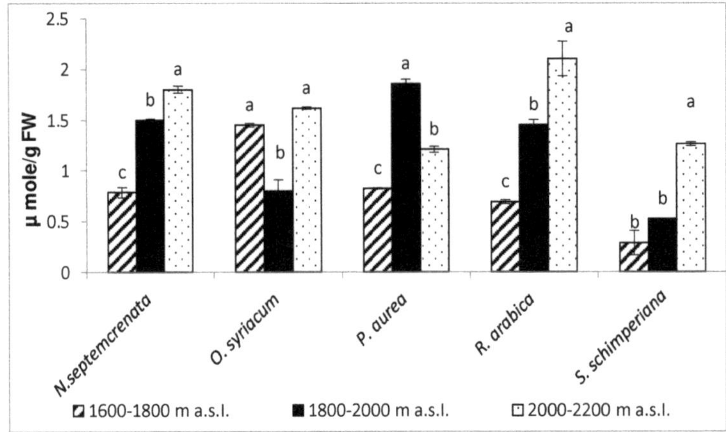

Figure 9. Change in contents of ascorbic acid of different five plant species grown at different altitudes of SKP mountains. Each value is the mean of three replicates ± SE. Bars with different letters are significantly different at $p \leq 0.05$.

3. Discussion

In mountainous environments, the altitudinal gradient is associated with a wide variation in environmental conditions, which affect plant distribution and population structure [42]. In high mountain regions, plants are challenged by unfavorable or even adverse abiotic environmental conditions that affect growth dynamics and threaten their existence. This is particularly the case for endemic endangered species [6]. With increasing altitude in the SKP mountains, there is a marked decrease in the distribution and intensity of the growth of plant species [43]. This may be due to the predominance of unfavorable climatic conditions at the highest region of the SKP mountains. The results obtained in the present work (Table 1) indicated that, with the increase in altitude in the SKP mountains, there is a decrease in soil pH and deficiency in macronutrients, which may be due to lower mineralization prevailing at low pH [44]. Other adverse abiotic factors of SKP as a highly elevated region include low temperature and deficiencies of oxygen [45,46], water precipitation, light intensity, and UV radiation [47,48].

Generally, in natural systems, a complex interplay between abiotic stressors and plant growth has resulted in several physiological traits by adaptation, acclimation, and speciation, which may differ between different plant species [49]. The photosynthetic process responds to the environment and maintains homeostasis by displaying several forms of adaptation [50]. Some secondary metabolites actively participate in this issue, particularly in the inhibition of chlorophyll photo-oxidation and accumulation of free radicals [51]. Moreover, as a derivative of photosynthetic products furnishes the secondary metabolic pathway with certain intermediates, this enhances photosynthetic performance via the positive feedback mechanism. The results obtained in the present work showed that with the increase in altitude there was a significant increase in the content of total soluble sugars (Figure 2) and total soluble proteins (Figure 4) in the plant species under investigation, with the highest content of total soluble sugars recorded in *Rosa Arabica* grown at 2000–2200 m a.s.l. In accordance with these results, Castrillo [52] found that higher altitudes triggered the accumulation of total soluble sugars in *Espeletia schultzii*. These data may reflect a higher photosynthetic performance of plants grown at the highest altitude. The data could be explained based on the increase in the concentration of HCO_3^- and Mg^{+2} in the soil of the highest altitude region (Table 1). Magnesium, as a component of the chlorophyll structure and a cofactor of some enzymes of the photosynthetic process, may enhance the photosynthetic process [53,54]. Moreover, there was a significant increase in the content of chlorophyll a, chlorophyll b, and carotenoids (Figure 1). Carotenoids, in addition to being active as light-harvesting pigments [55,56], act as a scavenger of singlet oxygen species and quench the triplet state of chlorophyll molecules [57].

Under abiotic stress, cellular homeostasis is disrupted, leading to the production of free radicals, which in turn can result in oxidative damage [58,59]. Moreover, in the high mountain region of SKP, among the more aggressive stressors are high light intensity and UV-B radiation [60]. Under these conditions, it is possible that electrons released from excited chlorophylls are transferred from photosystem I of the photosynthetic process to O_2 to form superoxide radicals, which initiate a chain of free radical liberation [61]. Hydrogen peroxide at low concentrations acts as a signaling molecule to induce the defense responses of plants under stresses but, at high levels, they may cause a substantial disturbance in the metabolism through damage of lipids of membranes and nucleic acids, conformational changes of enzymic proteins, destruction of thiol-containing compounds, etc. [62]. Soluble sugars were reported to be involved in defense mechanisms against stress via their efficiency in balancing ROS [63]. In addition, soluble sugars were reported to play a role in cold acclimation of plants [64]. SDS-PAGE of the protein extract of leaves of the five plant species under investigation is illustrated in Figure 5. The results refer to some differences in the pattern of the protein bands between different plant species and in the same species grown at different altitudes.

Virtually all organisms respond to environmental stress with the synthesis of a specific type of proteins [65]. However, in view of plant growth in SKP under a multitude of stressors, it would seem reasonable to detect the expression of a wide range of proteins. SDS-PAGE (Figure 5) reveals the presence of characteristic bands in the leaves of different plant species grown at the same altitude of the SKP mountains. The protein band with a molecular weight of 35 KDa, which was detected in various plant species irrespective of SKP altitude, is identified as one of the jasmonate-induced proteins that plays a crucial role in the defense against biotic and abiotic stresses [66]. Moreover, the protein with a molecular weight of 48 KDa, which has been identified to function as a proteinase inhibitors [67], was recorded in the five species grown at the three levels of the SKP mountains. In addition, the low molecular weight proteins having an antifungal activity (11 KDa) were detected in all plant species irrespective of the level of altitude of the SKP mountains. It must be stressed that the interpretations of these bands are speculative and require further in-depth investigation using advanced mass spectroscopic identification. Proline accumulation was reported to play a role in adaptation to stress or a consequence of stress [68]. It plays a role in redox homeostasis against ROS; it acts as an antioxidant [69]. In addition, proline acts as a molecular chaperone, capable of defending protein integrity from ROS [70].

A consequence of the abiotic stress prevailing in the SKP mountains, particularly at the highest altitude, is the qualitative and quantitative increase in a variety of SMs as flavonoids, phenolic compounds, alkaloids, carotenoids, steroids, tannins, and terpenoids [71,72]. In the present work, with the increase in altitude of the SKP mountains from 1600 to 2200 m a.s.l., there was a significant increase in the content of proline, total phenols, flavonoids, and tannins (Figures 3 and 6). In addition, phenolic antioxidants inhibit lipid peroxidation by trapping the free radicals, stabilize membranes by decreasing membrane fluidity, hinder the diffusion of free radicals, and restrict peroxidative reactions [73]. Through these interactions, the phenolic and flavonoid compounds increased the adaptation to abiotic oxidative stress. Tannins, on the other hand, have a crucial role against oxidative stress, particularly at high light intensity [74]. Tannins minimize oxidative damage via scavenging free radicals [75]. Tannins also bind to membranes, and the tannin-phospholipid complex may serve to moderate membrane morphology and permeability [76]. It is worth mentioning here that the plants with higher exposure to UV-B triggered phenylalanine ammonia-lyase. Phenylalanine ammonia-lyase is the key enzyme in the biosynthesis of the most important secondary metabolites [77], although these metabolites play a crucial role in plant adaptation to their adverse environmental conditions [78,79]. In addition to the adaptive function of SMs under adverse environments, it plays a crucial role in the maintenance of the stability of the primary metabolism, protecting it from disturbance induced by accumulation of specific metabolites via its conversion into secondary metabolites [80].

Plants successfully grown under oxidative stress actively initiate antioxidant systems, playing a role in their adaptation. Antioxidant defense systems can combat oxidative stress via scavenging of free radicals [81]. The results obtained (Figures 7–9) indicated that plant species grown at the relatively lower altitude (1600–1800 m a.s.l.), at which plants grow under relatively mild stress conditions, depend on scavenging free radicals on both the antioxidant enzymes and the antioxidant compounds [30,82]. A reverse pattern of change was observed in the activity level of antioxidant enzymes (catalase, superoxide dismutase, ascorbic acid peroxidase, and polyphenol peroxidase) in all species, with the exception of *R. arabica*, in which the increase in altitude mostly coincided with a corresponding increase in the activity level of antioxidant enzymes (Figure 8).

The modes of action of different antioxidants in the relief of oxidative stress vary [20,33,83]. Thus, plants that grows under stress conditions have many antioxidant molecules that represent the second line of defense against ROS, including ascorbic acid, carotenoids, glutathione, and phenolic compounds [84]. Ascorbic acid is considered the most potent antioxidant in plant tissues due to its capability of donating free electrons in many non-enzymatic and enzymatic reactions. Furthermore, ascorbic acid can scavenge $O_2^{\bullet-}$ and $OH^{\bullet}$ directly with greater ability to regenerate the oxidized carotenoids and, consequently, provide great protection to the cell membrane and minimize the oxidative damage synergically with other antioxidants [82].

Differences in environmental conditions (such as illumination, temperature, soil characteristics, and altitude) strongly contribute to the antioxidant activity and the amount of active ingredients in endemic medicinal plants [85]. In this study, many significant variations were detected in the antioxidant activity and chemical composition of five endemic targeted species collected from three different altitudes. It is clear that *R. arabica* depends mainly on increasing the activity of its antioxidant enzymes to adapt to high altitude, while the other investigated species tend to rely on antioxidant compounds as an adaptive response against high altitude to survive. Finally, the great variation in antioxidant activity and SM quantity and quality in these plants will possibly lead to considerable differences in their efficacy as herbal medicines [85].

4. Materials and Methods

4.1. Study Area

The current study was conducted in the south part of Sinai, specifically, in the mountainous region of SKP (Figure 10), which was declared a protectorate area by the Egyptian Environmental Affairs Agency (EEAA) in 1996. SKP is Egypt's fourth largest protectorate and is located between 28°30′ to 28°35′ N and 33°55′ to 34°30′ E. Its plateau altitude ranges between 1300 and more than 2600 m above sea level [43,86]. The area encompasses approximately 180 km² and is characterized by the presence of the highest rugged mountains in Egypt, namely Gebel Catherine (2624 m) and Gebel Mousa (2285 m), and the adjoining peaks [42]. This mountainous arid habitat supports an astounding biodiversity and a high share of rare and endemic plants because of its unique geological, morphological, and climatic aspects [43,87].

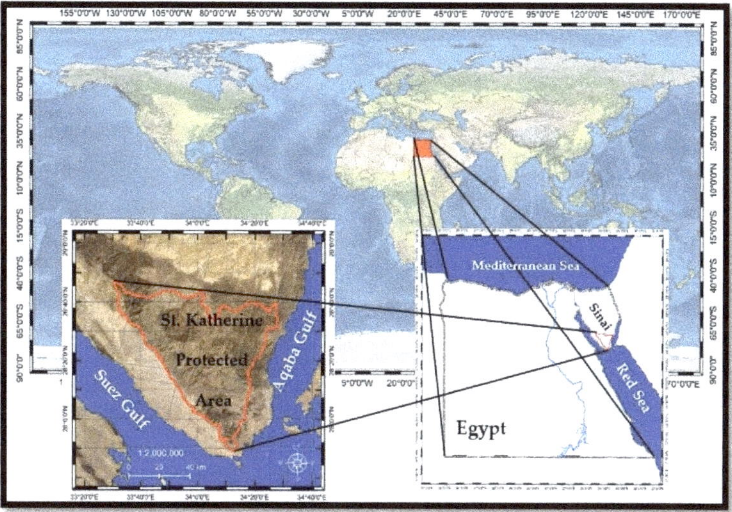

Figure 10. Study area map for Saint Katherine protected area, Sinai, Egypt (Using Landsat8 OLI 2019, Path/Row 178/39).

The plant materials were collected and field measurements carried out in April 2019 in mountainous habitats of SKP. The study area has been classified as one of the hyper-arid zones of the peninsula. SKP is the coolest area in Sinai owing to its high altitude [8,88]. The mean temperatures range from 5.4 to 25.2 °C, with the lowest temperature in January and February and the highest temperature in July and August [89]. Ayyad et al. [13] stated that high mountains in the SKP receive higher amounts of precipitation (100 mm/year) as rain and snow. The low altitude sites are climatically characterized by very dry summers with 5–30 mm precipitation per year. On the other hand, the high altitude district of South Sinai receives 35–50 mm of precipitation per year [90,91].

4.2. Target Species

Five endemic species were collected from three different sites with varying altitudes in the SKP mountains. For each species, five individual plants were taken and, for phytochemical and biochemical measurements, three replicates were taken from each. Details of the target species, including their scientific names with their families, altitude, and field photo are provided in Table 2. Samples from the shoot systems of the five plant species under investigation (namely, *Nepeta septemcrenata* Benth., *Origanum syriacum* subsp. *Sinaicum* (Boiss.) Greater and Burdet., *Phlomis aurea* Decene., *Rosa arabica* Crep., and *Silene schimperiana* Boiss.) were collected from three different altitudes of the SKP mountains (1600–1800, 1800–2000, and 2000–2200 m a.s.l.) and either kept frozen in a deep freezer (−20 °C) for extraction and estimation of enzyme, proline, photosynthetic pigments, total antioxidant capacity, ascorbic acid, malondialdehyde, and protein electrophoresis, or air-dried for extraction of carbohydrate and phenolic compounds. The identification of the five plant species was confirmed with the help of the Herbarium Section, Botany Department, Faculty of Science, Ain Shams University.

4.3. Soil Analysis

For each studied altitude, three soil samples were collected from profiles of 0–50 cm depth. Then, air-dried and thoroughly mixed together to form one composite sample. Textures were determined by sieving method to separate gravels, coarse sand, fine sand, silt, and clay. Determination of electric conductivity and pH was determined in soil–water (1:5) extracts using the potentiometric method [92,93]. Calcium and magnesium were determined volumetrically by the versene titration method described by Johnson and Ulrich [94]. Sodium and potassium were determined by flame photometry according to Shapiro and Brannock [95]. Estimation of chlorides was carried out by titration methods using 0.005N Silver Nitrate [93,96]. Sulphates were determined according to Bardsley and Lancaster [96].

4.4. Phytochemical Assay

4.4.1. Extraction and Estimation of Photosynthetic Pigments

The photosynthetic pigments including (Chl a, Chl b, and carotenoids) were extracted in 80% acetone and then estimated colorimetrically using Spectronic 601, Milton Roy company, USA according to the method described by Metzner et al. [97].

4.4.2. Extraction and Estimation of Total Soluble Sugars

Sugars were extracted according to Homme et al. [98], the air-dried tissue was boiled in 80% (*v/v*) ethanol, the filtrated extract was oven-dried at 60 °C, and then dissolved in water and was made to a known volume with water. The total soluble sugars contents were estimated with the anthrone reagent following the method of Loewus [99]. Finally, the concentration of soluble sugar was determined from the standard curve of glucose and calculated as µg glucose equivalent/g DW.

4.4.3. Extraction and Estimation of Total Phenolic Content

The total phenolic content was extracted and estimated by following the method adopted by Malik and Singh [100]. After extraction with ethanol (80%, *v/v*), the total phenolic content was estimated by Folin and Ciocalteau's reagent and the optical density of the reaction mixture was read at 750 nm. The concentrations were calculated from a standard curve of pyrogallol as gallic acid equivalents/g.

4.4.4. Extraction and Estimation of Total Flavonoids Content

The total flavonoids were extracted in methanol and estimated colorimetrically by using aluminum chloride based on the method described by Harborne [101]. The total flavonoid contents were calculated from a standard curve of quercetin and expressed as µg/g dry weight.

4.4.5. Extraction and Estimation of Tannins Content

Tannins were extracted and estimated as described by Ejikeme et al. [102]. In a conical flask 1 g of powdered plant tissue was added to 100 mL of distilled water. Then, boiled gently for 1 h on an electric hot plate and then filtered. The diluted extract (10 mL) was added to a conical flask containing 50 mL of distilled water, 5.0 mL of Folin–Denis reagent, and 10 mL of saturated Na_2CO_3 solution for color development. The mixture was left to react in a water bath at 25 °C for 30 min. Optical density was read at 700 nm and the concentration was calculated from a standard curve of tannic acid as following:

$$Tannic\ acid\ (mg/100\ g) = C \times extract\ volume \times 100 / Aliquot\ volume \times weight\ of\ sample \qquad (1)$$

where C is the concentration of tannic acid read off the graph.

Table 2. Families, scientific names, altitude, and field photos of the studied endemic species.

Family	Plant Species	Altitude (m a.s.l.)			Field Photo
		(1600–1800)	(1800–2000)	(2000–2200)	
Lamiaceae	*Nepeta septemcrenata* Benth.	1630	1945	2038	
	Origanum syriacum subsp. *Sinaicum* (Boiss.) Greater and Burdet.	1630	1825	2038	
	Phlomis aurea Decene.	1710	1825	2038	
Rosaceae	*Rosa arabica* Crep.	1750	1940	2150	
Caryophyllaceae	*Silene schimperiana* Boiss.	1750	1825	2110	

4.5. Biochemical Assay

4.5.1. Extraction and Estimation of Total Soluble Proteins

Total proteins were extracted by ground 0.5 g fresh tissue of leaves in 1 mL of phosphate buffer (0.1 M, pH 7.0) with a mortar and pestle and kept in ice. The protein concentration was estimated, and the absorbance was read at 595 nm on spectrophotometer based on the method described by Bradford [103].

4.5.2. Determination of Protein Banding Pattern

Total proteins were extracted from 0.5 g fresh tissue, the tissues were ground in liquid nitrogen. Then a few mL of tris buffer was added (1:2, tissue:buffer). The tris-HCl buffer contained 0.1 mM tris, pH 7.5, 4 mM B-mercaptoethanol, 0.1 mM EDTA-Na$_2$, 10 mM KCl, and 10 mM MgCl$_2$. The crude homogenate was centrifuged at 10,000× g for 20 min. The supernatant was used for gel analysis by SDS-polyacrylamide gel electrophoresis (SDS-PAGE) according to the method of Laemmli [104].

4.5.3. Estimation of Proline

Free proline was estimated by using ninhydrin reagent according to the method described by Bates et al. [105]. Proline concentration was measured from a standard curve of proline and expressed as μg/g fresh weight.

4.5.4. Extraction and Estimation of Malondialdehyde

Lipid peroxidation level was determined by measuring the amount of malondialdehyde (MDA) produced by the thiobarbituric acid reaction as described by Heath and Packer [106]. The crude extract was mixed with the same volume of a 0.5% thiobarbituric acid solution containing 20% trichloroacetic acid. The reaction mixture was incubated at 95 °C for 30 min and then cooled in an ice-bath. The mixture was centrifuged at 3000× g for 5 min. The absorbance of the supernatant was recorded at 532 and 600 nm. The MDA concentration was calculated by dividing the difference in absorbance (A_{532}–A_{600}) by its molar extinction coefficient (155 mM^{-1} cm^{-1}), and the results expressed as μmol g^{-1} fresh weight.

4.6. Enzyme Extraction and Assays

Extraction and Assaying Activity of Certain Enzymes

The method adopted in enzyme extraction was that described by Mukheriee and Choudhuri, [107]. A fresh tissue (250 mg) was frozen in liquid nitrogen and finely grounded by pestle in a chilled mortar. The frozen powder was added to 10 mL of 100 mM phosphate buffer (KH$_2$PO$_4$/K$_2$HPO$_4$, pH 6.8). The homogenates were centrifuged at 20,000× g for 20 min. The supernatant was made up to a known volume with the same buffer and used as enzyme preparation for assaying the activity of certain enzymes.

Superoxide dismutase activity was measured according to the method of Dhindsa et al. [108]. A total of 3 mL of assay mixture contained 13 mM methionine, 0.025 mM p–nitro blue tetrazolium chloride (NBT), 0.1 mM EDTA, 50 mM phosphate buffer (pH 7.8), 50 mM sodium bicarbonate, and 0.5 mL enzyme extract. The reaction was started by adding riboflavin (0.002 mM) and incubating the tubes below two fluorescent lamps (15 W) for 15 min. The reaction was stopped by switching off the light and covering the tubes with black cloth. The tubes without enzyme developed maximal colors. A no irradiated complete reaction mixture served as blank. The absorbance was measured at 560 nm using Spekol spectrocolourimeter VEB Carl Zeiss. The enzyme activity was calculated as unit/mg protein.

Catalase activity was assayed according to the method of Hermans et al. [109]. The reaction mixture with final volume of 10 mL, containing 40 μL enzyme extract, was added to 9.96 mL of H$_2$O$_2$ contained in phosphate buffer, pH 7.0 (0.16 mL of 30% H$_2$O$_2$ to 100 mL of 50 mM phosphate buffer). CAT activity was determined by measuring the rate of change of H$_2$O$_2$ absorbance in 60 s using a Spekol spectrocolourimeter VEB Carl Zeiss at 250 nm. The blank sample was made by using buffer instead of enzyme extract. The enzyme activity was calculated as mM of H$_2$O$_2$/g FW/min.

Peroxidase activity was assayed according to the method of Kar and Mishra [110] after slight modifications. A total of 5 mL of the reaction mixture contained 300 μM of phosphate buffer (pH 6.8),

50 μM catechol, 50 μM H_2O_2 was added to 1 mL of crude enzyme extract. After incubation at 25 °C for 5 min, 1 mL of 10% H_2SO_4 was added for stopping the reaction. The optical density was measured at 340 nm, and the activity was expressed as the amount of quinone/g fresh weight/min.

APX activity was estimated following the method of Koricheva et al. [111] after slight modifications. The reaction mixture (10 mL) contained 5.5 mL of 50 mM phosphate buffer (pH 7.0), 0.5 mL of the enzyme extract, 1 mL 20 mM H_2O_2, 1 mL 20 mM EDTA, and 2 mL of 20 mM ascorbic acid. After ascorbate oxidation the rate of decrease in absorbance was recorded at 290 nm using a UV spectrophotometer (Unicam Heλios Gamma and Delta). The enzyme activity was expressed as mM of ascorbate oxidized/g fresh weight/min.

4.7. Determination of Ascorbic Acid

Ascorbic acid was extracted estimated according to the methods of Kampfenkel et al. [112]. Fresh tissue (0.1 g) was homogenized in 1 mL 6% (*w/v*) trichloroacetic acid (TCA) solution and the homogenate was centrifuged at 12,000× *g* and 4 °C for 10 min. The supernatant was used for estimation of ascorbic acid.

4.8. Determination of Total Antioxidant Capacity

Total antioxidant capacity of the extract was evaluated by the phosphomolybdenum method as described by Prieto et al. [113]. A 0.3 mL extract was mixed with 3 mL of reagent solution (0.6 M sulfuric acid, 28 mM sodium phosphate, and 4 mM ammonium molybdate). The tubes containing the reaction mixture were incubated in water bath at 95 °C for 90 min. After cooling to room temperature, the absorbance of the mixture was measured at 695 nm against the blank. In the blank methanol (0.3 mL) was used in the place of extract.

4.9. Statistical Analysis

Analyses of variance (ANOVA) for all experimental results presented in this study were calculated using SPSS v20.0 (SPSS Inc., Chicago, USA) analyzing software. Statistical significances of the means were compared with Duncan's test at $p \leq 0.05$ levels, the standard error (SE) of the means were presented in tables, and figures are means ± SE ($n = 3$).

Author Contributions: Conceptualization, A.M.H., A.E., A.M.A., B.M.A. and H.M.H.; methodology, A.E., W.N.H. and H.M.H.; formal analysis, A.M.H.; and H.M.H. investigation, A.E., W.N.H.; resources, A.M.A., H.M.H., W.N.H. and B.M.A.; data curation, H.M.H., A.M.A. and W.N.H.; writing—original draft preparation, A.M.H., H.M.H., A.E., B.M.A.; writing—review and editing, A.M.H., H.M.H., A.E., W.N.H., A.M.A. and B.M.A. All authors have read and agreed to the published version of the manuscript.

Funding: The authors are grateful to the Deanship of Scientific Research, King Saud University for funding through Vice Deanship of Scientific Research Chairs.

Acknowledgments: The authors are grateful to the Deanship of Scientific Research, King Saud University for funding through Vice Deanship of Scientific Research Chairs.

Conflicts of Interest: The authors declare no conflict of interest.

References

1. Williams, S.E.; Williams, Y.M.; VanDerWal, J.; Isaac, J.L.; Shoo, L.P.; Johnson, C.N. Ecological specialization and population size in a biodiversity hotspot: How rare species avoid extinction. *Proc. Natl. Acad. Sci. USA* **2009**, *106*, 19737–19741. [CrossRef] [PubMed]
2. Alhaithloul, H.A.S. Impact of Combined Heat and Drought Stress on the Potential Growth Responses of the Desert Grass Artemisia sieberi alba: Relation to Biochemical and Molecular Adaptation. *Plants* **2019**, *8*, 416. [CrossRef] [PubMed]

3. Habib, N.; Ali, Q.; Ali, S.; Javed, M.T.; Zulqurnain Haider, M.; Perveen, R.; Shahid, M.R.; Rizwan, M.; Abdel-Daim, M.M.; Elkelish, A.; et al. Use of Nitric Oxide and Hydrogen Peroxide for Better Yield of Wheat (*Triticum aestivum* L.) under Water Deficit Conditions: Growth, Osmoregulation, and Antioxidative Defense Mechanism. *Plants* **2020**, *9*, 285. [CrossRef] [PubMed]

4. Grabherr, G.; Gottfried, M.; Pauli, H. Climate effects on mountain plants. *Nature* **1994**, *369*, 448. [CrossRef]

5. Moustafa-Farag, M.; Almoneafy, A.; Mahmoud, A.; Elkelish, A.; Arnao, M.B.; Li, L.; Ai, S. Melatonin and Its Protective Role against Biotic Stress Impacts on Plants. *Biomolecules* **2020**, *10*, 54. [CrossRef]

6. El-Keblawy, A.A.; Khedr, A.-H.A. Population structure and ecological role of *Moringa peregrina* (Forssk.) Fiori. at its northwestern range edge in the Hajar Mountains. *Plant Biosyst. Int. J. Deal. All Asp. Plant Biol.* **2017**, *151*, 29–38. [CrossRef]

7. Elkelish, A.A.; Alhaithloul, H.A.S.; Qari, S.H.; Soliman, M.H.; Hasanuzzaman, M. Pretreatment with Trichoderma harzianum alleviates waterlogging-induced growth alterations in tomato seedlings by modulating physiological, biochemical, and molecular mechanisms. *Environ. Exp. Bot.* **2019**, 103946. [CrossRef]

8. Moustafa, A.R.A.; Klopatek, J.M. Vegetation and landforms of the Saint Catherine area, southern Sinai, Egypt. *J. Arid Environ.* **1995**, *30*, 385–395. [CrossRef]

9. Abdel-Azeem, A.; Nada, A.A.; O'Donovan, A.; Kumar Thakur, V.; Elkelish, A. Mycogenic Silver Nanoparticles from Endophytic Trichoderma atroviride with Antimicrobial Activity. *J. Renew. Mater.* **2019**, *7*, 171–185. [CrossRef]

10. Ayyad, M.A.; Fakhry, A.M.; Moustafa, A.-R.A. Plant biodiversity in the Saint Catherine area of the Sinai peninsula, Egypt. *Biodivers. Conserv.* **2000**, *9*, 265–281. [CrossRef]

11. Tackholm, V. *Students' flora of Egypt 1974*, 2nd ed.; Cairo University Publishing: Beirut, Lebanon, 1974; 888p.

12. IUCN. The 100 Most Threatened Species. Are They Priceless or Worthless? Available online: https://www.iucn. org/content/100-most-threatened-species-are-they-priceless-or-worthless (accessed on 21 March 2020).

13. El-Esawi, M.A.; Alaraidh, I.A.; Alsahli, A.A.; Alzahrani, S.M.; Ali, H.M.; Alayafi, A.A.; Ahmad, M. Serratia liquefaciens KM4 Improves Salt Stress Tolerance in Maize by Regulating Redox Potential, Ion Homeostasis, Leaf Gas Exchange and Stress-Related Gene Expression. *Int. J. Mol. Sci.* **2018**, *19*, 3310. [CrossRef]

14. Shaltout, K.H.; El-Hamdi, K.H.; El-Masry, S.A.; Eid, E.M. Bedouin farms in the Saint Katherine mountainous area (South Sinai, Egypt). *J. Mt. Sci.* **2019**, *16*, 2232–2242. [CrossRef]

15. Boulos, L. *Flora of Egypt. Checklist*; Al-Hadara Publishing: Cairo, Egypt, 1995.

16. El-Demerdash, M. The ex Situ Conservation Technical Report on Propagation of Medicinal Plants; MPCP, The Egyptian Environmental Affairs Agency (EEAA): Cairo, Egypt. Available online: http://www.eeaa.gov. eg/en-us/topics/nature/biodiversity/plantgeneticresources/floraconservation.aspx (accessed on 1 April 2020).

17. Shaltout, K.H.; Ahmed, D.A.; Shabana, H.A. Population structure and dynamics of the endemic species Phlomis aurea Decne in different habitats in southern Sinai Peninsula, Egypt. *Glob. Ecol. Conserv.* **2015**, *4*, 505–515. [CrossRef]

18. Roupioz, L.; Jia, L.; Nerry, F.; Menenti, M. Estimation of Daily Solar Radiation Budget at Kilometer Resolution over the Tibetan Plateau by Integrating MODIS Data Products and a DEM. *Remote Sens.* **2016**, *8*, 504. [CrossRef]

19. Wonsick, M.M.; Pinker, R.T. The radiative environment of the Tibetan Plateau. *Int. J. Climatol.* **2014**, *34*, 2153–2162. [CrossRef]

20. Elkeilsh, A.; Awad, Y.M.; Soliman, M.H.; Abu-Elsaoud, A.; Abdelhamid, M.T.; El-Metwally, I.M. Exogenous application of β-sitosterol mediated growth and yield improvement in water-stressed wheat (*Triticum aestivum*) involves up-regulated antioxidant system. *J. Plant Res.* **2019**. [CrossRef] [PubMed]

21. Elkelish, A.A.; Soliman, M.H.; Alhaithloul, H.A.; El-Esawi, M.A. Selenium protects wheat seedlings against salt stress-mediated oxidative damage by up-regulating antioxidants and osmolytes metabolism. *Plant Physiol. Biochem.* **2019**. [CrossRef]

22. Abdelaal, K.A.; EL-Maghraby, L.M.; Elansary, H.; Hafez, Y.M.; Ibrahim, E.I.; El-Banna, M.; El-Esawi, M.; Elkelish, A. Treatment of Sweet Pepper with Stress Tolerance-Inducing Compounds Alleviates Salinity Stress Oxidative Damage by Mediating the Physio-Biochemical Activities and Antioxidant Systems. *Agronomy* **2019**, *10*, 26. [CrossRef]

23. Soliman, M.H.; Abdulmajeed, A.M.; Alhaithloul, H.; Alharbi, B.M.; El-Esawi, M.A.; Hasanuzzaman, M.; Elkelish, A. Saponin biopriming positively stimulates antioxidants defense, osmolytes metabolism and ionic status to confer salt stress tolerance in soybean. *Acta Physiol. Plant* **2020**, *42*, 114. [CrossRef]

24. El-Esawi, M.A.; Elkelish, A.; Soliman, M.; Elansary, H.O.; Zaid, A.; Wani, S.H. Serratia marcescens BM1 Enhances Cadmium Stress Tolerance and Phytoremediation Potential of Soybean Through Modulation of Osmolytes, Leaf Gas Exchange, Antioxidant Machinery, and Stress-Responsive Genes Expression. *Antioxidants* **2020**, *9*, 43. [CrossRef]

25. Soliman, M.H.; Alayafi, A.A.M.; El Kelish, A.A.; Abu-Elsaoud, A.M. Acetylsalicylic acid enhance tolerance of *Phaseolus vulgaris* L. to chilling stress, improving photosynthesis, antioxidants and expression of cold stress responsive genes. *Bot. Stud.* **2018**, *59*. [CrossRef] [PubMed]

26. Saleem, M.H.; Ali, S.; Rehman, M.; Rana, M.S.; Rizwan, M.; Kamran, M.; Imran, M.; Riaz, M.; Soliman, M.H.; Elkelish, A.; et al. Influence of phosphorus on copper phytoextraction via modulating cellular organelles in two jute (*Corchorus capsularis* L.) varieties grown in a copper mining soil of Hubei Province, China. *Chemosphere* **2020**, *248*, 126032. [CrossRef] [PubMed]

27. Batiha, G.E.-S.; Magdy Beshbishy, A.; Adeyemi, O.S.; Nadwa, E.H.; Rashwan, E.; Kadry, M.; Alkazmi, L.M.; Elkelish, A.A.; Igarashi, I. Phytochemical Screening and Antiprotozoal Effects of the Methanolic Berberis vulgaris and Acetonic Rhus coriaria Extracts. *Molecules* **2020**, *25*, 550. [CrossRef]

28. Rashad, Y.; Aseel, D.; Hammad, S.; Elkelish, A. Rhizophagus irregularis and Rhizoctonia solani Differentially Elicit Systemic Transcriptional Expression of Polyphenol Biosynthetic Pathways Genes in Sunflower. *Biomolecules* **2020**, *10*, 379. [CrossRef] [PubMed]

29. Julkunen-Tiitto, R.; Rousi, M.; Bryant, J.; Sorsa, S.; Keinänen, M.; Sikanen, H. Chemical diversity of several Betulaceae species: Comparison of phenolics and terpenoids in northern birch stems. *Trees* **1996**, *11*, 16–22. [CrossRef]

30. Gong, J.; Zhang, Z.; Zhang, C.; Zhang, J.; Ran, A. Ecophysiological Responses of Three Tree Species to a High-Altitude Environment in the Southeastern Tibetan Plateau. *Forests* **2018**, *9*, 48. [CrossRef]

31. Gea-Izquierdo, G.; Fonti, P.; Cherubini, P.; Martín-Benito, D.; Chaar, H.; Cañellas, I. Xylem hydraulic adjustment and growth response of *Quercus canariensis* Willd. to climatic variability. *Tree Physiol.* **2012**, *32*, 401–413. [CrossRef]

32. Polle, A.; Baumbusch, L.O.; Oschinski, C.; Eiblmeier, M.; Kuhlenkamp, V.; Vollrath, B.; Scholz, F.; Rennenberg, H. Growth and protection against oxidative stress in young clones and mature spruce trees (*Picea abies* L.) at high altitudes. *Oecologia* **1999**, *121*, 149–156. [CrossRef]

33. Elkelish, A.; Qari, S.H.; Mazrou, Y.S.A.; Abdelaal, K.A.A.; Hafez, Y.M.; Abu-Elsaoud, A.M.; Batiha, G.E.-S.; El-Esawi, M.A.; El Nahhas, N. Exogenous Ascorbic Acid Induced Chilling Tolerance in Tomato Plants Through Modulating Metabolism, Osmolytes, Antioxidants, and Transcriptional Regulation of Catalase and Heat Shock Proteins. *Plants* **2020**, *9*, 431. [CrossRef]

34. Soliman, M.; Elkelish, A.; Souad, T.; Alhaithloul, H.; Farooq, M. Brassinosteroid seed priming with nitrogen supplementation improves salt tolerance in soybean. *Physiol. Mol. Biol. Plants* **2020**. [CrossRef]

35. Conklin, P.L. Recent advances in the role and biosynthesis of ascorbic acid in plants. *Plant Cell Environ.* **2001**, *24*, 383–394. [CrossRef]

36. Soliman, M.; Alhaithloul, H.A.; Hakeem, K.R.; Alharbi, B.M.; El-Esawi, M.; Elkelish, A. Exogenous Nitric Oxide Mitigates Nickel-Induced Oxidative Damage in Eggplant by Upregulating Antioxidants, Osmolyte Metabolism, and Glyoxalase Systems. *Plants* **2019**, *8*, 562. [CrossRef]

37. Zamin, M.; Fahad, S.; Khattak, A.M.; Adnan, M.; Wahid, F.; Raza, A.; Wang, D.; Saud, S.; Noor, M.; Bakhat, H.F.; et al. Developing the first halophytic turfgrasses for the urban landscape from native Arabian desert grass. *Env. Sci. Pollut. Res.* **2019**. [CrossRef] [PubMed]

38. Streb, P.; Aubert, S.; Gout, E.; Bligny, R. Reversibility of cold- and light-stress tolerance and accompanying changes of metabolite and antioxidant levels in the two high mountain plant species *Soldanella alpina* and *Ranunculus glacialis*. *J. Exp. Bot.* **2003**, *54*, 405–418. [CrossRef]

39. Öncel, I.; Yurdakulol, E.; Keleş, Y.; Kurt, L.; Yıldız, A. Role of antioxidant defense system and biochemical adaptation on stress tolerance of high mountain and steppe plants. *Acta Oecologica* **2004**, *26*, 211–218. [CrossRef]

40. Hegazy, A.; Doust, J.L. *Plant Ecology in the Middle East*; Oxford University Press: Oxford, UK, 2016; ISBN 978-0-19-966081-0.

41. Helal, N.M.; Alharby, H.F.; Alharbi, B.M.; Bamagoos, A.A.; Hashim, A.M. *Thymelaea hirsuta* and *Echinops spinosus*: Xerophytic Plants with High Potential for First-Generation Biodiesel Production. *Sustainability* **2020**, *12*, 1137. [CrossRef]

42. Abd El-Wahab, R.; Zayed, A.E.-M.; Moustafa, A.E.-R.; Klopatek, J.; Helmy, M. Landforms, Vegetation, and Soil Quality in South Sinai, Egypt. *Catrina Int. J. Environ. Sci.* **2018**, *1*, 127–138.

43. Moustafa, A.E.-R.A.; Zaghloul, M.S. Environment and vegetation in the montane Saint Catherine area, south Sinai, Egypt. *J. Arid Environ.* **1996**, *34*, 331–349. [CrossRef]

44. Neina, D. The Role of Soil pH in Plant Nutrition and Soil Remediation. Available online: https://www.hindawi.com/journals/aess/2019/5794869/ (accessed on 12 April 2020).

45. Santiago, L.S. Use of Coarse Woody Debris by the Plant Community of a Hawaiian Montane Cloud Forest1. *Biotropica* **2000**, *32*, 633–641. [CrossRef]

46. Soethe, N.; Wilcke, W.; Homeier, J.; Lehmann, J.; Engels, C. Plant Growth along the Altitudinal —Role of Plant Nutritional Status, Fine Root Activity, and Soil Properties. In *Gradients in a Tropical Mountain Ecosystem of Ecuador*; Ecological Studies; Beck, E., Bendix, J., Kottke, I., Makeschin, F., Mosandl, R., Eds.; Springer: Berlin/Heidelberg, Germany, 2008; pp. 259–266. ISBN 978-3-540-73526-7.

47. Sharma, P.; Rana, J.C.; Devi, U.; Randhawa, S.S.; Kumar, R. Floristic Diversity and Distribution Pattern of Plant Communities along Altitudinal Gradient in Sangla Valley, Northwest Himalaya. Available online: https://www.hindawi.com/journals/tswj/2014/264878/ (accessed on 12 April 2020).

48. Chen, Y.; Zhang, X.; Guo, Q.; Cao, L.; Qin, Q.; Li, C.; Zhao, M.; Wang, W. Plant morphology, physiological characteristics, accumulation of secondary metabolites and antioxidant activities of *Prunella vulgaris* L. under UV solar exclusion. *Biol. Res.* **2019**, *52*. [CrossRef]

49. Pandey, P.; Irulappan, V.; Bagavathiannan, M.V.; Senthil-Kumar, M. Impact of Combined Abiotic and Biotic Stresses on Plant Growth and Avenues for Crop Improvement by Exploiting Physio-morphological Traits. *Front. Plant Sci.* **2017**, *8*. [CrossRef] [PubMed]

50. Lambers, H.; Oliveira, R.S. *Plant Physiological Ecology*; Springer Nature: Berlin/Heidelberg, Germany, 2019; ISBN 978-3-030-29639-1.

51. Guidi, L.; Tattini, M.; Landi, M. How Does Chloroplast Protect Chlorophyll Against Excessive Light? In *Chlorophyll*; Jacob-Lopes, E., Zepka, L.Q., Queiroz, M.I., Eds.; InTech: Vienna, Austria, 2017; ISBN 978-953-51-3107-6.

52. Castrillo, M. Photosynthesis in three altitudinal populations of the Andean plant *Espeletia schultzii* (Compositae). *Rev. Biol. Trop.* **2006**, *54*, 1143–1149. [CrossRef] [PubMed]

53. Guo, W.; Nazim, H.; Liang, Z.; Yang, D. Magnesium deficiency in plants: An urgent problem. *Crop J.* **2016**, *4*, 83–91. [CrossRef]

54. Elkelish, A.; Alnusaire, T.S.; Soliman, M.H.; Gowayed, S.; Senousy, H.H.; Fahad, S. Calcium availability regulates antioxidant system, physio-biochemical activities and alleviates salinity stress mediated oxidative damage in soybean seedlings. *J. Appl. Bot. Food Qual.* **2019**, *258–266*. [CrossRef]

55. D'angiolillo, F.; Mammano, M.M.; Fascella, G. Pigments, Polyphenols and Antioxidant Activity of Leaf Extracts from Four Wild Rose Species Grown in Sicily. *Not. Bot. Horti Agrobot. Cluj-Napoca* **2018**, *46*, 402–409. [CrossRef]

56. González, J.A.; Gallardo, M.G.; Boero, C.; Liberman Cruz, M.; Prado, F.E. Altitudinal and seasonal variation of protective and photosynthetic pigments in leaves of the world's highest elevation trees *Polylepis tarapacana* (Rosaceae). *Acta Oecologica* **2007**, *32*, 36–41. [CrossRef]

57. Havaux, M. Carotenoid oxidation products as stress signals in plants. *Plant J.* **2014**, *79*, 597–606. [CrossRef]

58. Kusvuran, S. Influence of drought stress on growth, ion accumulation and antioxidative enzymes in okra genotypes. *Int. J. Agric. Biol.* **2012**, *14*, 401–406.

59. Kusvuran, S.; Kiran, S.; Ellialtioglu, S.S. Antioxidant Enzyme Activities and Abiotic Stress Tolerance Relationship in Vegetable Crops. In *Abiotic and Biotic Stress in Plants—Recent Advances and Future Perspectives*; Shanker, A.K., Shanker, C., Eds.; InTech: Vienna, Austria, 2016; ISBN 978-953-51-2250-0.

60. El-Ghani, M.M.A.; Huerta-Martínez, F.M.; Hongyan, L.; Qureshi, R. *Plant Responses to Hyperarid Desert Environments*; Springer: Berlin/Heidelberg, Germany, 2017; ISBN 978-3-319-59135-3.

61. Logan, B.A. Reactive Oxygen Species and Photosynthesis. In *Antioxidants and Reactive Oxygen Species in Plants*; John Wiley & Sons, Ltd.: Hoboken, NJ, USA, 2007; pp. 250–267. ISBN 978-0-470-98856-5.

62. Sharma, P.; Dubey, R.S. Involvement of oxidative stress and role of antioxidative defense system in growing rice seedlings exposed to toxic concentrations of aluminum. *Plant Cell Rep.* **2007**, *26*, 2027–2038. [CrossRef]

63. Couée, I.; Sulmon, C.; Gouesbet, G.; El Amrani, A. Involvement of soluble sugars in reactive oxygen species balance and responses to oxidative stress in plants. *J. Exp. Bot.* **2006**, *57*, 449–459. [CrossRef]

64. Li, S.; Yang, Y.; Zhang, Q.; Liu, N.; Xu, Q.; Hu, L. Differential physiological and metabolic response to low temperature in two zoysiagrass genotypes native to high and low latitude. *PLoS ONE* **2018**, *13*, e0198885. [CrossRef] [PubMed]

65. Gidalevitz, T.; Prahlad, V.; Morimoto, R.I. The Stress of Protein Misfolding: From Single Cells to Multicellular Organisms. *Cold Spring Harb Perspect. Biol.* **2011**, *3*. [CrossRef] [PubMed]

66. Andresen, I.; Becker, W.; Schlüter, K.; Burges, J.; Parthier, B.; Apel, K. The identification of leaf thionin as one of the main jasmonate-induced proteins of barley (*Hordeum vulgare*). *Plant Mol. Biol.* **1992**, *19*, 193–204. [CrossRef] [PubMed]

67. Hemantaranjan, A. *Advances in Plant Physiology*; Scientific Publishers: Oxford, UK, 2013; Volume 14, ISBN 978-93-86237-59-0.

68. Ashraf, M.; Foolad, M.R. Roles of glycine betaine and proline in improving plant abiotic stress resistance. *Environ. Exp. Bot.* **2007**, *59*, 206–216. [CrossRef]

69. Szabados, L.; Savouré, A. Proline: A multifunctional amino acid. *Trends Plant Sci.* **2010**, *15*, 89–97. [CrossRef]

70. Liang, X.; Zhang, L.; Natarajan, S.K.; Becker, D.F. Proline mechanisms of stress survival. *Antioxid. Redox Signal.* **2013**, *19*, 998–1011. [CrossRef]

71. ElKelish, A.E.; Winkler, J.B.; Lang, H.; Holzinger, A.; Behrendt, H.; Durner, J.; Kanter, U.; Ernst, D. *Effects of ozone, CO$_2$ and drought stress on the growth and pollen production of common ragweed (Ambrosia artemisiifolia)*; Julius Kühn Institut, Bundesforschungsinstitut für Kulturpflanzen: Quedlinburg, Germany, 2014; pp. 139–147.

72. Cirak, C.; Radusiene, J.; Jakstas, V.; Ivanauskas, L.; Seyis, F.; Yayla, F. Altitudinal changes in secondary metabolite contents of *Hypericum androsaemum* and *Hypericum polyphyllum*. *Biochem. Syst. Ecol.* **2017**, *70*, 108–115. [CrossRef]

73. Watson, R.R. *Polyphenols in Plants: Isolation, Purification and Extract Preparation*; Academic Press: New York, NY, USA, 2014; ISBN 978-0-12-398491-3.

74. Bartwal, A.; Mall, R.; Lohani, P.; Guru, S.K.; Arora, S. Role of Secondary Metabolites and Brassinosteroids in Plant Defense Against Environmental Stresses. *J. Plant Growth Regul.* **2013**, *32*, 216–232. [CrossRef]

75. Gourlay, G.; Constabel, C.P. Condensed tannins are inducible antioxidants and protect hybrid poplar against oxidative stress. *Tree Physiol.* **2019**, *39*, 345–355. [CrossRef]

76. Hasanuzzaman, M.; Hossain, M.A.; da Silva, J.A.T.; Fujita, M. Plant Response and Tolerance to Abiotic Oxidative Stress: Antioxidant Defense Is a Key Factor. In *Crop Stress and Its Management: Perspectives and Strategies*; Venkateswarlu, B., Shanker, A.K., Shanker, C., Maheswari, M., Eds.; Springer: Dordrecht, The Netherlands, 2012; pp. 261–315. ISBN 978-94-007-2220-0.

77. Dai, L.-P.; Xiong, Z.-T.; Huang, Y.; Li, M.-J. Cadmium-induced changes in pigments, total phenolics, and phenylalanine ammonia-lyase activity in fronds of *Azolla imbricata*. *Env. Toxicol.* **2006**, *21*, 505–512. [CrossRef]

78. Morales, F.; Ancín, M.; Fakhet, D.; González-Torralba, J.; Gámez, A.L.; Seminario, A.; Soba, D.; Ben Mariem, S.; Garriga, M.; Aranjuelo, I. Photosynthetic Metabolism under Stressful Growth Conditions as a Bases for Crop Breeding and Yield Improvement. *Plants* **2020**, *9*, 88. [CrossRef] [PubMed]

79. Sai, K.; Thapa, R.; Devkota, H.P.; Joshi, K.R. Phytochemical Screening, Free Radical Scavenging and α-Amylase Inhibitory Activities of Selected Medicinal Plants from Western Nepal. *Medicines* **2019**, *6*, 70. [CrossRef] [PubMed]

80. Yang, L.; Wen, K.-S.; Ruan, X.; Zhao, Y.-X.; Wei, F.; Wang, Q. Response of Plant Secondary Metabolites to Environmental Factors. *Molecules* **2018**, *23*, 762. [CrossRef] [PubMed]

81. Sharma, P.; Jha, A.B.; Dubey, R.S.; Pessarakli, M. Reactive Oxygen Species, Oxidative Damage, and Antioxidative Defense Mechanism in Plants under Stressful Conditions. Available online: https://www.hindawi.com/journals/jb/2012/217037/ (accessed on 12 April 2020).

82. Hasanuzzaman, M.; Bhuyan, M.H.M.B.; Anee, T.I.; Parvin, K.; Nahar, K.; Mahmud, J.A.; Fujita, M. Regulation of Ascorbate-Glutathione Pathway in Mitigating Oxidative Damage in Plants under Abiotic Stress. *Antioxidants* **2019**, *8*, 384. [CrossRef]

83. El-Esawi, M.A. *Phytohormones: Signaling Mechanisms and Crosstalk in Plant Development and Stress Responses*; BoD—Books on Demand: Norderstedt, Germany, 2017; ISBN 978-953-51-3411-4.

84. Gill, S.S.; Tuteja, N. Reactive oxygen species and antioxidant machinery in abiotic stress tolerance in crop plants. *Plant Physiol. Biochem.* **2010**, *48*, 909–930. [CrossRef]

85. Liu, W.; Yin, D.; Li, N.; Hou, X.; Wang, D.; Li, D.; Liu, J. Influence of Environmental Factors on the Active Substance Production and Antioxidant Activity in *Potentilla fruticosa* L. and Its Quality Assessment. *Sci. Rep.* **2016**, *6*. [CrossRef]

86. Moustafa, A.G. Long Term Monitoring of Rosa arabica Populations as a Threatened Species in South Sinai, Egypt. Available online: https://www.semanticscholar.org/paper/Long-Term-Monitoring-of-Rosa-arabica-Populations-as-Moustafa/a9209f600f5eef6928a0df021c7866cea10018f1 (accessed on 2 April 2020).

87. El-Wahab, R.A.; Zaghloul, M.S.; Moustafa, A.A. Conservation of Medicinal plants in St. Catherine Protectorate, South Sinai, Egypt. In Proceedings of the First International Conference on Strategy of Egyptian Herbaria, Dokki, Egypt, 9–11 March 2004; pp. 231–251.

88. Danin, A. Desert Rocks—A Habitat Which Supports Many Species That Were New to Science in the Last 40 Years. *Turk. J. Bot.* **2008**, *32*, 459–464.

89. Fakhry, A.M.; El-Keblawy, A.; Shabana, H.A.; Gamal, I.E.; Shalouf, A. Microhabitats Affect Population Size and Plant Vigor of Three Critically Endangered Endemic Plants in Southern Sinai Mountains, Egypt. *Land* **2019**, *8*, 86. [CrossRef]

90. Moustafa, A.E.-R.A.; Zayed, A. Effect of environmental factors on the flora of alluvial fans in southern Sinai. *J. Arid Environ.* **1996**, *32*, 431–443. [CrossRef]

91. Omar, K.A. Ecological and Climatic Attribute Analysis for Egyptian *Hypericum sinaicum*. *Am. J. Life Sci.* **2014**, *2*, 369. [CrossRef]

92. Dewis, J.; Freitas, F. *Physical and Chemical Methods of Soil and Water Analysis*; Food and Agriculture Organization of the United Nations: Rome, Italy, 1984; ISBN 978-92-5-102010-4.

93. Blume, H.-P. Page, A.L., R. H. Miller and D.R. Keeney (Ed., 1982): *Methods of soil analysis*; 2. Chemical and microbiological properties, 2. Aufl. 1184 S., American Soc. of Agronomy (Publ.), Madison, Wisconsin, USA, gebunden 36 Dollar. *Zeitschrift für Pflanzenernährung und Bodenkunde* **1985**, *148*, 363–364. [CrossRef]

94. Johnson, C.M.; Ulrich, A. *Analytical Methods for Use in Plant Analysis*; University of California: Oakland, CA, USA, 1959.

95. Shapiro, L.; Brannock, W.W. *A Field Method for the Determination of Calcium and Magnesium in Limestone and Dolomite*; Open-File Report; U.S. Geological Survey: Reston, VA, USA, 1957; Volume 57–99, p. 9.

96. Black, C.A.; American Society of Agronomy; American Society for Testing and Materials. *Methods of soil Analysis Part 1*; American Society of Agronomy: Crop Science Society of America: Soil Science Society of America: Madison, WI, USA, 1965.

97. Metzner, H.; Rau, H.; Senger, H. Untersuchungen zur Synchronisierbarkeit einzelner Pigmentmangel-Mutanten von Chlorella. *Planta* **1965**, *65*, 186–194. [CrossRef]

98. Prud'homme, M.-P.; Gonzalez, B.; Billard, J.-P.; Boucaud, J. Carbohydrate Content, Fructan and Sucrose Enzyme Activities in Roots, Stubble and Leaves of Ryegrass (*Lolium perenne* L.) as Affected by Source/Sink Modification after Cutting. *J. Plant Physiol.* **1992**, *140*, 282–291. [CrossRef]

99. Loewus, F.A. Improvement in Anthrone Method for Determination of Carbohydrates. *Anal. Chem.* **1952**, *24*, 219. [CrossRef]

100. Malik, C.P.; Singh, M.B. *Plant Enzymology and Histo-Enzymology: A Text Manual*; Kalyani Publishers: Delhi, India, 1980.

101. Harborne, A.J. *Phytochemical Methods A Guide to Modern Techniques of Plant Analysis*, 3rd ed.; Springer: Amsterdam, The Netherlands, 1998; ISBN 978-0-412-57260-9.

102. Ejikeme, C.M.; Ezeonu, C.S.; Eboatu, A.N. Determination of Physical and Phytochemical Constituents of Some Tropical Timbers Indigenous to Niger Delta Area of Nigeria. *Eur. Sci. J. ESJ* **2014**, *10*, 247–270.

103. Bradford, M.M. A rapid and sensitive method for the quantitation of microgram quantities of protein utilizing the principle of protein-dye binding. *Anal. Biochem.* **1976**, *72*, 248–254. [CrossRef]

104. Laemmli, U.K. Cleavage of Structural Proteins during the Assembly of the Head of Bacteriophage T4. *Nature* **1970**, *227*, 680–685. [CrossRef]

105. Bates, L.S.; Waldren, R.P.; Teare, I.D. Rapid determination of free proline for water-stress studies. *Plant Soil* **1973**, *39*, 205–207. [CrossRef]

106. Heath, R.L.; Packer, L. Photoperoxidation in isolated chloroplasts: I. Kinetics and stoichiometry of fatty acid peroxidation. *Arch. Biochem. Biophys.* **1968**, *125*, 189–198. [CrossRef]

107. Mukherjee, S.P.; Choudhuri, M.A. Implications of water stress-induced changes in the levels of endogenous ascorbic acid and hydrogen peroxide in Vigna seedlings. *Physiol. Plant.* **1983**, *58*, 166–170. [CrossRef]

108. Dhindsa, R.S.; Plumb-Dhindsa, P.; Thorpe, T.A. Leaf Senescence: Correlated with Increased Levels of Membrane Permeability and Lipid Peroxidation, and Decreased Levels of Superoxide Dismutase and Catalase. *J. Exp. Bot.* **1981**, *32*, 93–101. [CrossRef]

109. Hermans, C.; Conn, S.J.; Chen, J.; Xiao, Q.; Verbruggen, N. An update on magnesium homeostasis mechanisms in plants. *Metallomics* **2013**, *5*, 1170–1183. [CrossRef] [PubMed]

110. Kar, M.; Mishra, D. Catalase, Peroxidase, and Polyphenoloxidase Activities during Rice Leaf Senescence. *Plant Physiol.* **1976**, *57*, 315–319. [CrossRef]

111. Koricheva, J.; Roy, S.; Vranjic, J.A.; Haukioja, E.; Hughes, P.R.; Hänninen, O. Antioxidant responses to simulated acid rain and heavy metal deposition in birch seedlings. *Env. Pollut.* **1997**, *95*, 249–258. [CrossRef]

112. Kampfenkel, K.; Montagu, M.V.; Inze, D. Effects of Iron Excess on Nicotiana plumbaginifolia Plants (Implications to Oxidative Stress). *Plant Physiol.* **1995**, *107*, 725–735. [CrossRef]

113. Prieto, P.; Pineda, M.; Aguilar, M. Spectrophotometric quantitation of antioxidant capacity through the formation of a phosphomolybdenum complex: Specific application to the determination of vitamin E. *Anal. Biochem.* **1999**, *269*, 337–341. [CrossRef]

Article

Persistence of the Effects of Se-Fertilization in Olive Trees over Time, Monitored with the Cytosolic Ca^{2+} and with the Germination of Pollen

Alberto Marco Del Pino [1], Luca Regni [1,*], Roberto D'Amato [1], Alessandro Di Michele [2], Primo Proietti [1,*] and Carlo Alberto Palmerini [1]

[1] Department of Agricultural, Food and Environmental Sciences (DSA3), University of Perugia, Borgo XX Giugno 74, 06121 Perugia, Italy; alberto.delpino@unipg.it (A.M.D.P.); roberto.damato@unipg.it (R.D.); carlo.palmerini@unipg.it (C.A.P.)

[2] Department of Physics and Geology, University of Perugia, Via Pascoli, 06123 Perugia, Italy; alessandro.dimichele@unipg.it

* Correspondence: luca.regni@unipg.it (L.R.); primo.proietti@unipg.it (P.P.)

Abstract: Selenium (Se) is an important micronutrient for living organisms, since it is involved in several physiological and metabolic processes. Biofortification with Se increases the nutritional and qualitative values of foods in Se-deficient regions and increases tolerance to oxidative stress in olive trees. Many studies have shown that Se, in addition to improving the qualitative and nutritional properties of EVO oil, also improves the plant's response to abiotic stress. This study addressed this issue by monitoring the effects of Se on cytosolic Ca^{2+} and on the germination of olive pollen grains in oxidative stress. The olive trees subjected to treatment with Na-selenate in the field produced pollen with a Se content 6–8 times higher than the controls, even after 20 months from the treatment. Moreover, part of the micronutrient was organic in selenium methionine. The higher selenium content did not produce toxic effects in the pollen, rather it antagonized the undesirable effects of oxidative stress in the parameters under study. The persistence of the beneficial effects of selenium observed over time in pollens, in addition to bringing out an undisputed adaptability of olive trees to the micronutrient, suggested the opportunity to reduce the number of treatments in the field.

Keywords: *Olea europaea* L.; selenium biofortification; oxidative stress; olive pollen; cytosolic Ca^{2+}

Citation: Del Pino, A.M.; Regni, L.; D'Amato, R.; Di Michele, A.; Proietti, P.; Palmerini, C.A. Persistence of the Effects of Se-Fertilization in Olive Trees over Time, Monitored with the Cytosolic Ca^{2+} and with the Germination of Pollen. *Plants* **2021**, *10*, 2290. https://doi.org/10.3390/plants10112290

Academic Editors: Masayuki Fujita and Mirza Hasanuzzaman

Received: 7 September 2021
Accepted: 20 October 2021
Published: 25 October 2021

Publisher's Note: MDPI stays neutral with regard to jurisdictional claims in published maps and institutional affiliations.

1. Introduction

Selenium (Se) is an essential microelement normally present in humans and its endogenous levels fluctuate among populations of different geographic areas and are influenced by environmental factors [1,2]. The microelement, despite being used for many years in the prevention of many diseases, is used with caution as a food additive due to its toxicity at high concentrations [1]. Se in inorganic and organic forms is absorbed by the small intestine and distributed to various tissues, and enters as selenium–cysteine and selenium–methionine in proteins and participates in important biological processes [3,4]. The therapeutic role of selenium was identified since 1957 by Wrobel, through the observation that at low doses it can prevent liver necrosis in rats [3]. Subsequently, many studies have showed beneficial effects of selenium in processes such as immuno-endocrine, metabolic processes and in the maintenance of cellular homeostasis [4,5]. Moreover, the Se biofortification is considered as an agronomic-based strategy, utilized by farmers to produce Se-enriched food products which may help reduce dietary deficiencies in Se-deficient regions such as the Mediterranean Basin [6–11]. In recent years, the Se biofortification has shown beneficial effects in plants by increasing the antioxidant defense against reactive oxygen species (ROS) in vegetative growth and in the response to environmental stress [12–16]. ROS, normally produced at low concentrations, participate in membrane signals, reproduction and pollen—stigma recognition [17–19]. ROS become toxic at high concentrations, induce oxidative stress and deregulate molecular signals including cytosolic Ca^{2+} [20–24]. Several studies conducted on the cytosolic Ca^{2+} of olive pollen, among the molecular signalling networks, resulted in a reliable experimental model [25,26]. The levels of Ca^{2+} are closely related to the cytosolic concentration of the ions, which changes over time are possible to trace through the marking of pollen with the FURA-2AM probe [25–27]. Furthermore, the fluctuations of cytosolic Ca^{2+} have an important role in the germination of pollen and in the growth of the pollen tube, even if up to now the interactions between the two events are little known. [25,26]. The objective of this work is to evaluate the half-life of selenium in olive trees and the persistence of the beneficial effects in oxidative stress in order to reduce the number of treatments in the field with Na-selenate. This objective was verified by monitoring the fluctuations of cytosolic Ca^{2+} and the germination rate of olive pollen in oxidative stress.

2. Results

2.1. Scanning Electron Microscopy Images of Olive Pollen Grains

Pollen grains collected after 8 and 20 months from untreated (C_8, C_{20}) and Se-fertilized (T_8, T_{20}) olive trees were analyzed by scanning field emission electron microscopy (SEM). Pollen images from untreated plants (A) and, Se-fertilized (B) are shown in Figure 1. The individual granules of the two populations did not show differences in size and shape, but showed some qualitative differences in morphology: an angular structure instead of a smooth one, a violation of the pattern and thickness of the cuticle (Figure 1). Furthermore, the population of pollen grains from Se-fertilized olive trees (T_8, T_{20}) showed a lower aggregation capacity than that of untreated plants (C_8, C_{20}) in both harvesting periods.

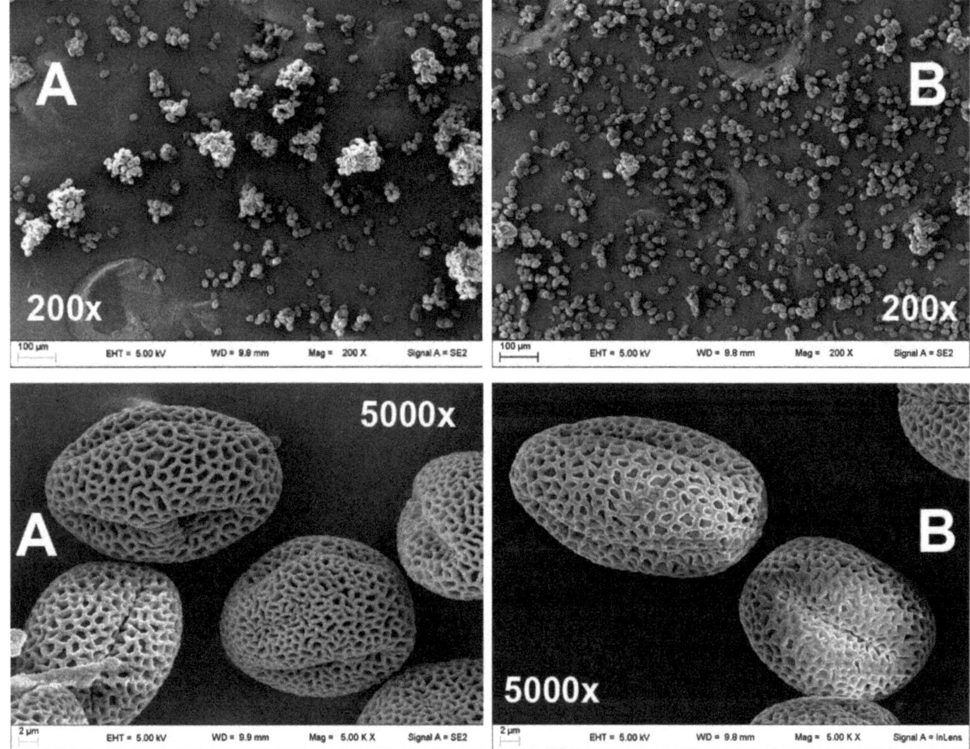

Figure 1. SEM images (200× and 5000×) of olive pollen from control plants (**A**) and from Se-fertilized plants (**B**).

2.2. Speciation of Selenium in Olive Pollen

Olive pollen grains collected after 8 and 20 months from untreated (C_8 and C_{20}) and Se-fertilized (T_8 and T_{20}) plants were analyzed by ICPMS HPLC. The analyses showed that the total selenium content (Se-tot.) in the pollens of the Se-fertilized plants was higher than that of the untreated plants (Table 1).

Table 1. Speciation of Se in olive pollen grains collected after 8 and 20 months from untreated (C_8 and C_{20}) and Se-fertilized (T_8 and T_{20}) olive trees in the field.

	MeSeCys ppb	Se Met ppb	Se (IV) ppb	Se (VI) ppb	Se Total ppb
C_8	23 ± 3 a	858.2 ± 15 a	300.5 ± 8 a	1725.0 ± 32 a	3210 ± 30 a
C_{20}	41 ± 4 a	778.7 ± 14 a	460.7 ± 9 b	1883.2 ± 33 a	3600 ± 29 a
T_8	153 ± 10 b	4895.2 ± 23 c	218.4 ± 7 a	$15,860.0 \pm 34$ b	$28,370 \pm 51$ b
T_{20}	211.8 ± 12 b	2541.2 ± 22 b	582.4 ± 11 b	$14,563.0 \pm 31$ b	$17,900 \pm 41$ b

Means in each column followed by the different letter are significantly different at $p < 0.05$.

The ratio [Se-tot. (T_8/C_8)] was 8.8 in pollen collected after 8 months and dropped to 5.0 [Se-tot. (T_{20}/C_{20})] in the pollen collected after 20 months from the field treatment.

Part of the Na-selenate of the treatment was transformed in selenium–methionine (Se-met) as the predominant species. The Se-met decreased over time and the ratio [Se-met (T_8)/Se tot. (T_8)] of 17.2% dropped to 14.2% [Se-met (T_{20})/Se-tot. (T_{20})].

Selenium (VI), the predominant species of inorganic selenium, remained fairly stable. The ratio [Se (VI) (T_8)/Se-tot. (T_8)] was 55.9% and increased to 81.3% [Se (VI) (T_{20})/Se-tot. (T_{20})] after 20 months (Table 1).

2.3. The Cytosolic Ca^{2+} Tested in Pollen during Oxidative Stress

The labelling of olive pollen grains with the FURA-2AM fluorescent probe allowed to determine the variations over time of cytosolic pollen calcium ($\Delta[Ca^{2+}]_{cp}$) in oxidative stress. The experiment was conducted in two phases, initially in the absence of Ca^{2+} in the incubation medium, then after 200 s, $CaCl_2$ (1 mM) was added. The two phases of the measurement made it possible to differentiate the fluctuations of the cytosolic Ca^{2+} from those deriving from the entry of Ca^{2+} from the extracellular medium.

The $[Ca^{2+}]_{cp}$ of pollen from untreated plants (C_8 and C_{20}) was perturbed by oxidative stress induced in vitro with H_2O_2 (0.1–5.0 mM), while that of Se-fertilized plants was not. In particular, pollens from untreated trees (C_8 and C_{20}) showed an increase in cytosolic Ca^{2+} proportional to the H_2O_2 used, while those from Se-fertilized trees (T_8 and T_{20}) did not show any changes in $[Ca^{2+}]_{cp}$ at the same concentrations of hydrogen peroxide (Figure 2A).

The addition of $CaCl_2$ (1 mM) determined an increase in $[Ca^{2+}]_{cp}$ only in the pollen of untreated plants (C_8 and C_{20}) (Figure 2B).

2.4. The Cytosolic Ca^{2+} Tested in Olive Pollen in the Presence of the Extracts of the Germinative Apexes

Extracts of olive vegetative apexes collected from untreated (EC_8 and EC_{20}) and Se-fertilized (ET_8 and ET_{20}) plants were tested for cytosolic Ca^{2+} of control olive pollen. An aliquot (1 mg) of extract was added to the incubation medium containing $CaCl_2$ (2 mM). All the extracts of the germinative apexes (EC_8, EC_{20}, ET_8, ET_{20}) determined a marked decrease in $[Ca^{2+}]_{cp}$ only for the duration of 100 s. The addition of H_2O_2 (0.25–5 mM) in the incubation medium, after 100 s from the reestablishment of Ca^{2+} homeostasis, determined a dose-dependent increase in $[Ca^{2+}]_{cp}$ (Figure 3). The $[Ca^{2+}]_{cp}$ was not perturbed by the hydrogen peroxide, if plant extracts of the ET_8 or ET_{20} were present in the incubation medium (Figure 3).

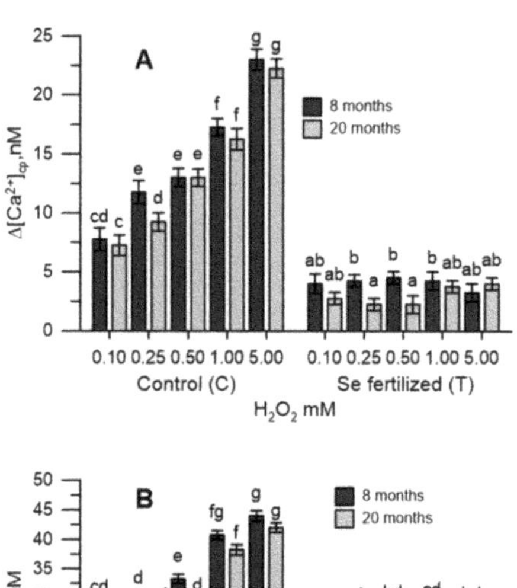

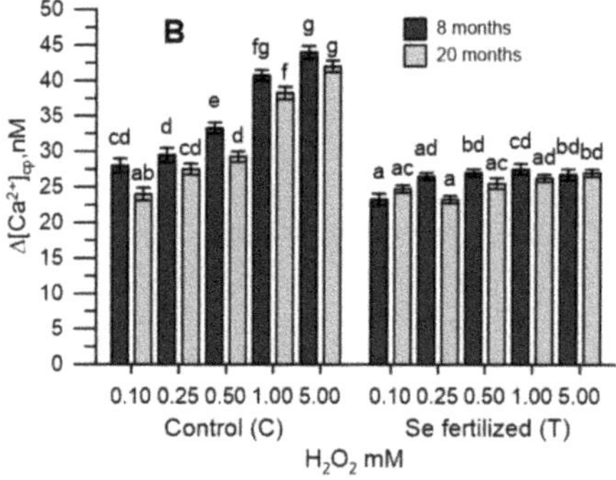

Figure 2. Effect of H_2O_2 (0.1 to 5.0 mM) in the $[Ca^{2+}]_{cp}$ of pollen grains from untreated (C_8 and C_{20}) and Se-fertilized plants (T_8 and T_{20}). Measurements were performed in the absence (**A**) and in the presence (**B**) of $CaCl_2$ (1 mM) in the incubation medium. Data are expressed as means $\pm$ SEM from 4 independent tests. Different letters show significant difference at $p < 0.05$.

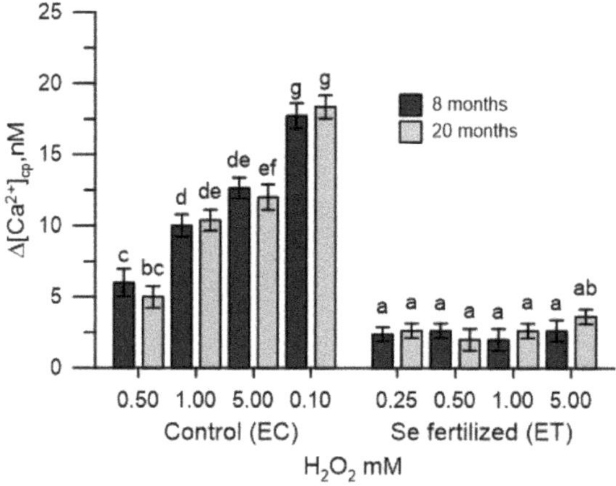

Figure 3. Effects of H_2O_2 (0.25 to 5.0 mM) in the cytosolic Ca^{2+} of olive pollen, in the presence of the extracts of the germinative apexes from untreated plants (EC_8, EC_{20}) and Se-fertilized (ET_8, ET_{20}). Data are expressed as means $\pm$ SEM from 4 independent tests. Different letters show significant difference at $p < 0.05$.

Therefore, although all the extracts of the germinative apexes from untreated or Se-fertilized plants had a marked Ca^{2+} chelating activity (data not shown), the effect in $[Ca^{2+}]_{cp}$ of oxidative stress was only manifested with extracts from untreated plants (EC_8 and EC_{20}). The germinative extracts of the Se-fertilized plants (ET_8 and ET_{20}) had a total selenium content 4–5 times higher than that of the untreated plants (EC_8 and EC_{20}).

2.5. Germination of Olive Pollen Grains in Oxidative Stress

Pollen grains collected from untreated (C_8 and C_{20}) and Se-fertilized (T_8 and T_{20}) plants were incubated for germination in the presence of H_2O_2 (1 mM and 5 mM). Hydrogen peroxide reduced germination differently and, respectively (Figure 4): with H_2O_2 (1 mM): 65% C_8 and 46% T_8, 55% C_{20} and 35% T_{20}; with H_2O_2 (5 mM): 93% C_8 and 74% T_8, 90% C_{20} and 76% T_{20}. The results showed that the pollen grains from Se-fertilized plants are less sensitive to the effects of hydrogen peroxide in germination at both times examined and at the same concentrations of hydrogen peroxide (Figure 4).

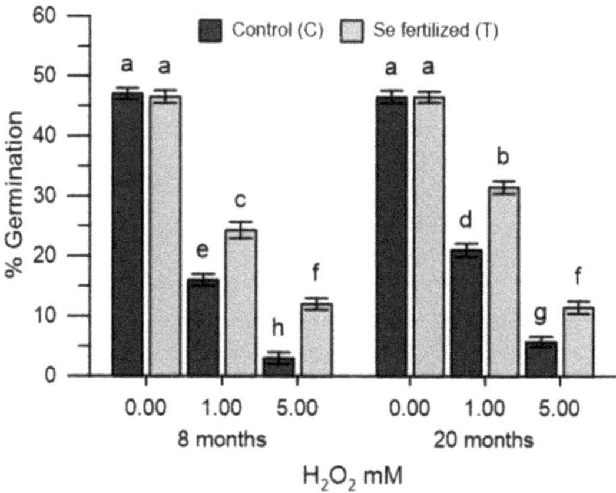

Figure 4. Effects of H_2O_2 (1 and 5 mM) in the germination of olive pollen from untreated (C_8 and C_{20}) and Se-fertilized plants (T_8 and T_{20}). Data are expressed as means $\pm$ SEM from 4 independent tests. Different letters show significant difference at $p < 0.05$.

3. Discussion

In this study, the determination of the half-life of Se in pollen collected 8 and 20 months after field fertilization of olive trees with Na-selenate was present in the pollen of selenium in organic (Se-met) and inorganic form, confirming what was observed in previous works [18]. Here, it emerged, for the first time, that the absorbed selenium remained in the olive pollen even after 20 months from the treatment. A modest decay was evidenced in the Se-met and in the total Se, while the inorganic species (SeVI) did not show variations over time. It is plausible that the biological self has a more dynamic metabolic turnover in olive trees. Previous studies have shown the beneficial effects of selenium in increasing the antioxidant defense against reactive oxygen species (ROS) in vegetative growth and in the response to environmental stress [14–16].

The determination of the variations produced in the cytosolic Ca^{2+} and in the germination were easy and quick to perform, allowing the monitoring of the oxidative stress onset and the effectiveness of the antioxidant measures proposed, as described in previous papers [25,26]. The pollen, labelled with the FURA 2AM probe, was used as an experimental model, allowing to easily trace the dynamic changes of the cytosolic Ca^{2+}. In this study, oxidative stress was induced in vitro with H_2O_2 in olive pollen and the effects were tested in cytosolic Ca^{2+} pollen. The experimental protocol used was conducted in the absence (Ca^{2+} free) and in the presence of Ca^{2+} in the incubation medium. This allowed to differentiate, if the variations of the cytosolic Ca^{2+} caused by the oxidative stress in the pollens, resulted from the release of the ion from the internal stores or from the Ca^{2+} entry from the extracellular medium. In Ca^{2+}-free conditions, H_2O_2 stress of the pollen internal stores caused the release of stored Ca^{2+} and the increase in cytosolic Ca^{2+}. The addition of $CaCl_2$ to the incubation medium resulted also in an increase in cytosolic Ca^{2+}, but this was secondary to Ca^{2+} depletion. H_2O_2 was dependent on pollen internal stores. At both times examined, the homeostasis of Ca^{2+} in the pollen of non-treated plants in the field was altered by oxidative stress, while that of the pollen of Se-fertilized plants was not. It is presumed that the beneficial effects produced by selenium allowed the prevention of oxidative stress in pollen internal stores. A similar protective effect of selenium in cytosolic Ca^{2+}, also occurred in pollens incubated with the extracts of the Se-enriched germinal apexes, collected from the same Se-fertilized olive trees. Under these experimental conditions, H_2O_2 did not cause changes in the levels of cytosolic Ca^{2+} and selenium also did not interfere with the Ca^{2+}-chelating properties of the extracts of the germinating apexes (data not shown). Given the short execution times of the measurement of cytosolic Ca^{2+}, it is possible to exclude, at least in vitro, the implications of any metabolic mechanism and to conclude that selenium over time can substantially act as a simple ROS scavenger, which ultimately prevents the ROS-mediated dysfunction in Ca^{2+} signals. The effects of selenium were also manifested in pollen germination, which was markedly reduced by H_2O_2. The pollen collected after 8 and 20 months from the Se-fertilized plants showed a germination of 63–65% higher with 1 mM H_2O_2 than that of the pollen of untreated plants. This result is particularly important for agricultural productivity, based on multiple abiotic factors that can lead to excess ROS formation [28]. The data obtained fits in with what was asserted by some authors, who consider abiotic stresses of different nature responsible for an excessive accumulation of ROS and, consequently, for the sterility of pollen [29].

The SEM images of the two pollen populations did not show significant differences in size and shape, but qualitative differences emerge in the surface structure. The latter, probably, is responsible for the lower aggregation capacity of pollen. In spite of this, the Se-fertilization, despite resulting in an increase in the selenium content of about 6–8 times and of the changes in the morphology of the pollen, did not influence the germination rates in the absence of oxidative stress over time. The olive trees therefore showed an undisputed ability to adapt to selenium in the long term, thus excluding any toxicity problems.

Studies conducted previously have shown that Se-fertilization, in addition to improving the qualitative and nutritional properties of EVO oil, allowed an increase in the response of the plant to abiotic stress [14–16].

This study, in addition to showing the beneficial effects of selenium in cytosolic Ca^{2+} and in the germination of pollen in oxidative stress, adds, for the first time, that these effects persist over time, even after 20 months from selenium fertilization of plants and suggests the reduction in treatments in the field in order to pursue the goal of precision agriculture.

4. Materials and Methods

4.1. Reagents

FURA-2AM (FURA-2-pentakis (acetoxymethyl) ester), PBS (Phosphate-Buffered Saline), Triton X-100, EGTA (ethylene glycol-bis (β-aminoethyl ether), sodium elenite (Na_2SeO_4), selenium methionine, hydrogen peroxide (H_2O_2), sodium chloride (NaCl), potassium chloride (KCl), magnesium chloride ($MgCl_2$), glucose, Hepes, and dimethyl sulfoxide (DMSO), were acquired from Sigma-Aldrich (St. Louis, MO, USA). Any other chemicals and reagents (reagent grade) were of the highest quality, and obtained from reputable commercial sources.

4.2. Plant Material, Growing Conditions and Pollen and Vegetative Apexes Collection

This study was conducted in 2017–2019 on trees of *Olea europaea* L., cultivar Leccino, grown in a thirty-year-old olive grove near Perugia (Central Italy, 42°57′39.2″ N, 12°25′02.5″ E). The soil is clay loam and the trees are trained to the vase system (with a trunk 1 m high and 3–4 main branches) with a planting distances of 5×6 m. The area has a semi-continental climate. The average temperature difference between the coldest (January) and hottest (July) months is 19–20 °C (with an average diurnal thermal range of 10–11 °C and an average annual air temperature of 13–14 °C). The maximum and minimum temperatures are 36 °C and -7 °C, respectively. The annual average precipitation is about 800 mm, distributed mostly in the autumn, winter and spring. The olive grove is considered to be representative of many intensively managed olive groves in central Italy. In the rainfed olive grove, an area away from the margins with uniform exposure, slope and chemical and physical soil characteristics was selected. Within the selected area, 20 trees (average height 3.5 m) were selected, and among them, at the end of September 10, trees with homogeneous size and yield were treated with Se (100 mg L^{-1}) while another 10 trees with homogeneous size and yield similar to those treated with Se were treated with water and wetting agent only (control). Between a treated tree and a control, there were three trees that received no treatment. The Se dosage was established based on previous studies [7,10]. This solution was obtained by dissolving sodium selenate (SeO_4^{2-}) in water. For each treatment, 0.5% of the Albamilagro wetting agent (Albamilagro International S.p.A., Parabiago, MI, Italy) was added. Each plant was treated with 10 L Se solution. At the base of the tree a filter paper impermeable in the side in contact with the soil was put to prevent the solution from dripping onto the soil. On the other hand, twenty randomly selected 'control' trees were sprayed with the same technique, but with a water solution containing only the wetting agent. All trees reached the 1st stage of flowering in 2018 and 2019 in the last days of May. The olive phenology assessment of flowering beginning was established when the pollen was freely released by shaking the anthers of different branches, located at different heights on the tree and with different exposures [14]. At the beginning of the flowering phase, three branches for each tree (treated and control) were bagged using white double-layer paper bags (0.65×0.35 m) in order to collect the pollen. The bags were placed in the southeast portion of the canopy. In each tree, the bags were placed on branches of similar vigor at the apical, medial and basal positions of the part of the canopy considered. The branches had 70–80 inflorescences each. At the end of the flowering phase, the bags were removed and the pollen was filtered with a cell strainer (40 μm). At the same time from the same trees, 4 g/tree of vegetative apexes were collected. In each year (2018 and 2019), the pollen and vegetative apexes were collected from the same treated and control trees. Therefore, in summary, the pollen collection was carried out after 8 and 20 months (C_8 and C_{20}) from untreated olive trees and after 8 and 20 months (T_8 and T_{20}) from Se-fertilized olive trees.

4.3. Extracts of Vegetative Apexes of Olive Trees

The vegetative apexes were collected from untreated and Se-fertilized olive trees in the field after 8 (EC_8 and ET_8) and 20 months (EC_{20} and ET_{20}) from treatment. A sample (2 g) of the collected vegetative apexes was extracted three times with 20 mL of methanol, dried and then resuspended in 10 mL of methanol. Aliquots of the extract were used to test the variations of cytosolic Ca^{2+} in olive pollen.

4.4. Determination of Total Selenium in Olive Pollens and Vegetative Apexes

Measurements of total selenium content in olive pollen were performed using defrozen and dry samples, respectively. Samples of pollen (0.5 g sample^{-1}) were microwave digested (ETHOS One high-performance microwave digestion system; Milestone Inc., Sorisole, Bergamo, Italy) with 8 mL of ultrapure concentrated nitric acid (65% w/w) and 2 mL of hydrogen peroxide (30% w/w). The heating program for the digestion procedure was 30 min with power of 1000 W and 200 °C. After cooling down, the digests were diluted with water up to 20 mL, then passed through 0.45 μm filters. The analysis were conducted using a graphite furnace atomic absorption spectrophotometry, Shimadzu AA-6800 apparatus (GF-AAS; GFA-EX7, Shimadzu Corp., Tokyo, Japan) with deuterium lamp background correction and a matrix modifier ($Pd(NO_3)_2$, 0.5 mol L^{-1} in HNO_3). All analyses were carried out in triplicate.

4.5. Se Speciation with HPLC ICPMS

Defrozen pollen material (0.25 g) was mixed with 10 mL of solution and 2.0 mg mL^{-1} of protease. Samples were sonicated with an ultrasound probe for 2 min and stirred in a water bath at 37 °C for 4 h. Then, samples were cooled at room temperature and centrifuged for 10 min at 9000 rpm. The supernatant was filtered through 0.22 μm Millex GV filters (Millipore Corporation, Billerica, MA, USA). The standards solutions (1, 5, 10, and 20 μg L^{-1}) for inorganic (i.e., selenite, SeO$_3$$^{-2}$ and selenate, SeO$_4$$^{-2}$) and organic (i.e., selenocystine, (SeCys2); Se-(methyl) selenocysteine, (SeMeSeCys); selenomethionine, (SeMet) were employed. Se forms were prepared with ultrapure (18.2 MΩ cm) water. Speciation of Se was performed by HPLC (Agilent 1100, Agilent Technologies, Santa Clara, CA, USA) using an anion exchange column (Hamilton, PRP-X100, 250 × 4.6 mm^2, 5 μm particle size). The mobile phase was made by ammonium acetate with gradient elution. The analytical method and instrumental conditions were previously described in [9].

4.6. Measurement of Cytosolic Ca^{2+}

Intracellular calcium levels were determined spectrofluorometrically using FURA-2AM the probe [27]. 100 mg of olive pollens were suspended in 10 mL PBS and hydrated for 3 days. Hydrated pollens were harvested by centrifugation at 1000× g × 4 min and then resuspended in 2 mL Ca^{2+}-free HBSS buffer (120 mM NaCl, 5.0 mM KCl, MgCl2 1 mM, 5 mM glucose, 25 mM Hepes, pH 7.4). Pollen suspensions were incubated in the dark with FURA-2AM (2 μL of a 2 mM solution in DMSO) for 120 min, after which samples were centrifuged at 1000× g × 4 min. Pollens were then harvested and suspended in ~10 mL of Ca^{2+}-free HBSS containing 0.1 mM EGTA, which was included to rule out or, at least, minimize a potential background due to contaminating ions (so as to obtain a suspension of 1 × 106 of pollen granules hydrated per mL). Oxidative stress was induced by adding hydrogen peroxide to the suspended pollen. Effects on cytosolic Ca^{2+} were evaluated after 100 s. Fluorescence was measured in a Perkin-Elmer LS 50 B spectrofluorometer (ex. 340 and 380 nm, em. 510 nm) (Figure 1), set with a 10 nm and a 7.5 nm slit width in the excitation and emission windows, respectively. Fluorometric readings were normally taken after 300–350 s. When required, samples of pollen, CaCl$_2$, H$_2$O$_2$, Na$_2$SeO$_4$ and aliquot extracts of vegetative apexes were added for specific purposes, as described in the Results section. Cytosolic calcium concentrations ([Ca^{2+}]c) were calculated as shown by Grynkiewicz [27].

4.7. Pollen Germination

Olive pollen grains used in the experimentation are collected in the field from treated trees (Se-enriched) and control trees. Freshly collected pollen samples were rehydrated by incubation in a humid chamber at room temperature for 30 min [30] and then transferred to culture plate (6-well culture plate (1.0 mg of pollen per plate) containing 3 mL of an agar-solidified growing medium: agar 1%, sucrose 10%, boric acid (H$_3$BO$_3$) 100 ppm and calcium chloride (CaCl$_2$) 1 mM, at pH 5.5 [31]. Subsequently, with the aid of a brush, a uniform distribution was obtained on the surface of the medium. Oxidative stress was induced by adding hydrogen peroxide to the suspended pollen. Effects on germination were assessed at the end of the incubation period. Pollen grains were then incubated for 24–48 h in a growth chamber at 25 °C. The number of germinated and non-germinated pollen grains were determined with the aid of a microscope with a 10× objective lens. Germination rate were determined using two replicates of 100 grain. Grains were considered germinated if the size of the pollen tube was greater than the diameter of the grain [31]. Experiments were conducted in a completely randomized design with four replications.

4.8. Statistical Analysis

Statistical tests were performed using Graph Pad Prism 6.03 software for Windows (La Jolla, CA, USA). Tests for variance assumptions were conducted (homogeneity of variance by Levene's test, normal distribution by the D'Agostino-Pearson omnibus normality test).

Results obtained are expressed as mean values $\pm$ standard error of the mean (SEM). Significance of differences were analyzed by Fisher's least significant differences test, after the analysis of variance according to the split plot in time design. Differences with $p < 0.05$ were considered statistically significant.

5. Conclusions

This study shows that through the determination of cytosolic Ca^{2+} and pollen germination, quick and easy measurements are possible to monitor the onset of oxidative stress and the effectiveness of any antioxidant measure adopted. Furthermore, the field treatment with selenium maintains its effects even after 20 months, suggesting that it is possible to reduce the amount of selenium fertilization in a precision agriculture perspective.

Author Contributions: Conceptualization, A.M.D.P., L.R., P.P., C.A.P.; methodology, A.M.D.P., L.R., R.D., A.D.M., P.P., C.A.P.; investigation, A.M.D.P., L.R., R.D., A.D.M.; data curation, A.M.D.P.; writing—original draft preparation, A.M.D.P., L.R., R.D., A.D.M., P.P., C.A.P.; writing—review and editing, L.R., P.P.; supervision, P.P., C.A.P. All authors have read and agreed to the published version of the manuscript.

Funding: This research received no external funding.

Institutional Review Board Statement: Not applicable.

Informed Consent Statement: Not applicable.

Data Availability Statement: All raw or derived data supporting the findings of this study are available from the corresponding author, upon request.

Acknowledgments: We are grateful to Mirco Boco, Giorgio Sisani and Andrea Sforna for their technical support.

Conflicts of Interest: The authors declare no conflict of interest.

References

1. Rayman, M.P. Selenium and human health. *Lancet* **2012**, *379*, 1256–1268. [CrossRef]
2. Park, K.; Rimm, E.; Siscovick, D.; Spiegelman, D.; Morris, J.S.; Mozaffarian, D. Demographic and lifestyle factors and selenium levels in men and women in the U.S. *Nutr. Res. Pract.* **2011**, *5*, 357–364. [CrossRef] [PubMed]
3. Wrobel, K.; Power, R.; Toborek, M. Biological activity of selenium: Revisited. *IUBMB Life* **2016**, *68*, 97–105. [CrossRef] [PubMed]
4. Wang, N.; Tan, H.-Y.; Li, S.; Xu, Y.; Guo, W.; Feng, Y. Supplementation of Micronutrient Selenium in Metabolic Diseases: Its Role as an Antioxidant. *Oxidative Med. Cell. Longev.* **2017**, *2017*, 7478523. [CrossRef]
5. Wei, J.; Zeng, C.; Gong, Q.Y.; Li, X.X.; Lei, G.H.; Yang, T.B. Associations between dietary antioxidant intake and metabolic syndrome. *PLoS ONE* **2015**, *10*, e0130876. [CrossRef] [PubMed]
6. D'Amato, R.; Proietti, P.; Nasini, L.; Del Buono, D.; Tedeschini, E.; Businelli, D. Increase in the selenium content of extra virgin olive oil: Quantitative and qualitative implications. *Grasas Aceites* **2014**, *65*, e025. [CrossRef]
7. D'Amato, R.; Proietti, P.; Onofri, A.; Regni, L.; Esposto, S.; Servili, M.; Businelli, D.; Selvaggini, R. Biofortification (Se): Does it increase the content of phenolic compounds in virgin olive oil (VOO)? *PLoS ONE* **2017**, *12*, e0176580. [CrossRef]
8. D'Amato, R.; De Feudis, M.; Hasuoka, P.E.; Regni, L.; Pacheco, P.H.; Onofri, A.; Businelli, D.; Proietti, P. The selenium supplementation influences olive tree production and oil stability against oxidation and can alleviate the water deficiency effects. *Front. Plant Sci.* **2018**, *9*, 1191. [CrossRef]
9. Fontanella, M.C.; D'Amato, R.; Regni, L.; Proietti, P.; Beone, G.M.; Businelli, D. Selenium speciation profiles in biofortified sangiovese wine. *J. Trace Elem. Med. Biol.* **2017**, *43*, 87–92. [CrossRef]
10. D'Amato, R.; Regni, L.; Falcinelli, B.; Mattioli, S.; Benincasa, P.; Dal Bosco, A.; Pacheco, P.; Proietti, P.; Troni, E.; Santi, C.; et al. Current Knowledge on Selenium Biofortification to Improve the Nutraceutical Profile of Food: A Comprehensive Review. *J. Agric. Food Chem.* **2020**, *68*, 4075–4097. [CrossRef] [PubMed]
11. D'Amato, R.; Petrelli, M.; Proietti, P.; Onofri, A.; Regni, L.; Perugini, D.; Businelli, D. Determination of changes in the concentration and distribution of elements within olive drupes (cv. Leccino) from Se biofortified plants, using laser ablation inductively coupled plasma mass spectrometry. *J. Sci. Food Agric.* **2018**, *98*, 4971–4977. [CrossRef] [PubMed]
12. Regni, L.; Palmerini, C.A.; Del Pino, A.M.; Businelli, D.; D'Amato, R.; Mairech, H.; Marmottini, F.; Micheli, M.; Pacheco, P.H.; Proietti, P. Effects of selenium supplementation on olive under salt stress conditions. *Sci. Hortic.* **2021**, *278*, 109866. [CrossRef]
13. Proietti, P.; Nasini, L.; Del Buono, D.; D'Amato, R.; Tedeschini, E.; Businelli, D. Selenium protects olive (*Olea europaea* L.) from drought stress. *Sci. Hortic.* **2013**, *164*, 165–171. [CrossRef]

14. Tedeschini, E.; Proietti, P.; Timorato, V.; D'Amato, R.; Nasini, L.; Del Buono, D.; Businelli, D.; Frenguelli, G. Selenium as stressor and antioxidant affects pollen performance in *Olea europaea*. *Flora* **2015**, *215*, 16–22. [CrossRef]

15. Lyons, G.H.; Genc, Y.; Soole, K.; Stangoulis, J.C.R.; Liu, F.; Graham, R.D. Selenium increases seed production in Brassica. *Plant Soil* **2009**, *318*, 73–80. [CrossRef]

16. Prins, C.N.; Hantzis, L.J.; Quinn, C.F.; Pilon-Smits, E.A.H. Effects of selenium accumulation on reproductive functions in Brassica juncea and Stanleya pinnata. *J. Exp. Bot.* **2011**, *62*, 5633–5640. [CrossRef]

17. Kwak, J.M.; Nguyen, V.; Schroeder, J.I. The Role of Reactive Oxygen Species in Hormonal Responses. *Plant Physiol.* **2006**, *141*, 323–329. [CrossRef]

18. Hancock, J.T.; Desikan, R.; Neill, S.I. Role of reactive oxygen species in cell signalling pathways. *Biochem. Soc. Trans.* **2001**, *29*, 345–350. [CrossRef]

19. Laloi, C.; Apel, K.; Danon, A. Reactive oxygen signalling: The latest news. *Curr. Opin. Plant Biol.* **2004**, *7*, 323–328. [CrossRef] [PubMed]

20. Görlach, A.; Bertram, K.; Hudecova, S.; Krizanova, O. Calcium and ROS: A mutual interplay. *Redox Biol.* **2015**, *6*, 260–271. [CrossRef] [PubMed]

21. Yan, Y.; Wei, C.L.; Zhang, W.R.; Cheng, H.P.; Liu, J. Cross-talk between calcium and reactive oxygen species signaling. *Acta Pharmacol. Sin.* **2006**, *27*, 821–826. [CrossRef]

22. Brini, M.; Calì, T.; Ottolini, D.; Carofoli, E. Intracellular calcium homeostasis and signaling. In *Metallomics and the Cell*; Banci, L., Ed.; Springer: Dordrecht, The Netherlands, 2013; Volume 12, pp. 119–168. [CrossRef]

23. Orrenius, S.; Gogvadze, V.; Zhivotovsky, B. Calcium and mitochondria in the regulation of cell death. *Biochem. Biophys. Res. Commun.* **2015**, *460*, 72–81. [CrossRef]

24. Proietti, P.; Trabalza Marinucci, M.; Del Pino, A.M.; D'Amato, R.; Regni, L.; Acuti, G.; Chiaradia, E.; Palmerini, C.A. Selenium maintains Ca^{2+} homeostasis in sheep lymphocytes challenged by oxidative stress. *PLoS ONE* **2018**, *13*, e0201523. [CrossRef]

25. Del Pino, A.M.; Guiducci, M.; D'Amato, R.; Di Michele, A.; Tosti, G.; Datti, A.; Palmerini, C.A. Selenium maintains cytosolic Ca^{2+} homeostasis and preserves germination rates of maize pollen under H$_2$O$_2$-induced oxidative stress. *Sci. Rep.* **2019**, *9*, 1350. [CrossRef]

26. Del Pino, A.M.; Regni, L.; D'Amato, R.; Tedeschini, E.; Businelli, D.; Proietti, P.; Palmerini, C.A. Selenium-Enriched Pollen Grains of *Olea europaea* L.: Ca^{2+} Signaling and Germination Under Oxidative Stress. *Front. Plant Sci.* **2019**, *10*, 1611. [CrossRef]

27. Grynkiewicz, G.; Poenie, M.; Tsien, R.Y. A new generation of Ca^{2+} indicators with greatly improved fluorescence properties. *J. Biol. Chem.* **1985**, *260*, 3440–3450. [CrossRef]

28. Fahad, S.; Bajwa, A.A.; Nazir, U.; Shakeel, A.A.; Farooq, A.; Zohaib, A.; Sadia, S.; Nasim, W.; Adkins, S.; Saud, S.; et al. Crop Production under Drought and Heat Stress: Plant Responses and Management Options. *Front. Plant Sci.* **2017**, *8*, 1147. [CrossRef]

29. Lazzaro, M.D.; Cardenas, L.; Bhatt, A.P.; Justus, C.D.; Phillips, M.S.; Holdaway-Clarke, T.L.; Hepler, P.K. Calcium gradients in conifer pollen tubes; dynamic properties differ from those seen in angiosperms. *J. Exp. Bot.* **2005**, *56*, 2619–2628. [CrossRef] [PubMed]

30. Rejo'n, J.D.; Zienkiewicz, A.; Rodríguez-García, M.I.; Castro, A.J. Profiling and functional classification of esterases in olive (*Olea europaea*) pollen during germination. *Ann. Bot.* **2012**, *110*, 1035–1045. [CrossRef] [PubMed]

31. Martins, E.S.; Davide, L.M.C.; Miranda, G.J.; Barizon, J.O.; de Assis, F., Jr.; de Carvalho, R.P.; Gonçalves, M.C. In vitro pollen viability of maize cultivars at different times of collection. *Ciênc. Rural* **2017**, *47*, e20151077. [CrossRef]

plants

Article

Supplemental Selenium and Boron Mitigate Salt-Induced Oxidative Damages in *Glycine max* L.

Mira Rahman [1], Khussboo Rahman [1], Khadeja Sultana Sathi [1], Md. Mahabub Alam [1], Kamrun Nahar [2], Masayuki Fujita [3,*] and Mirza Hasanuzzaman [1,*]

1   Department of Agronomy, Faculty of Agriculture, Sher-e-Bangla Agricultural University, Dhaka 1207, Bangladesh; mirarahman73@gmail.com (M.R.); khussboorahman1305594@sau.edu.bd (K.R.); sathikhadeja1405945@sau.edu.bd (K.S.S.); shamim1983@sau.edu.bd (M.M.A.)
2   Department of Agricultural Botany, Faculty of Agriculture, Sher-e-Bangla Agricultural University, Dhaka 1207, Bangladesh; knahar84@yahoo.com
3   Laboratory of Plant Stress Responses, Faculty of Agriculture, Kagawa University, Miki-cho, Kita-gun, Kagawa 761-0795, Japan
*   Correspondence: fujita@ag.kagawa-u.ac.jp (M.F.); mhzsauag@yahoo.com (M.H.)

Citation: Rahman, M.; Rahman, K.; Sathi, K.S.; Alam, M.M.; Nahar, K.; Fujita, M.; Hasanuzzaman, M. Supplemental Selenium and Boron Mitigate Salt-Induced Oxidative Damages in *Glycine max* L.. *Plants* **2021**, *10*, 2224. https://doi.org/10.3390/plants10102224

Academic Editors: Magda Pál and Fabrizio Araniti

Received: 26 September 2021
Accepted: 18 October 2021
Published: 19 October 2021

Publisher's Note: MDPI stays neutral with regard to jurisdictional claims in published maps and institutional affiliations.

Abstract: The present investigation was executed with an aim to evaluate the role of exogenous selenium (Se) and boron (B) in mitigating different levels of salt stress by enhancing the reactive oxygen species (ROS) scavenging, antioxidant defense and glyoxalase systems in soybean. Plants were treated with 0, 150, 300 and 450 mM NaCl at 20 days after sowing (DAS). Foliar application of Se (50 μM Na$_2$SeO$_4$) and B (1 mM H$_3$BO$_3$) was accomplished individually and in combined (Se+B) at three-day intervals, at 16, 20, 24 and 28 DAS under non-saline and saline conditions. Salt stress adversely affected the growth parameters. In salt-treated plants, proline content and oxidative stress indicators such as malondialdehyde (MDA) content and hydrogen peroxide (H$_2$O$_2$) content were increased with the increment of salt concentration but the relative water content decreased. Due to salt stress catalase (CAT), monodehydroascorbate reductase (MDHAR), dehydroascorbate reductase (DHAR), glyoxalase I (Gly I) and glyoxalase II (Gly II) activity decreased. However, the activity of ascorbate peroxidase (APX), glutathione reductase (GR), glutathione peroxidase (GPX), glutathione *S*-transferase (GST) and peroxidase (POD) increased under salt stress. On the contrary, supplementation of Se, B and Se+B enhanced the activities of APX, MDHAR, DHAR, GR, CAT, GPX, GST, POD, Gly I and Gly II which consequently diminished the H$_2$O$_2$ content and MDA content under salt stress, and also improved the growth parameters. The results reflected that exogenous Se, B and Se+B enhanced the enzymatic activity of the antioxidant defense system as well as the glyoxalase systems under different levels of salt stress, ultimately alleviated the salt-induced oxidative stress, among them Se+B was more effective than a single treatment.

Keywords: abiotic stress; AsA-GSH pathway; methylglyoxal; micronutrient; osmoregulation; reactive oxygen species; trace elements

1. Introduction

Abiotic stress or environmental stress is not a sole entity. Under the umbrella of abiotic stress, it comprises all types of hostile environmental conditions that a plant may face in nature [1]. Salt stress is a major abiotic stress. Salinity threatens the productivity of plants by negatively affecting the biochemical, physiological and molecular features of the plants [2]. Because of inappropriate management and climate change, the saline-affected area has been increasing more than before in arid, semi-arid and coastal areas, along with other types of land [3]. Worldwide, about 20–50% of irrigated land areas are affected by salt [3]. The alarming issue is by 2050 up to 50% of agricultural land is expected to be affected by salinity [2].

In the soil solution, sodium chloride (NaCl) and sodium sulfate (Na_2SO_4) are the most available soluble salts. An increase in salinity level, in most of the cases, indicates mainly an increase in Na^+ and Cl^- concentration. Both Na^+ and Cl^- ions produce critical conditions for plant survival, but between them, Cl^- is more dangerous [4]. Salinity primarily creates osmotic stress and ionic toxicity. Osmotic stress occurs due to the accumulation of a higher concentration of salt ions in the root zone. Osmotic stress hinders the uptake of water and nutrient of the plants and ultimately causes stomatal closure, reduction in cell expansion and division. In the later stage, a higher accumulation of salt inside the cells and tissues induces ionic toxicity, disruption of ion homeostasis, alteration of cellular functions, premature senescence, and in extreme condition plant death. Salinity-induced osmotic and ionic stresses are responsible for the overproduction of reactive oxygen species (ROS) [5]. Therefore, an excess concentration of ROS in plants induces deleterious oxidative stress, which causes oxidation of plant cell components (lipid, protein, nucleic acid, etc.), along with the cell organelles and membranes which also disrupts the redox homeostasis [6]. The overproduction of ROS such as superoxide radical ($O_2^{\bullet-}$), hydrogen peroxide (H_2O_2), hydroxyl radical ($^{\bullet}OH$) etc. is needed to be stopped to protect the plants from oxidative damage and to regulate the proper physiological and biochemical activities. Plants maintain the balance between formation and detoxification of ROS by an antioxidant defense system [7]. Multiple enzymatic components of antioxidant defense system like catalase (CAT), ascorbate peroxidase (APX), monodehydroascorbate reductase (MDHAR), dehydroascorbate reductase (DHAR), glutathione reductase (GR), glutathione peroxidase (GPX), glutathione S-transferase (GST), and peroxidase (POD) coordinately act to control the ROS and ROS-induced oxidative stress, along with non-enzymatic components [8].

Many investigations have been carried out and attempts have been taken to mitigate the hazardous effect of salt stress on crops [9–11]. After a plethora of investigations, selenium (Se) a beneficial trace element (at lower concentrations) was found to be an effective one in improving growth, antioxidant defense and inducing tolerance mechanisms against salt stress [9–11]. Selenium acts as a plant growth regulator, stress modulator, antioxidant agent at lower concentrations but at higher concentrations, it is phytotoxic and may act as a pro-oxidant [12]. Supplementation of Se mitigates salt stress by reducing Na^+ accumulation in plant parts, Na^+ compartmentalization, upregulating Na^+ and Cl^- ions transporter genes, chelation and boosting of the antioxidant defense system. Selenium protects plants from oxidative stress by triggering the detoxification of ROS, which is overgenerated due to salt stress [9,10,12]. Boron (B) is an essential micronutrient that actively participates in the crop growth and development process. It is associated with respiration, transportation of water, protein synthesis, sugar transport, RNA metabolism and plant hormones. Importantly, it maintains the structural integrity of bio-membranes [13]. Boron is involved with lignin synthesizing, strengthening the cell walls, and indirectly protecting cell membranes in plants [14]. Supplementation of B increased the pectin (19%) and hemicellulose (50%) content of the cell wall under oxidative stress [15]. Exogenous application of B decreased the Cl^- content in sugar beet under salt stress by upregulating the transportation of Cl^- ions [16]. Supplementation of B declined lipid peroxidation (indicated by malondialdehyde, MDA content) and hydrogen peroxide (H_2O_2) content under oxidative stress. Moreover, the application of B increased the antioxidant defense and decreased oxidative stress in different crops [15,17]. However, among the different combinations of treatments with or without Fe, Se+B treatment alone showed the maximum relative water content (RWC) and the minimum relative water loss, and also improved the growth parameters along with other treatments [18].

Soybean (*Glycine max* L.) is a widely cultivated legume around the world because of its versatile uses and economic importance. It is a prominent source of proteins and edible oil, it has valuable uses as food, feed and oilseed crop [19]. In soybean plants, salinity creates oxidative, osmotic and ionic stress [20]. Salt-induced osmotic, ionic and oxidative stress hampers the growth, biomass production and crucial physiological activities of soybean, ultimately productivity is negatively affected [20,21].

In previous investigations, the positive effect of Se and B in improving the salt tolerance of plants were studied in the laboratory and hydroponic conditions, mostly. However, the interaction role of Se and B, along with individual applications was rarely studied. Moreover, the role of B in methylglyoxal (MG) detoxification was hardly studied. The present investigation was conducted to study the injurious effect of the different levels of salt stress on soybean and also to study the roles of Se, B and Se+B in mitigating salt-induced oxidative stress, and enhancing the salt stress tolerance in soybean by upregulating the antioxidant defense system and MG detoxification system.

2. Results

2.1. Growth Parameters

Upon exposure to 150 mM NaCl stress, the reduction in plant height was not significant but at 300 and 450 mM NaCl, stress plant height was reduced by 35 and 55%, respectively, in comparison to untreated control (without salt and Se, B and Se+B treatment). However, supplementation of Se and Se+B increased the plant height under all salt treatments. At 150 and 300 mM NaCl stress B alone did not increase the plant height significantly, compared to a respective only salt-treated plant without Se, B and Se+B spray. Under mild, moderate and severe salinity Se+B spray increased the plant height by 17, 30 and 39%, respectively, compared to corresponding salt-treatment alone (150, 300 and 450 mM NaCl-treated plant without Se, B and Se+B spray; Figure 1A).

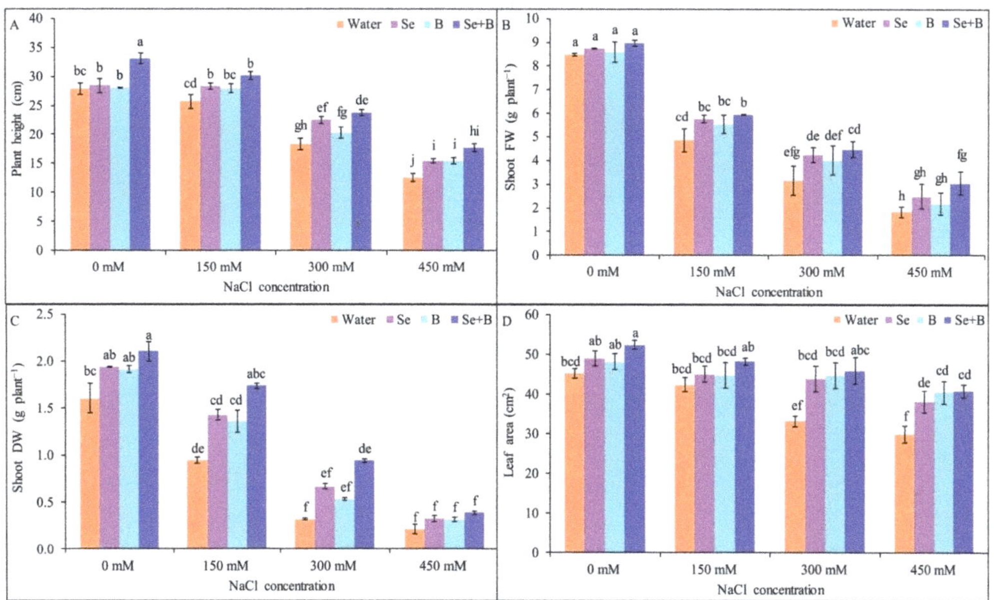

Figure 1. Effect of exogenous Se, B and Se+B on plant height (**A**), shoot FW (**B**), shoot DW (**C**) and leaf area (**D**) in soybean at 0, 150, 300 and 450 mM NaCl-induced salt stress. Here, Se, B and Se+B indicate 50 μM Na$_2$SeO$_4$, 1 mM H$_3$BO$_3$ and 50 μM Na$_2$SeO$_4$ + 1 mM H$_3$BO$_3$, respectively. Values in a column with different letters are significantly different at $p \leq 0.05$ applying Fisher's LSD test.

Shoot fresh weight (FW) was declined by 43, 63 and 78% at 150, 300 and 450 mM NaCl stress, in comparison to untreated control. Rather than single supplementation, Se+B combinedly increased the shoot FW significantly under salt stress. Application of Se+B spray increased the shoot FW by 22, 41 and 66% at 150, 300 and 450 mM NaCl stress, respectively, in comparison to corresponding only NaCl-treated plants (Figure 1B).

Shoot dry weight (DW) was decreased by 41% at 150 mM NaCl stress, compared to untreated control. Moreover, at 300 and 450 mM NaCl stress the reduction in shoot DW plant^{-1} was statistically similar, in comparison to untreated control. However, Se and B alone did not increase the shoot DW significantly under salt stress. Combinedly Se+B increased the shoot DW plant^{-1} significantly by 85 and 93% at 150 and 300 mM NaCl stress but at 450 mM NaCl stress the increment of shoot DW was not significant, respectively, compared to respective salt treatment alone (Figure 1C).

In 150 mM NaCl-stressed plants, reduction in leaf area was not significant, compared to untreated control. Moreover, at 300 mM and 450 mM NaCl stress decrement of leaf area was statistically similar. Single supplementation of Se and B showed statistically similar results to Se+B and increased the leaf area at 300 and 450 mM NaCl stress but not at 150 mM NaCl stress. However, due to Se+B spray at 300 and 450 mM NaCl stress, leaf area was increased by 39 and 37%, respectively, compared to salt-treated plant alone (without Se, B and Se+B spray; Figure 1D).

2.2. Leaf Relative Water Content

Due to imposition of 150 mM NaCl stress leaf RWC was decreased by 19%, and at 300 and 450 mM NaCl stress the reduction in leaf RWC was statistically similar, respectively, in comparison to untreated control (without salt and Se, B and Se+B treatment). On the contrary, foliar application of Se, B and Se+B increased the leaf RWC and showed a statistically similar increment of leaf RWC under different levels of NaCl stress, compared to only salt-treated plants. Under mild, moderate and severe salinity Se+B spray increased the leaf RWC by 18, 21 and 20%, respectively, compared to corresponding salt-treatment alone (150, 300 and 450 mM NaCl-treated plant without Se, B and Se+B spray; Figure 2A).

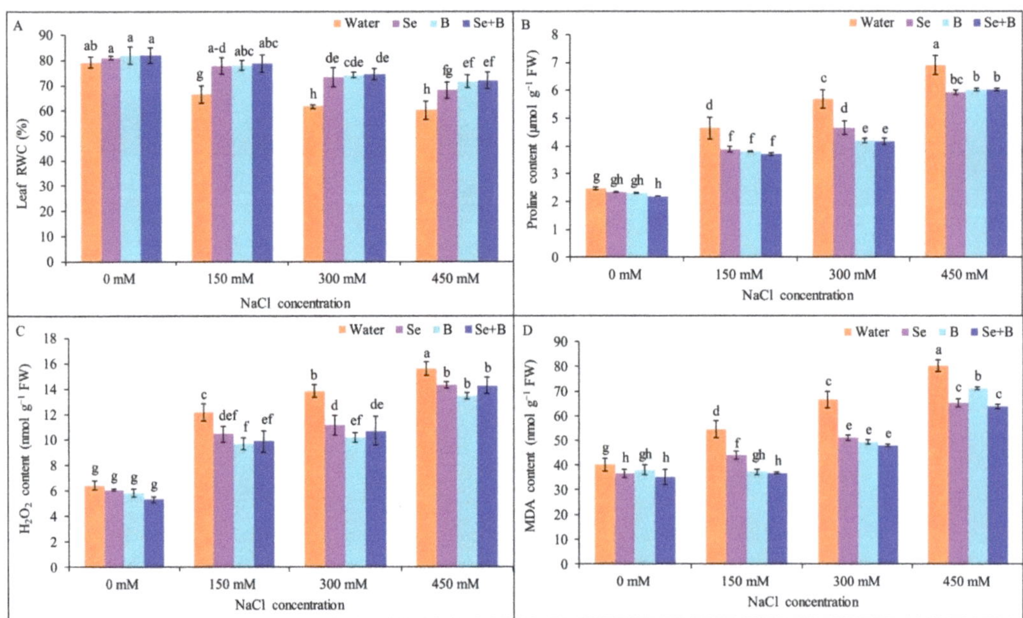

Figure 2. Effect of exogenous Se, B and Se+B on leaf RWC (**A**), Pro content (**B**), H$_2$O$_2$ content (**C**) and MDA content (**D**) in soybean at 0, 150, 300 and 450 mM NaCl-induced salt stress. Here, Se, B and Se+B indicate 50 μM Na$_2$SeO$_4$, 1 mM H$_3$BO$_3$ and 50 μM Na$_2$SeO$_4$ + 1 mM H$_3$BO$_3$, respectively. Values in a column with different letters are significantly different at $p \leq 0.05$ applying Fisher's LSD test.

2.3. Proline Content

A higher amount of proline (Pro) accumulation was observed in 150, 300 and 450 mM NaCl-stressed plants; 88, 131 and 180%, respectively, in comparison to untreated control. On the other hand, foliar-applied Se, B and Se+B diminished the Pro content in saline condition only. Moreover, the single and combined application of Se and B showed a statistically similar result in decreasing Pro content under salt stress. Application of Se+B decreased the Pro content by 20, 19 and 12% at 150, 300 and 450 mM NaCl stress, respectively, compared to corresponding only NaCl-treated plants (150, 300 and 450 mM NaCl-treated plants without Se, B and Se+B treatments; Figure 2B).

2.4. H_2O_2 Content

In salt-stressed plants, H_2O_2 content was increased by 40, 93 and 131% under mild, moderate and severe salinity, respectively, in comparison to untreated control plants. However, Se and B alone and combinedly showed a statistically similar result in the case of decreasing the H_2O_2 content under different levels of salt stress. Moreover, H_2O_2 content decreased due to Se+B spray by 17, 21 and 24% under mild, moderate and severe salinity, respectively, compared to corresponding only salt-treated plants (Figure 2C).

2.5. Lipid Peroxidation (MDA Content)

Salt-induced oxidative stress is responsible for lipid peroxidation. In order to estimate the lipid peroxidation level, MDA content is measured as a major indicator.

When subjected to salt stress MDA content was increased with the increase in the NaCl concentration, in a dose-dependent manner. Under mild, moderate and severe salinity MDA content was increased sharply by 112, 142 and 172%, respectively, compared to untreated control. On the contrary, under saline conditions, exogenous Se, B and Se+B diminished the MDA content. However, MDA content was reduced due to Se+B application by 19, 23 and 9% under mild, moderate and severe salinity, compared to respective salt treatment alone (Figure 2D).

2.6. Activities of Antioxidant Enzymes

The activity of APX was increased by 45, 58 and 85% under mild, moderate and severe salinity, respectively, compared to untreated control (without salt and Se, B and Se+B treatment). Moreover, foliar application of Se, B and Se+B further increased the APX activity in plants under different levels of salinity, compared to only salt-treated plants. Supplementation of Se+B enhanced the activity of APX at 150 mM NaCl stress which was statistically similar to the result of Se and B alone. However, at 300 and 450 mM NaCl stress, Se+B amplified the APX activity more than the individual supplementation of Se and B, and increased the APX activity by 19 and 19%, respectively, in comparison to corresponding salt treatment alone (Figure 3A).

In response to 150, 300 and 450 mM NaCl stress, MDHAR activity was declined by 27, 42 and 54%, respectively, in comparison to untreated control. On the contrary, Se, B and Se+B treatments increased the MDHAR activity under salt stress, compared to corresponding salt treatment alone. Among them Se+B showed more increment in MDHAR activity than single supplementation. However, at 150, 300 and 450 mM NaCl stress exogenous Se+B enhanced the MDHAR activity by 43, 45 and 36%, respectively, in comparison to respective only salt-treated plants (Figure 3B).

The DHAR activity was decreased by 12, 22 and 36% at 150, 300 and 450 mM NaCl, respectively, in comparison to untreated control. However, DHAR activity was enhanced under salt stress because of Se, B and Se+B spray, compared to the respective only salt treatment (without Se, B and Se+B spray). A statistically similar enhancement was observed in the DHAR activity at 150 mM NaCl stress due to the application of Se, B and Se+B spray, compared to 150 mM NaCl stress (without Se, B and Se+B spray). The combined application of Se+B increased the DHAR activity by 21 and 16%, respectively, at 300 and 450 mM NaCl, in comparison to the corresponding salt treatment alone which was followed by individual application of Se and B (Figure 3C).

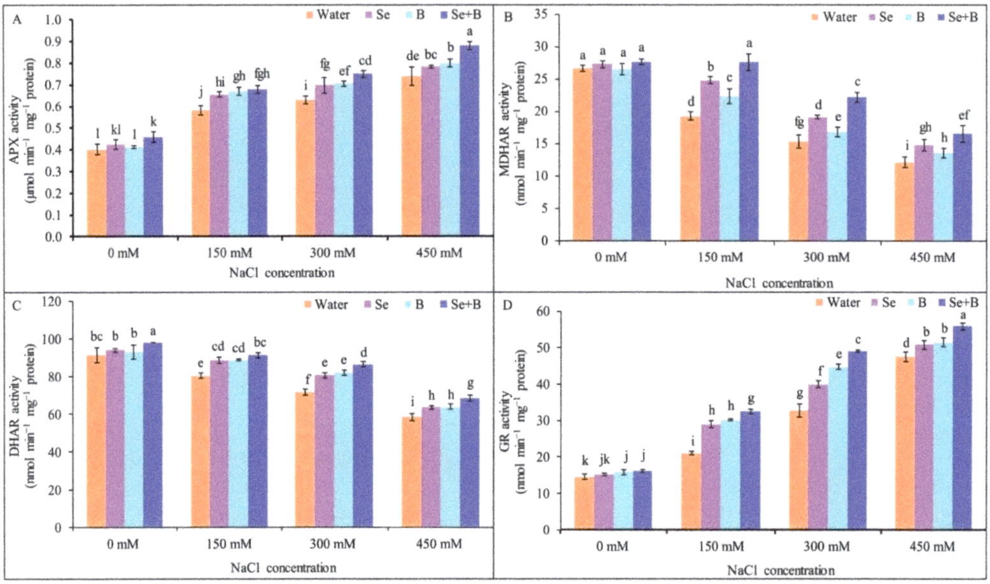

Figure 3. Effect of exogenous Se, B and Se+B on APX activity (**A**), MDHAR activity (**B**), DHAR activity (**C**) and GR activity (**D**) in soybean at 0, 150, 300 and 450 mM NaCl-induced salt stress. Here, Se, B and Se+B indicate 50 μM Na_2SeO_4, 1 mM H_3BO_3 and 50 μM Na_2SeO_4 + 1 mM H_3BO_3, respectively. Values in a column with different letters are significantly different at $p \leq 0.05$ applying Fisher's LSD test.

When exposed to 150, 300 and 450 mM NaCl, GR activity was increased by 45, 126 and 228%, respectively, compared to untreated control. However, exogenous supplementation of Se, B and Se+B further increased the GR activity at 150, 300 and 450 mM NaCl, compared to salt treatment alone. In combined Se+B increased the GR activity more than spraying with of Se or B alone, and due to exogenous Se+B the GR activity was accelerated by 54, 50 and 18% at 150, 300 and 450 mM NaCl, respectively, in comparison to respective only salt-treated plants (Figure 3D).

All salt treatments caused a substantial reduction in the CAT activity, in comparison to untreated control. Under mild, moderate and severe salinity CAT activity was decreased by 33, 43 and 56%, respectively, compared to untreated control. On the contrary, foliar application of Se, B and Se+B enhanced the CAT activity under salt stress. Among the foliar supplementations, Se+B showed a higher increment in CAT activity than Se or B alone. Moreover, application of Se+B increased the CAT activity by 38, 25 and 27% under mild, moderate and severe salinity, respectively, in comparison to respective salt treatment alone (Figure 4A).

In NaCl-treated plants, GPX activity was enhanced by 33, 108 and 125%, respectively, in comparison to untreated control. Moreover, foliar supplementation of Se, B and Se+B further stimulated the GPX activity under different levels of salinity. Among the treatments, under salt stress Se+B increased the GPX activity more than Se or B alone. Comparing with corresponding salt-treated plants (without Se, B and Se+B spray), GPX activity was increased by 66, 20 and 29%, respectively, in Se+B-treated plants under mild, moderate and severe salinity (Figure 4B).

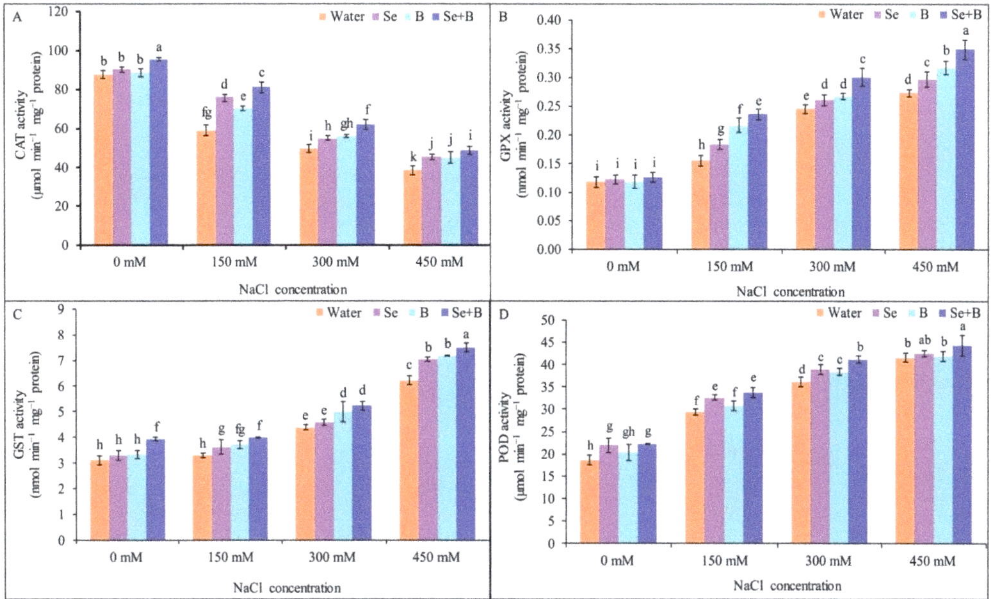

Figure 4. Effect of exogenous Se, B and Se+B on CAT activity (**A**), GPX activity (**B**), GST activity (**C**) and POD activity (**D**) in soybean at 0, 150, 300 and 450 mM NaCl-induced salt stress. Here, Se, B and Se+B indicate 50 μM Na_2SeO_4, 1 mM H_3BO_3 and 50 μM Na_2SeO_4 + 1 mM H_3BO_3, respectively. Values in a column with different letters are significantly different at $p \leq 0.05$ applying Fisher's LSD test.

Between untreated control plants and 150 mM salt-treated plants, no significant difference in GST activity was recorded. Moreover, at 300 and 450 mM salt stress GST activity was increased by 41 and 111%, in comparison to untreated control. However, at 150, 300 and 450 mM salt stress, application of Se+B further accelerated the GST activity by 21, 19 and 21%, respectively, compared to the respective only salt treatment. Although Se+B showed higher GST activity than an individual spray, but at 150 and 300 mM NaCl stress, B spray showed a similar increase in GST activity as Se+B spray, compared to only salt treatment (Figure 4C).

When plants were exposed to 150, 300 and 450 mM salt stress, POD activity was increased by 58, 95 and 124%, respectively, in comparison to untreated control. Application of Se+B increased the POD activity by 15, 14 and 7% at 150, 300 and 450 mM salt stress, respectively, compared to respective salt treatment alone. Moreover, no significant increment in POD activity was observed at 150 and 450 mM NaCl stress because of single supplementation of B. Single supplementation of Se caused a statistically similar increment of POD activity as Se+B at 150 and 450 mM NaCl stress, compared to corresponding salt treatment alone (Figure 4D).

2.7. Activities of Glyoxalase Enzymes

The activity of Gly I and Gly II showed reverse relation with the salt concentration, Gly I and Gly II activity decreased with the gradual increase in the salt concentration.

Under mild, moderate and severe salinity Gly I activity was declined by 26, 46 and 56%, respectively, in comparison to untreated control. On the contrary, foliar application of Se and B alone significantly increased the Gly I activity under mild and moderate salinity but not at severe salinity. Moreover, combined application of Se+B enhanced the Gly I activity by 31, 29 and 18% under mild, moderate and severe salinity, respectively, in comparison to corresponding salt treatment alone, followed by an individual spray of Se and B (Figure 5A).

In response to mild, moderate and severe salinity the activity of Gly II was declined by 24, 46 and 59% in 150, 300 and 450 mM NaCl-treated plants, respectively, in comparison to untreated control (without salt and Se, B and Se+B treatment). At 150 and 300 mM NaCl stress foliar application of Se+B increased the Gly II activity by 24 and 42%, respectively, compared to respective salt-treatment alone. Moreover, at 450 mM salt stress B alone and combined supplementation of Se+B showed a similar result, in the case of increasing the Gly II activity, respectively, in comparison to corresponding only salt-treated plant (Figure 5B).

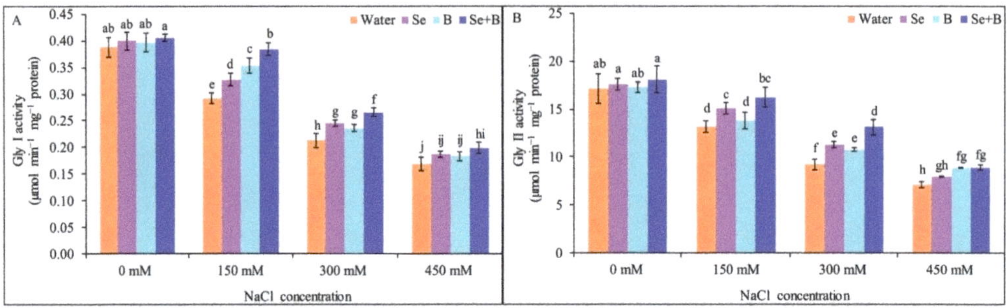

Figure 5. Effect of exogenous Se, B and Se+B on glyoxalase system: Gly I activity (**A**) and Gly II activity (**B**) in soybean at 0, 150, 300 and 450 mM NaCl-induced salt stress. Here, Se, B and Se+B indicate 50 µM Na_2SeO_4, 1 mM H_3BO_3 and 50 µM Na_2SeO_4 + 1 mM H_3BO_3, respectively. Values in a column with different letters are significantly different at $p \leq 0.05$ applying Fisher's LSD test.

3. Discussion

Salt stress has a detrimental effect on the growth, development and physiological activities of soybean [20,21]. In combating salt stress, exogenous protectants (micronutrients, trace elements, osmoprotectants, phytohormones, polyamines and antioxidants) showed promising results [4]. Supplementation of beneficial trace elements like Se improves antioxidant defense, and stress tolerance in plants [12,22]. The application of micronutrients increases plant stress tolerance by upregulating physiological activities and by improving the antioxidant defense system [23].

In response to salt stress shoot growth stunts [5]. Akram et al. [24] reported salt-induced plant height reduction in different genotypes of soybean. Salinity decreased the plant height, shoot FW and shoot DW in soybean along with other growth parameters [25]. Wu et al. [26] investigated that salt stress reduced the leaf area. Due to NaCl stress, plant height, DW and leaf area were reduced in two soybean cultivars [27]. Plant height, plant FW, plant DW and leaf area were decreased in NaCl-stressed soybean plants [28]. As salt stress imposes both ionic and osmotic stress in the plant, it negatively affects the cell division and cell elongation process as well as normal cell functioning, ultimately hinders the plant growth [4]. Foliar application of Se, at low concentration, enhanced the plant height of wheat under salt stress [29]. Application of Se increased the growth and shoot DW under salt stress [9]. Foliar applied Se enhanced the leaf area of cowpea under salt stress [30]. Boron is a crucial factor for the processes like respiration, synthesis of proteins, transportation of sugars and carbohydrate metabolism. Importantly, growth hormone-like IAA is also associated with B [13]. Ullah et al. [31] found that B aided in cell division of the actively growing region, especially the region near the shoot and root tips. However, Se+B spray reverted the negative effect of salt on growth parameters, in the present study.

Higher NaCl concentration in the soil solution of the root zone decreases the water potential and also hinders the water uptake through the root [5]. Moreover, a higher accumulation of Na^+ and Cl^- inside the cell also restricts the water uptake. Therefore, the plant faces physiological drought (osmotic stress) under salt stress [5]. Many researchers investigated that salt stress reduced the RWC in different crops [10,29,30]. Due to NaCl-induced osmotic stress, the RWC of the soybean plants decreased, in the current study. However, supplementation of Se improved the RWC of plants, in previous studies [10,29,30]. Upon exposure to 300 mM NaCl stress foliar-applied B reduced the concentration of Cl^- ion in xylem sap, thus improved the water uptake [16].

Higher accumulation of osmolytes occurs to protect the plant cells from NaCl-induced dehydration [32]. Salt stress-induced higher accumulation of Pro was observed in soybean and other crops [20,33,34]. With the increment of the NaCl concentration, Pro accumulation increased in NaCl-stressed soybean plants of the present study to protect the cell from osmotic stress by maintaining the osmotic pressure of the cell. Several previously published research articles denoted that Se and B have the potential to regulate the accumulation of osmoprotectants in plants. Selenium is involved in regulating osmoprotectants and secondary metabolites [12]. Selenium supplementation regulated the Pro content and osmotic pressure which helped in enhancing the water translocation to the shoot and ultimately increased RWC under salt stress [10]. Supplementation of B reduced the leaf Pro content under salt stress and enhanced the glycine betaine content [35]. In the present study, B improved the water uptake (RWC increased) but decreased Pro content. Selenium and B may have a role in the enhancement of the other osmolytes rather than Pro like glycine betaine, glutathione, soluble sugars [35–37] which acted to confer osmoprotection and improved the water content. Similarly, to our findings, supplementation of Se+B decreased the Pro content and increased the RWC under NaCl stress in stevia [18]. Boron-induced regulation of pyrroline-5-carboxylate synthetase (*P5CS*) and proline dehydrogenase (*PDH*) genes and the activities of P5CS and PDH were involved in modulating Pro accumulation in *Brassica napus* [38]. Selenium applications regulated transcript levels of genes *P5CS2* and *PDH* in fragrant rice [39]. This evidence strongly supports the osmoprotective roles of B and Se.

Salt-induced osmotic stress, ionic stress, nutrient imbalance, disrupted ion homeostasis and excessive generation of ROS, altogether threaten the antioxidant defense system of plants. However, salt-induced oxidative stress deteriorates the cell membrane, cellular organelles, cellular components (proteins, lipids, carbohydrates, etc.) and nucleic acids (DNA and RNA), and may also cause programmed cell death [2]. As H_2O_2 is a ROS, overproduction of H_2O_2 results in higher lipid peroxidation which is indicated by higher MDA content [22]. In previous investigations, salt stress was responsible for the overgeneration of H_2O_2 and MDA content in different soybean genotypes [20,24,40]. In the present study, H_2O_2 content, as well as MDA content, increased with the increase in the NaCl concentrations. Supplementation of Se protected plants from salt-induced oxidative stress by triggering the detoxification of ROS, which were generated due to salt stress [11]. Moreover, by upregulating the antioxidant defense system, Se diminished the MDA content and membrane damage under salt stress [9]. Boron plays a protective role against excessive ROS by maintaining membrane integrity, metabolic activity and enzymatic activity [13]. Supplementation of B decreased the H_2O_2 content and as well as MDA content under salt stress [41]. Foliar applied B upregulated the Cl^- transportation and protected plants from salt-induced oxidative stress [16]. However, Se and B alone and in combined alleviated the salt-induced oxidative stress, in the present investigation which was indicated by the reduced oxidative stress indicators (H_2O_2 and MDA content) under salt stress due to Se, B and Se+B spray.

In the antioxidant defense system four key enzymes viz. APX, MDHAR, DHAR and GR of the ascorbate-glutathione (AsA-GSH) pathway are crucial for enhancing stress tolerance as well as in minimizing the stress-induced oxidative damages, by detoxification of the ROS [8]. Moreover, APX is involved in direct ROS scavenging. In chloroplast, APX is the only ROS scavenger enzyme, as CAT is absent in chloroplast. Moreover, APX and GR are the major ROS detoxifier and also maintain redox homeostasis [7]. In soybean plants, the APX and GR activity was further increased with the increase in the salt concentration but MDHAR and DHAR activity were decreased under salt stress, in the present study. Application of Se increased the APX, MDHAR, DHAR and GR activity under salt stress [22]. Supplementation of B under salt stress increased the enzymatic activity [41]. The combined application of Se+B was more effective than individual Se and B spray in enhancing the activity of MDHAR and GR under salt stress.

Catalase produces H_2O and O_2 by direct dismutation of H_2O_2. Thus, CAT plays a major role in ROS detoxification under abiotic stress [7]. In previous studies, several scientists observed lower CAT activity in response to different levels of salt stress [33,42]. Similarly, in the present study, lower CAT activity was observed with the increment of the salt concentration, compared to untreated control. Previously, in various crops supplementation of Se at lower concentration increased the CAT activity under salt stress [10,29,43,44]. Application of B also increased the CAT activity under salt stress [41]. However, exogenous application of Se alone and with B, boosted the CAT activity in the salt-treated plant. Higher CAT activity along with lower H_2O_2 content under salt stress due to Se, B and Se+B spray indicated that Se and B alone and combinedly enhanced the antioxidant defense against salt-induced oxidative stress by increasing the ROS scavenging. The GPX enzymes are pivotal for cell protection and detoxification under oxidative stress. Selenium enhances the activity of GPX enzymes as it is a cofactor of GPX enzymes [45,46]. Along with peroxide breakdown, hormone biosynthesis and stress signaling GST is involved in amplifying the activity of GPX. The GPX enzymes scavenge H_2O_2 by utilizing GSH [7]. Upon exposure to salt stress, increased activity of GPX and GST was reported by several researchers [22,34,47]. However, GST and GPX activity increased in response to salt stress, in the present study. Foliar applied Se increased the GST and GPX activity in rapeseed under salt stress [22]. Application of B at high concentration increased the GPX activity [48]. In the current study, POD activity was increased under different levels of salinity, compared to untreated control. Likewise, in oat seedlings, the activity of POD was increased at 100 mM salt stress, in comparison to control [49]. The activity of POD was increased under salt stress in common beans [43]. Similarly, the POD activity was enhanced in maize under salinity [44]. Moreover, the application of Se at low concentration further increased the POD activity under salt-induced oxidative stress [44]. Under oxidative stress application of B also increased the POD activity [17].

Methylglyoxal is a by-product of the glyoxalase system which is a highly reactive and cytotoxic compound. The enzymes of the glyoxalase system, viz. Gly I and Gly II transform MG into nontoxic compounds. In the glyoxalase system, MG detoxification occurs in two steps. First, Gly I converts MG into *S*-D-lactoyl-glutathione by utilizing GSH. In last step, *S*-D-lactoyl-glutathione is converted into D-lactate by Gly II [8]. The activity of Gly I and Gly II decreased upon exposure to salt stress, in previous investigations [22,42]. Lower Gly I and Gly II activity under salt stress was observed, in our study which indicated that MG detoxification was not sufficient enough under salt stress. However, Se supplementation enhanced MG detoxification by amplifying the Gly I and Gly II activity under NaCl stress [22]. In the present study, in comparison to only salt-treated plants at 150 and 300 mM NaCl stress Se+B resulted in higher Gly I and Gly II activity than Se or B alone which means the higher MG detoxification under salt stress.

In summary, upon exposure to different levels of NaCl-induced salt stress leaf RWC decreased along with growth parameters, but Pro content and oxidative stress indicators (H_2O_2 and MDA content) increased. On the contrary, Se and B alone and combined improved the growth parameters, leaf RWC and decreased the ROS accumulation which was clear by the reduced H_2O_2 and MDA content (lipid peroxidation), and also decreased the Pro content when subjected to salt stress. All NaCl treatments decreased the MDHAR, DHAR and CAT activity and enhanced the APX, GR, GPX, GST and POD activity. However, exogenous Se, B and Se+B spray enhanced the enzymatic activities (APX, MDHAR, DHAR, GR, CAT, GPX, GST, POD) as a part of the antioxidant defense system under salt stress. Along with other enzymatic activities, the activity of Gly I and Gly II were also increased due to single and combined application of Se and B in salt-treated plants. However, it is clear from the results of the present study that foliar-applied Se, B and Se+B mitigated the salt stress by modulating the antioxidant defense system, ROS metabolism and glyoxalase system (Figure 6).

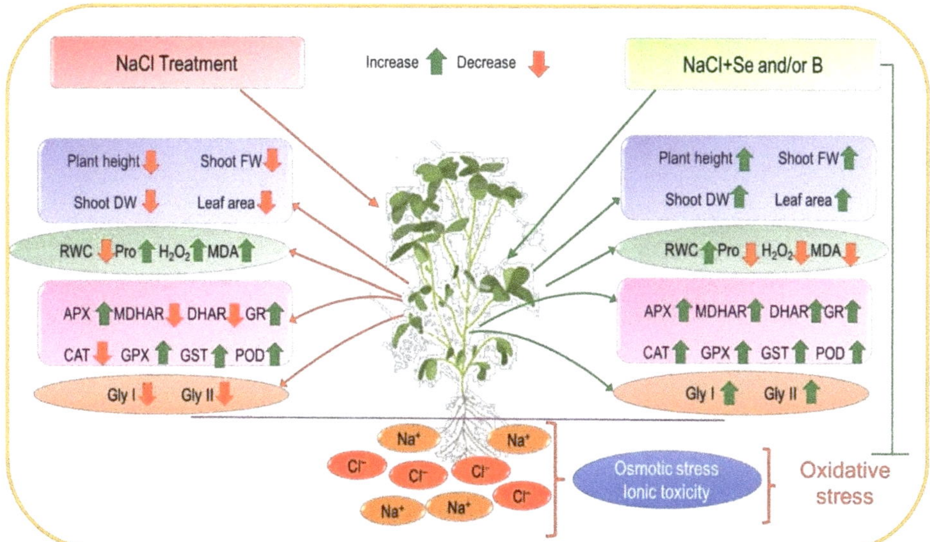

Figure 6. Schematic presentation of the protective roles of exogenous Se and B under salt stress.

4. Materials and Methods

4.1. Plant Material and Treatments

Healthy, matured, well-dried and uniform soybean (*Glycine max* cv. BINA Soybean-5) seeds were sown in plastic pots (14 L). Organic manure, urea, triple super phosphate and muriate of potash were applied as basal dose without B. At 20 days after sowing (DAS) along with control (0 mM NaCl), mild, moderate and severe salinity was imposed on plants by treating with 150, 300 and 450 mM NaCl, respectively. Single and combined supplementation of Se (50 µM Na_2SeO_4) and B (1 mM H_3BO_3) was accomplished at 16, 20, 24 and 28 DAS under control and saline conditions. The experiment was laid out in completely randomized design (CRD) with three replications.

4.2. Growth Parameters

For measuring plant height five plants were selected from each replication, and height was taken from the ground level to the tip of the plant. The average height of five plants was considered as the height of the plants for each replication and expressed as cm.

From each replication randomly three sample plants were uprooted, then roots were separated and shoots were weighed in a balance and after that averaged the weight to measure shoot FW plant^{-1}.

After measuring shoot FW, shoot samples were oven-dried at 80 °C for 48 h, then weighed. The average DW was measured and considered as the shoot DW plant^{-1}.

For leaf area measurement, leaf images were taken by a digital camera and the area was calculated using Image-J software v. 1.8.0 [50].

4.3. Leaf Relative Water Content and Proline Content

Leaf relative water content was measured according to Barrs and Weatherly [51]. After collecting fresh leaves, FW was measured and then dipped into distilled water for 8 h. After that, turgid weight (TW) was measured and followed by oven drying for 48 h at 80 °C to measure DW. The RWC was calculated from the following equation,

$$RWC\ (\%) = (FW - DW)/(TW - DW) \times 100$$

Proline content was measured from leaf sample according to Bates et al. [52]. Leaf samples were homogenized by using sulfosalicylic acid, followed by centrifuging at $11,500 \times g$. Then glacial acetic acid and acid ninhydrin was added with the supernatant. After cooling the mixture, toluene was added to separate ninhydrin pro-complex and then optical density of the chromophore was observed spectrophotometrically at 520 nm. Finally, Pro content was measured by comparing with a standard curve of known concentration of Pro. The Pro content was expressed as μmol g^{-1} FW.

4.4. Determination of H_2O_2 Content and Lipid Peroxidation

According to Yang et al. [53], leaf samples of 0.5 g were homogenized by trichloroacetic acid (TCA) and centrifuged at $11,500 \times g$. Then, supernatant was mixed with potassium-phosphate (K-P) buffer (pH 7.0) and potassium iodide (KI). After that, H_2O_2 content was determined spectrophotometrically by measuring the optical absorption of supernatant at 390 nm by using an extinction coefficient of 0.28 μM^{-1} cm^{-1} which was expressed as nmol g^{-1} FW.

By following the method of Heath and Packer [54] lipid peroxidation was assayed as MDA content which was measured spectrophotometrically at 532 and 600 nm on the basis of thiobarbituric acid reactive substances (TBARS) production by using an extinction coefficient of 155 mM^{-1} cm^{-1} and expressed as nmol g^{-1} FW.

4.5. Protein Determination

Protein content was estimated by following the method of Bradford [55] in which a standard curve was prepared from a known concentration of bovine serum albumin (BSA) and used to determine the protein content.

4.6. Determination of Enzyme Activities

The APX (EC: 1.11.1.11) activity was estimated spectrophotometrically at 290 nm according to the method of Nakano and Asada, [56]. The solution of K-P buffer (pH 7.0), reduced ascorbate (AsA), H_2O_2, ethylenediaminetetraacetic acid (EDTA) and extract for enzyme were used during the process where 2.8 mM^{-1} cm^{-1} was the extinction coefficient and it was expressed as μmol min^{-1} mg^{-1} protein.

The MDHAR (EC: 1.6.5.4) activity was determined from the reaction mixture consisting of Tris-HCl buffer (pH 7.5), AsA, ascorbate oxidase (AO), nicotinamide adenine dinucleotide phosphate (NADPH) along with enzyme and 6.2 mM^{-1} cm^{-1} was the extinction coefficient [57]. The activity of MDHAR is expressed as nmol min^{-1} mg^{-1} protein.

The DHAR (EC: 1.8.5.1) activity was measured spectrophotometrically by following the method of Nahar et al. [57] at 265 nm. The reaction mixture consisted of K–P buffer (pH 7.0), reduced glutathione (GSH), dehydroascorbate (DHA), EDTA and enzyme solution. Extinction coefficient 14 mM^{-1} cm^{-1} was used during the enzyme calculation and DHAR activity was expressed as nmol min^{-1} mg^{-1} protein.

The GR (EC: 1.6.4.2) activity was determined from the reaction of the reaction mixture which consists of K-P buffer (pH 7.0), oxidized glutathione (GSSG), EDTA, NADPH and enzyme solution at 340 nm using 6.2 mM^{-1} cm^{-1} as the extinction coefficient [22]. The GR activity was expressed as nmol min^{-1} mg^{-1} protein.

The CAT (EC: 1.11.1.6) activity was estimated according to Hasanuzzaman et al. [22] at 240 nm and read from a reaction mixture of K-P buffer (pH 7.0), H_2O_2 and enzyme solution and computed using 39.4 M^{-1} cm^{-1} as the extinction coefficient and expressed as μmol min^{-1} mg^{-1} protein.

The activity of GST (EC: 2.5.1.18) was measured from the reaction mixture of GSH, 1-chloro-2, 4-dinitrobenzene (CDNB), Tris-HCl buffer (pH 6.5) and enzyme extract, using 9.6 mM^{-1} cm^{-1} as the extinction coefficient [22]. The GST activity was expressed as nmol min^{-1} mg^{-1} protein.

The GPX (EC: 1.11.1.9) activity was read from the absorbance of the reaction between K-P buffer (pH 7.0), GSH, EDTA, sodium azide (NaN$_3$), NADPH, GR, H_2O_2 and enzyme at

340 nm. Extinction coefficient 6.62 mM^{-1} cm^{-1} was used during the calculation of enzyme activity [57]. The GPX activity was expressed as nmol min^{-1} mg^{-1} protein.

The POD activity (EC: 1.11.1.7) was determined according to Hemeda and Klein [58], at 470 nm guaiacol oxidation was observed spectrophotometrically. The reaction buffer contained sodium phosphate buffer (pH 6.0), methoxyphenol and 30% H$_2$O$_2$. The activity was computed by using an extinction coefficient of 6.58 mM^{-1} cm^{-1} and it was expressed as μmol min^{-1} mg^{-1} protein.

The Gly I (EC: 4.4.1.5) activity was read from the reaction of the mixture which consists of K-P buffer (pH 7.0), GSH, magnesium sulfate and MG at 240 nm by using 3.37 mM^{-1} cm^{-1} as the extinction coefficient [22]. The Gly I activity was expressed as μmol min^{-1} mg^{-1} protein.

The Gly II (EC: 3.1.2.6) activity was measured from the reaction of enzyme, Tris-HCl buffer (pH 7.2), 5, 5′-dithiobis-2-nitrobenzoic acid (DTNB) and *S*-D-lactoylglutathione at 412 nm by using an extinction coefficient of 13.6 mM^{-1} cm^{-1} [22]. The Gly II activity was expressed as μmol min^{-1} mg^{-1} protein.

4.7. Statistical Analysis

All data of three replications were statistically analyzed by using CoStat v.6. 400 [59]. Data were analyzed in one-way analysis of variance (ANOVA). The mean difference was compared by Fisher's least significant difference (LSD) test with the 5% level of significance.

5. Conclusions

The results of the present study clearly reveal that single and combined supplementation of Se and B mitigated the deleterious effect of salt stress and salt-induced oxidative stress in soybean, and also improved salt tolerance. However, exogenous Se, B and Se+B declined the salt-induced oxidative stress in soybean by upregulating the enzymatic activities of the antioxidant defense system and glyoxalase system, ultimately by diminishing the ROS accumulation which was indicated by the reduced H$_2$O$_2$ content and MDA content. Moreover, growth parameters and leaf RWC were also improved due to supplementation of Se+B under salt stress. Though in the present study, Se and B alone showed similar results to the combined application, but it was evident that the combined application of Se+B was more effective than a single application to confer the different levels of salt stress. Therefore, further investigations should be performed to ascertain the doses of Se, B and Se+B under different levels of salinity to mitigate the salt stress, as a high concentration of Se and B is phytotoxic. The role of B in MG detoxification has rarely been studied, and the mechanism is not well known yet; thus, it needs advanced future investigations.

Author Contributions: Conceptualization, M.H.; methodology, M.F., M.H.; software, M.H.; formal analysis, M.M.A., M.H.; investigation, M.R., K.R., K.S.S.; resources, M.F., M.H.; data curation, M.M.A.; writing–original draft preparation, M.R., writing–review and editing, K.N., M.F., M.H.; visualization, M.H.; supervision, M.H.; project demonstration, M.H.; funding acquisition, M.H. All authors have read and agreed to the published version of the manuscript.

Funding: This research was funded by Sher-e-Bangla Agricultural University Research System.

Institutional Review Board Statement: Not applicable.

Informed Consent Statement: Not applicable.

Data Availability Statement: All data are presented in this manuscript.

Acknowledgments: We are thankful to Maliha Rahman Falguni, Naznin Ahmed, Nazmin Sultana, Azaj Mahmud, Ayesha Siddika and Tonusree Saha for their heartiest help to execute the experiment. We acknowledge Rezaul Karim, Bangladesh Agricultural Development Corporation, for providing us high-quality seeds of soybean.

Conflicts of Interest: Mirza Hasanuzzaman, who is the coauthor of this article, is a current Editorial Board member of Plants. This fact did not affect the peer-review process.

Abbreviation

AO—ascorbate oxidase; APX—ascorbate peroxidase; AsA—ascorbate; BSA—bovine serum albumin; CAT—catalase; CDNB—1-chloro-2, 4-dinitrobenzene; DHA—dehydroascorbate; DHAR—dehydroascorbate reductase; DTNB—5, 5′-dithiobis (2-nitrobenzoic acid); EDTA—ethylenediamine tetraacetic acid; Gly I—glyoxylase I; Gly II—glyoxalase II; GPX—glutathione peroxidase; GR—glutathione reductase; GSH—glutathione; GSSG—oxidized glutathione; GST—glutathione *S*-transferase; KI—potassium iodide; MDA—malondialdehyde; MDHAR—monodehydroascorbate reductase; MG—methylglyoxal; NADPH—nicotinamide adenine dinucleotide phosphate; POD—peroxidase; Pro—proline; ROS—reactive oxygen species; TBARS—thiobarbituric acid reactive substances; TCA—trichloroacetic acid.

References

1. Bechtold, U.; Field, B. Molecular mechanisms controlling plant growth during abiotic stress. *J. Exp. Bot.* **2018**, *69*, 2753–2758. [CrossRef] [PubMed]
2. Kumar, A.; Singh, S.; Gaurav, A.K.; Srivastava, S.; Verma, J.P. Plant growth-promoting bacteria: Biological tools for the mitigation of salinity stress in plants. *Front. Microbiol.* **2020**, *11*, 1216. [CrossRef] [PubMed]
3. Food and Agriculture Organization of the United Nations. 2021. Available online: http://www.fao.org/global-soil-partnership/resources/highlights/detail/en/c/1412475/ (accessed on 18 July 2021).
4. Hasanuzzaman, M.; Nahar, K.; Fujita, M. Plant response to salt stress and role of exogenous protectants to mitigate salt-induced damages. In *Ecophysiology and Responses of Plants Under Salt Stress*; Ahmed, P., Azooz, M.M., Prasad, M.N.V., Eds.; Springer: New York, NY, USA, 2013; pp. 25–87.
5. Munns, R.; Tester, M. Mechanisms of salinity tolerance. *Annu. Rev. Plant Biol.* **2008**, *59*, 651–681. [CrossRef] [PubMed]
6. Sachdev, S.; Ansari, S.A.; Ansari, M.I.; Fujita, M.; Hasanuzzaman, M. Abiotic stress and reactive oxygen species: Generation, signaling, and defense mechanisms. *Antioxidants* **2021**, *10*, 277. [CrossRef]
7. Hasanuzzaman, M.; Bhuyan, M.H.M.B.; Zulfiqar, F.; Raza, A.; Mohsin, S.M.; Mahmud, J.A.; Fujita, M.; Fotopoulos, V. Reactive oxygen species and antioxidant defense in plants under abiotic stress: Revisiting the crucial role of a universal defense regulator. *Antioxidants* **2020**, *9*, 681. [CrossRef]
8. Hasanuzzaman, M.; Bhuyan, M.H.M.B.; Anee, T.I.; Parvin, K.; Nahar, K.; Mahmud, J.A.; Fujita, M. Regulation of ascorbate-glutathione pathway in mitigating oxidative damage in plants under abiotic stress. *Antioxidants* **2019**, *8*, 384. [CrossRef]
9. Jiang, C.; Zu, C.; Lu, D.; Zheng, Q.; Shen, J.; Wang, H.; Li, D. Effect of exogenous selenium supply on photosynthesis, Na$^+$ accumulation and antioxidative capacity of maize (*Zea may* L.) under salinity stress. *Sci. Rep.* **2017**, *7*, 42039. [CrossRef]
10. Subramanyam, K.; Du Laing, G.; Van Damme, E.J.M. Sodium selenate treatment using a combination of seed priming and foliar spray alleviates salinity stress in rice. *Front. Plant Sci.* **2019**, *10*, 116. [CrossRef]
11. Hassan, M.J.; Raza, M.A.; Khan, I.; Meraj, T.A.; Ahmed, M.; Shah, G.A.; Ansar, M.; Awan, S.A.; Khan, N.; Iqbal, N. Selenium and salt interactions in black gram (*Vigna mungo* L): Ion Uptake, antioxidant defense system, and photochemistry efficiency. *Plants* **2020**, *9*, 467. [CrossRef]
12. Hasanuzzaman, M.; Bhuyan, M.H.M.B.; Raza, A.; Hawrylak-Nowak, B.; Matraszek-Gawron, R.; Mahmud, J.A.; Nahar, K.; Fujita, M. Selenium in plants: Boon or bane? *Environ. Exp. Bot.* **2020**, *178*, 104170. [CrossRef]
13. Hansch, R.; Mendel., R.R. Physiological functions of mineral macronutrients (Cu, Zn, Mn, Fe, Ni, Mo, B, Cl). *Curr. Opin. Plant Biol.* **2009**, *12*, 259–266. [CrossRef]
14. Osakabe, Y.; Osakabe, K.; Shinozaki, K.; Tran, L.S.P. Response of plants to water stress. *Front. Plant Sci.* **2014**, *5*, 86. [CrossRef]
15. Riaz, M.; Kamran, M.; Fang, Y.; Yang, G.; Rizwan, M.; Ali, S.; Zhou, Y.; Wang, Q.; Deng, L.; Wang, Y.; et al. Boron supply alleviates cadmium toxicity in rice (*Oryza sativa* L.) by enhancing cadmium adsorption on cell wall and triggering antioxidant defense system in roots. *Chemosphere* **2021**, *266*, 128938. [CrossRef]
16. Dong, X.; Sun, L.; Guo, J.; Liu, L.; Han, G.; Wang, B. Exogenous boron alleviates growth inhibition by NaCl stress by reducing Cl$^-$ uptake in sugar beet (*Beta vulgaris*). *Plant Soil* **2021**, *464*, 423–439. [CrossRef]
17. Wu, X.; Song, H.; Guan, C.; Zhang, Z. Boron mitigates cadmium toxicity to rapeseed (*Brassica napus*) shoots by relieving oxidative stress and enhancing cadmium chelation onto cell walls. *Environ. Pollut.* **2020**, *263*, 114546. [CrossRef]
18. Shahverdi, M.A.; Omidi, H.; Damalas, C.A. Foliar fertilization with micronutrients improves *Stevia rebaudiana* tolerance to salinity stress by improving root characteristics. *Braz. J. Bot.* **2020**, *43*, 55–65. [CrossRef]
19. Liu, S.; Zhang, M.; Feng, F.; Tian, Z. Toward a "Green Revolution" for Soybean. *Mol. Plant.* **2020**, *13*, 688–697. [CrossRef]
20. Farhangi-Abriz, S.; Ghassemi-Golezani, K. How can salicylic acid and jasmonic acid mitigate salt toxicity in soybean plants? *Ecotoxicol. Environ. Saf.* **2018**, *147*, 1010–1016. [CrossRef]
21. Soliman, M.; Elkelish, A.; Souad, T.; Alhaithloul, H.; Farooq, M. Brassinosteroid seed priming with nitrogen supplementation improves salt tolerance in soybean. *Physiol. Mol. Biol. Plants* **2020**, *26*, 501–511. [CrossRef]
22. Hasanuzzaman, M.; Hossain, M.A.; Fujita, M. Selenium-induced up-regulation of the antioxidant defense and methylglyoxal detoxification system reduces salinity-induced damage in rapeseed seedlings. *Biol. Trace Elem. Res.* **2011**, *143*, 1704–1721. [CrossRef]

23. Noreen, S.; Fatima, Z.; Ahmad, S.; Ashraf, M. Foliar application of micronutrients in mitigating abiotic stress in crop plants. In *Plant Nutrients and Abiotic Stress Tolerance*; Hasanuzzaman, M., Fujita, M., Oku, H., Nahar, K., Hawrylak-Nowak, B., Eds.; Springer: Singapore, 2018; pp. 95–117.

24. Akram, S.; Siddiqui, M.N.; Hussain, B.M.N.; Bari, M.A.A.; Mostofa, M.G.; Hossaain, M.A.; Tran, L.S.P. Exogenous glutathione modulates salinity tolerance of soybean [*Glycine max* (L.) Merrill] at reproductive stage. *J. Plant Growth Regul.* **2017**, *36*, 877–888. [CrossRef]

25. Osman, M.S.; Badawy, A.A.; Osman, A.I.; Abdel Latef, A.A.H. Ameliorative impact of an extract of the halophyte *Arthrocnemum macrostachyum* on growth and biochemical parameters of soybean under salinity stress. *J. Plant Growth Regul.* **2021**, *40*, 1245–1256. [CrossRef]

26. Wu, Y.; Jin, X.; Liao, W.; Hu, L.; Dawuda, M.M.; Zhao, X.; Yu, J. 5-Aminolevulinic acid (ALA) alleviated salinity stress in cucumber seedlings by enhancing chlorophyll synthesis pathway. *Front. Plant Sci.* **2018**, *9*, 635. [CrossRef]

27. Wei, P.; Chen, D.; Jing, R.; Zhao, C.; Yu, B. Ameliorative effects of foliar methanol spraying on salt injury to soybean seedlings differing in salt tolerance. *Plant Growth Regul.* **2015**, *75*, 133–141. [CrossRef]

28. Baghel, L.; Kataria, S.; Guruprasad, K.N. Static magnetic field treatment of seeds improves carbon and nitrogen metabolism under salinity stress in soybean. *Bioelectromagnetics* **2016**, *37*, 455–470. [CrossRef]

29. Elkelish, A.A.; Soliman, M.H.; Alhaithloul, H.A.; El-Esawi, M.A. Selenium protects wheat seedlings against salt stress-mediated oxidative damage by up-regulating antioxidants and osmolytes metabolism. *Plant Physiol. Biochem.* **2019**, *137*, 144–153. [CrossRef]

30. Manaf, H.H. Beneficial effects of exogenous selenium, glycine betaine and seaweed extract on salt stressed cowpea plant. *Ann. Agric. Sci.* **2016**, *61*, 41–48. [CrossRef]

31. Ullah, S.; Khan, A.S.; Malik, A.U.; Afzal, I.; Shahid, M.; Razzaq, K. Foliar application of boron influences the leaf mineral status, vegetative and reproductive growth, and yield and fruit quality of 'kinnow' mandarin (*Citrus reticulata* Blanco.). *J. Plant Nutr.* **2012**, *35*, 2067–2079. [CrossRef]

32. Amist, N.; Singh, N.B. Responses of enzymes involved in proline biosynthesis and degradation in wheat seedlings under stress. *Allelopathy J.* **2017**, *42*, 195–206. [CrossRef]

33. Parvin, K.; Hasanuzzaman, M.; Bhuyan, M.H.M.B.; Nahar, K.; Mohsin, S.M.; Fujita, M. Comparative physiological and biochemical changes in tomato (*Solanum lycopersicum* L.) under salt stress and recovery: Role of antioxidant defense and glyoxalase systems. *Antioxidants* **2019**, *8*, 350. [CrossRef]

34. Parvin, K.; Nahar, K.; Hasanuzzaman, M.; Bhuyan, M.H.M.B.; Mohsin, S.M.; Fujita, M. Exogenous vanillic acid enhances salt tolerance of tomato: Insight into plant antioxidant defense and glyoxalase systems. *Plant Physiol. Biochem.* **2020**, *150*, 109–120. [CrossRef] [PubMed]

35. Karimi, S.; Tavallali, V.; Wirthensohn, M. Boron amendment improves water relations and performance of *Pistacia vera* under salt stress. *Sci. Hortic.* **2018**, *241*, 252–259. [CrossRef]

36. Dehnavi, M.M.; Yadavi, A.; Merajipoor, M. Physiological responses of sesame (*Sesamum indicum* L.) to foliar application of boron and zinc under drought stress conditions. *J. Plant Proc. Funct.* **2017**, *6*, 27–36.

37. Shah, W.H.; Aadil, R.; Inayatullah, T.; Rehman, R.U. Exogenously applied selenium (Se) mitigates the impact of salt stress in *Setaria italica* L. and *Panicum miliaceum* L. *Nucleus.* **2020**, *63*, 327–339. [CrossRef]

38. Yan, L.; Li, S.; Riaz, M.; Jiang, C. Proline metabolism and biosynthesis behave differently in response to boron-deficiency and toxicity in *Brassica napus*. *Plant Physiol. Biochem.* **2021**, *167*, 529–540. [CrossRef]

39. Luo, H.; He, L.; Lai, R.; Liu, J.; Xing, P.; Tang, X. Selenium applications enhance 2-acetyl-1-pyrroline biosynthesis and yield formation of fragrant rice. *Agron. J.* **2021**, *113*, 250–260. [CrossRef]

40. Kataria, S.; Baghel, L.; Jain, M.; Guruprasad, K.N. Magnetopriming regulates antioxidant defense system in soybean against salt stress. *Biocatal. Agric. Biotechnol.* **2019**, *18*, 101090. [CrossRef]

41. Tavallali, V.; Karimi, S.; Espargham, O. Boron enhances antioxidative defense in the leaves of salt-affected *Pistacia vera* seedlings. *Hort. J.* **2018**, *87*, 55–62. [CrossRef]

42. Mahmud, J.A.; Hasanuzzaman, M.; Khan, M.I.R.; Nahar, K.; Fujita, M. β-Aminobutyric acid pretreatment confers salt stress tolerance in *Brassica napus* L. by modulating reactive oxygen species metabolism and methylglyoxal detoxification. *Plants* **2020**, *9*, 241. [CrossRef]

43. Moussa, H.R.; Hassen, A.M. Selenium affects physiological responses of *Phaseolus vulgaris* in response to salt level. *Int. J. Veg. Sci.* **2018**, *24*, 236–253. [CrossRef]

44. Ashraf, M.A.; Akbar, A.; Parveen, A.; Rasheed, R.; Hussain, I.; Iqbal, M. Phenological application of selenium differentially improves growth, oxidative defense and ion homeostasis in maize under salinity stress. *Plant Physiol. Biochem.* **2018**, *123*, 268–280. [CrossRef]

45. Brigelius-Flohe, R.; Maiorino, M. Glutathione peroxidases. *Biochim. Biophys. Acta.* **2013**, *1830*, 3289–3303. [CrossRef]

46. Bermingham, E.N.; Hesketh, J.E.; Sinclair, B.R.; Koolaard, J.P.; Roy, N.C. Selenium-enriched foods are more effective at increasing glutathione peroxidase (GPx) activity compared with selenomethionine: A meta-analysis. *Nutrients* **2014**, *6*, 4002–4031. [CrossRef]

47. Nahar, K.; Hasanuzzaman, M.; Alam, M.M.; Fujita, M. Roles of exogenous glutathione in antioxidant defense system and methylglyoxal detoxification during salt stress in mung bean. *Biol. Plant.* **2015**, *59*, 745–756. [CrossRef]

48. Dilek Tepe, H.; Aydemir, T. Effect of boron on antioxidant response of two lentil (*Lens culinaris*) cultivars. *Commun. Soil Sci. Plant Anal.* **2017**, *48*, 1881–1894. [CrossRef]

49. Sapre, S.; Gontia-Mishra, I.; Tiwari, S. *Klebsiella* sp. confers enhanced tolerance to salinity and plant growth promotion in oat seedlings (*Avena sativa*). *Microbiol. Res.* **2018**, *206*, 25–32. [CrossRef]
50. Ahmad, S.; Ali, H.; Rehman, A.U.; Khan, R.J.Z.; Ahmad, W.; Fatima, Z.; Abbas, G.; Irfan, M.; Ali, H.; Khan, M.A.; et al. Measuring leaf area of winter cereals by different techniques: A comparison. *Pak. J. Life Soc. Sci.* **2015**, *13*, 117–125.
51. Barrs, H.D.; Weatherley, P.E. A re-examination of the relative turgidity technique for estimating water deficits in leaves. *Aust. J. Biol. Sci.* **1962**, *15*, 413–428. [CrossRef]
52. Bates, L.S.; Waldren, R.P.; Teari, D. Rapid determination of free proline for water stress studies. *Plant Soil* **1973**, *39*, 205–207. [CrossRef]
53. Yang, S.; Wang, L.; Li, S. Ultraviolet-B irradiation-induced freezing tolerance in relation to antioxidant system in winter wheat (*Triticum aestivum* L.) leaves. *Environ. Exp. Bot.* **2007**, *60*, 300–307. [CrossRef]
54. Heath, R.L.; Packer, L. Photo peroxidation in isolated chloroplast: I. Kinetics and stoichiometry of fatty acid peroxidation. *Arch. Biochem. Biophys.* **1968**, *125*, 189–198. [CrossRef]
55. Bradford, M.M. A rapid and sensitive method for the quantitation of microgram quantities of protein utilizing the principle of protein-dye binding. *Anal. Biochem.* **1976**, *72*, 248–254. [CrossRef]
56. Nakano, Y.; Asada, K. Hydrogen peroxide is scavenged by ascorbate specific peroxidase in spinach chloroplasts. *Plant Cell Physiol.* **1981**, *22*, 867–880.
57. Nahar, K.; Hasanuzzaman, M.; Alam, M.M.; Rahman, A.; Suzuki, T.; Fujita, M. Polyamine and nitric oxide crosstalk: Antagonistic effects on cadmium toxicity in mung bean plants through upregulating the metal detoxification, antioxidant defense and methylglyoxal detoxification systems. *Ecotoxicol. Environ. Saf.* **2016**, *126*, 245–255. [CrossRef]
58. Hemeda, H.M.; Klein, B.P. Effects of naturally occurring antioxidants on peroxidase activity of vegetable extracts. *J. Food Sci.* **1990**, *55*, 184–185. [CrossRef]
59. CoStat-Statistical Software. version 6.400; CoHort Software: Monterey, CA, USA, 2008.

Article

Gibberellic Acid and Jasmonic Acid Improve Salt Tolerance in Summer Squash by Modulating Some Physiological Parameters Symptomatic for Oxidative Stress and Mineral Nutrition

Mashael M. Al-harthi [1], Sameera O. Bafeel [2] and Manal El-Zohri [2,3,*]

[1] Department of Biology, Faculty of Applied Science, Umm Al-Qura University, Makkah 21955, Saudi Arabia; mmharthy@uqu.edu.sa
[2] Department of Biological Sciences, Faculty of Science, King Abdulaziz University, Jeddah 21488, Saudi Arabia; sbafil@kau.edu.sa
[3] Department of Botany and Microbiology, Faculty of Science, Assiut University, Assiut 71516, Egypt
* Correspondence: melzohri@kau.edu.sa or mnzohri@aun.edu.eg; Tel.: +966-557-168-898

Citation: Al-harthi, M.M.; Bafeel, S.O.; El-Zohri, M. Gibberellic Acid and Jasmonic Acid Improve Salt Tolerance in Summer Squash by Modulating Some Physiological Parameters Symptomatic for Oxidative Stress and Mineral Nutrition. *Plants* **2021**, *10*, 2768. https://doi.org/10.3390/plants10122768

Academic Editors: Mirza Hasanuzzaman and Masayuki Fujita

Received: 10 November 2021
Accepted: 12 December 2021
Published: 15 December 2021

Publisher's Note: MDPI stays neutral with regard to jurisdictional claims in published maps and institutional affiliations.

Abstract: Gibberellic acid (GA) and jasmonic acid (JA) are considered to be endogenous regulators that play a vital role in regulating plant responses to stress conditions. This study investigated the ameliorative role of GA, JA, and the GA + JA mixture in mitigating the detrimental effect of salinity on the summer squash plant. In order to explore the physiological mechanisms of salt stress alleviation carried out by exogenous GA and JA, seed priming with 1.5 mM GA, 0.005 mM JA, and their mixture was performed; then the germinated summer squash seedlings were exposed to 50 mM NaCl. The results showed that a 50 mM NaCl treatment significantly reduced shoot and root fresh and dry weight, water content (%), the concentration of carotenoid (Car), nucleic acids, K^+, and Mg^{++}, the K^+/Na^+ ratio, and the activity of catalase (CAT) and ascorbate peroxidase (APX), while it increased the concentration of proline, thiobarbituric acid reactive substances (TBARS), Na^+, and Cl^- in summer squash plants, when compared with the control. However, seed priming with GA, JA and the GA + JA mixture significantly improved summer squash salt tolerance by reducing the concentration of Na^+ and Cl^-, TBARS, and the Chl a/b ratio and by increasing the activity of superoxide dismutase, CAT, and APX, the quantities of K^+ and Mg^{++}, the K^+/Na^+ ratio, and the quantities of RNA, DNA, chlorophyll b, and Car, which, in turn, ameliorated the growth of salinized plants. These findings suggest that GA and JA are able to efficiently defend summer squash plants from salinity destruction by adjusting nutrient uptake and increasing the activity of antioxidant enzymes in order to decrease reactive oxygen species accumulation due to salinity stress; these findings offer a practical intervention for summer squash cultivation in salt-affected soils. Synergistic effects of the GA and JA combination were not clearly observed, and JA alleviated most of the studied traits associated with salinity stress induced in summer squash more efficiently than GA or the GA + JA mixture.

Keywords: phytohormones; salinity; *Cucurbita pepo* (L.); mineral uptake; proline; lipid peroxidation; antioxidant enzymes; nucleic acids

1. Introduction

Salinity stress results in extensive crop damage worldwide; most field crops are salt sensitive, and the problem is expected to increase in the coming decades [1]. Salinity is regarded as the most limiting and damaging of the factors that restrict crop growth, yield, and productivity [2]. Salt-affected soils negatively affect plants in various ways including through water stress, ion toxicity, nutritional imbalance, oxidative stress owing to the formation of reactive oxygen species, alterations to metabolic processes, reductions in photosynthesis rate, membrane damage, declines in cell division and expansion, and genetic disorders [3,4]. Plants have evolved many mechanisms to alleviate the adverse effects of salt stress, including plasticity in changing their morphological patterns, the accumulation of compatible solutes to preserve cell water content and prevent ultrastructural destruction, ion-homeostasis, enhanced water-use efficiency, improved photosynthesis activity, the detoxification of ROS via the activation of antioxidant systems, and the stimulation of plant hormones [5].

It is fundamentally important to recognize how plants perceive stress signals and respond to various environmental stress factors [6]. Plant hormones are active members of the signal cascade involved in the generation of plant responses to stress conditions [7]. The exogenous application of biological growth-promoting substances is a promising sustainable strategy to encourage plant growth and yield and to reinforce the plant's capacity to alleviate stress conditions [8]. Phytohormones are endogenously produced organic substances essential for regulating plant growth and productivity. Many phytohormones including, abscisic acid (ABA), gibberellins, ethylene, salicylic acid (SA), and jasmonic acid (JA) seem to be critical modules of complex signaling networks and have been integrated into current models of stress response [9]. Hence, they play an important role in prompting plant tolerance to various stress conditions.

Numerous works have proven the potential of gibberellic acid (GA) as a classical plant hormone. Gibberellins (GAs) are phytohormones that regulate numerous metabolic pathways, the activity of many enzymes, and gene expression; therefore, they play a vital role in seed germination and seedling growth, stem and root elongation, leaf expansion, and flowering [10]. The exogenous application of GA3 enhances stomatal conductance, water-use efficiency, photosynthesis activity, ion uptake, and the balance of other phytohormones [11]. Furthermore, GA3 maximizes the antioxidant capacity and osmoprotectants, while it minimizes lipid peroxidation, in order to alleviate the drastic effects of environmental stress [12]. Jasmonic acid, which is affiliated with the new phytohormone group, is a key regulator that plays a major role in plant growth and development and in the mitigation of both biotic and abiotic stress conditions [13]. Understanding of the complexity of the signaling network in which JA is involved is just developing [14]. Jasmonates exert their effects by orchestrating large-scale changes in gene expression [15,16].

Some investigations have shown that there is a relation between GAs and JAs under both usual and adverse conditions. Achard et al. [17], find that GA signaling declines under cold stress, causing plant growth reduction, while DELLA proteins accumulate. Wingler et al. [18] concluded that DELLA proteins can bind to JAZ proteins, causing the stimulation of jasmonate-responsive genes. This outcome reveals JA–GA crosstalk under environmental stress. Like other plant hormones, they exhibit stimulating and inhibiting activities in planta, and synergistic or antagonistic effects with respect to other plant hormones are well known [19].

Summer squash (*Cucurbita pepo* L.) is a popular, widely used crop worldwide, which belongs to family Cucurbitaceae. It grows under tropical and subtropical conditions during summer [20]. It is significant not only because of its utilization as human food but also because of its use as a medicinal plant due to its high content of zinc and antioxidants [21]. Summer squash is a sensitive crop for abiotic stress conditions including salinity [22]. Therefore, it has been chosen as a case study plant in this investigation of how tolerance to salinity stress is improved by seed priming with GA and JA. Several studies have shown that exogenous GA can mitigate toxic effects and improve the vegetative and reproductive responses of many plants under salinity stress [10–12]. Nevertheless, there is not a lot of data regarding summer squash and JA.

It is convenient to describe the role played by each phytohormone in salinity tolerance, as if each one worked separately; however, it is well known that phytohormone responses include extensive crosstalk and regulatory interfaces. In some cases, enhancing the signal of several phytohormones is needed for the activation of one stress tolerance gene. In other cases, the induction of one hormone repudiates the action of another. Still another form of crosstalk occurs when one hormone enhances or suppresses the biosynthesis or action of another. Unravelling these interfaces is one of the most urgent issues in plant stress physiology. Therefore, this present study was conducted to explore the interactive responses of salinity stressed summer squash plants to seed priming with GA and JA, as well as the GA + JA mixture, in order to evaluate whether the combination of GA and JA has synergistic effects in alleviating salinity stress in summer squash.

2. Results

2.1. Biomass and Water Content Percentage

The shoot and root biomass of summer squash plants significantly reduced in response to saline conditions. All investigated hormonal treatments significantly enhanced shoot and root fresh and dry weight under saline conditions. The highest values of shoot and root fresh and dry weight under saline conditions were observed for JA treatment (Figure 1). Seed priming with JA caused a significant enhancement in summer squash shoot and root fresh weights under saline conditions, by about 126% and 376%, respectively, as compared to their corresponding controls (Figure 1A,B). GA and the GA + JA mixture enhanced the shoot fresh weight to comparable values of about 107% higher than the saline control (Figure 1A). Under non-saline conditions, GA and JA significantly enhanced shoot fresh weights by about 17% and 30%, respectively, as compared to their corresponding controls (Figure 1A). Under saline conditions, JA caused a significant enhancement in summer squash shoot and root dry weights of about 157% and 155%, respectively, as compared to their corresponding controls. GA and the GA + JA mixture enhanced dry weight under saline conditions in shoots by about 130% and 141% as compared to the control, and in roots by about 114% and 123% as compared to the control (Figure 1C,D). The two-way ANOVA indicated highly significant ($p < 0.001$) effects of both salinity and the salinity $\times$ hormones interaction on shoot and root FW and DW. Hormonal treatments showed highly significant ($p < 0.01$) and significant ($p < 0.05$) effects on shoot and root FW, respectively, but no significant effects were shown for individual hormonal treatments on shoot and root DW (Table 1).

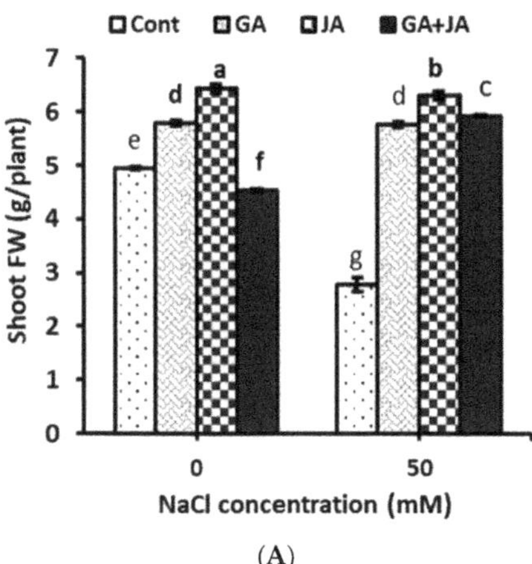

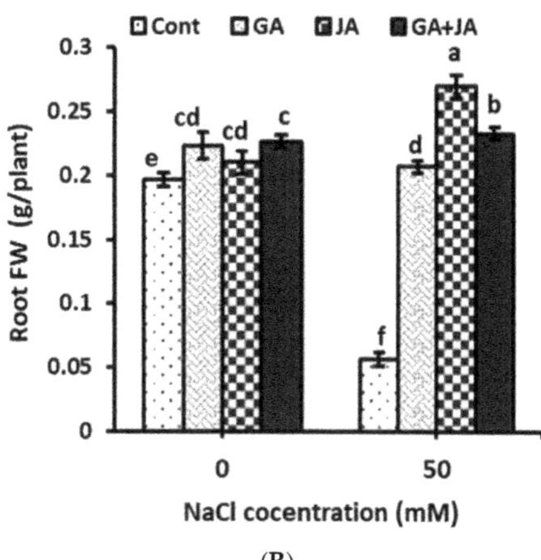

Figure 1. *Cont.*

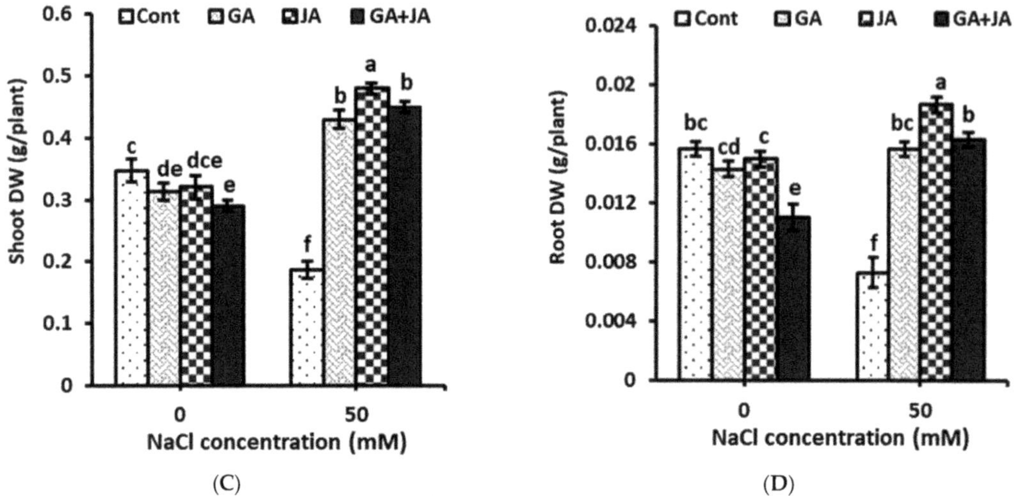

Figure 1. Biomass—(**A**) shoot fresh weight (FW); (**B**) root FW; (**C**) shoot dry weight (DW); and (**D**) root DW—of non-salinized and salinized summer squash plants, as affected by seed priming with 1.5 mM GA, 0.005 mM JA, and mixture of them. Data represent mean of three replicates ($n = 3$) with error bars indicating standard error of the mean. Bars carrying different lowercase letters are significantly different at $p < 0.05$ according to Duncan's multiple range test. p values for two-way ANOVA are reported in Table 1.

Table 1. The results of two-way ANOVA performed using salinity stress (0, 50 mM NaCl), hormonal treatments (GA, JA, GA + JA) and Salinity × Hormones as sources of variation.

Parameters	Significance of Sources of Variation		
	Salinity (S)	Hormones (H)	S × H
Shoot FW	***	**	***
Root FW	***	*	***
Shoot DW	***	ns	***
Root DW	***	ns	***
Shoot WC	*	ns	**
Root WC	*	ns	**
Proline in shoot	***	**	**
Proline in root	***	**	*
TBARS in shoot	***	*	*
TBARS in root	***	***	***
SOD activity in shoot	ns	***	ns
SOD activity in root	*	***	ns
CAT activity in shoot	***	ns	***
CAT activity in root	***	*	***
APX activity in shoot	***	ns	***
APX activity in root	***	ns	***
RNA in shoot	***	ns	***
RNA in root	***	ns	***
DNA in shoot	***	*	***
DNA in root	***	*	***

The stars indicate significant differences (* $p < 0.05$; ** $p < 0.01$; *** $p < 0.001$); ns, not significant.

The shoot and root water content (WC) percentage was reduced significantly with salinity stress (Figure 2). None of the hormonal treatments, when applied to non-salinized plants, showed any significant effect on WC. However, when salinized plants were treated with all hormonal treatments, their shoot and root WC increased by 20% and 30%, respectively, higher than stressed untreated plants. The two-way ANOVA showed significant effects ($p < 0.05$) of salinity stress and highly significant effects ($p < 0.01$) of the salinity × hormones interaction on shoot and root WC, while no significant effects were shown for individual hormonal treatments (Table 1).

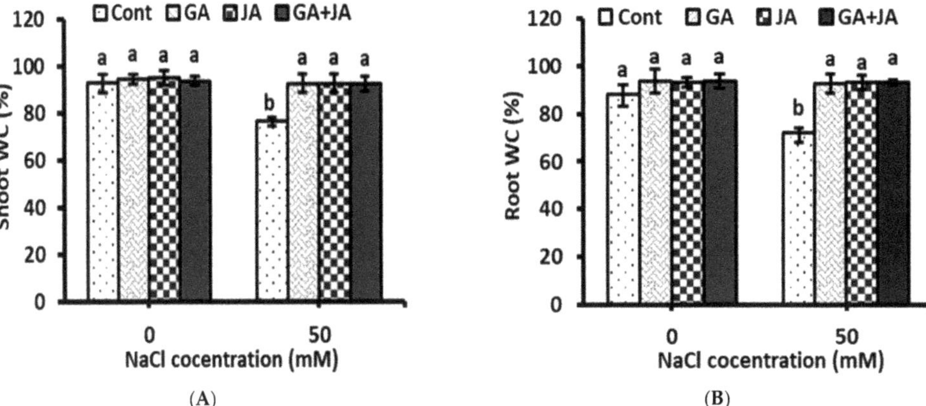

Figure 2. Percentage of water content (WC) in (**A**) shoot and (**B**) root of non-salinized and salinized summer squash plants, as affected by seed priming with 1.5 mM GA, 0.005 mM JA, and mixture of them. Data represent mean of three replicates ($n = 3$) with error bars indicating standard error of the mean. Bars carrying different lowercase letters are significantly different at $p < 0.05$ according to Duncan's multiple range test. p values for two-way ANOVA are reported in Table 1.

2.2. Plant Pigments

Salinity stress did not affect Chlorophyll a (Chl a), while it significantly enhanced Chlorophyll b (Chl b) and reduced the Chl a/b ratio and Carotenoid (Car) concentrations. All hormonal treatments significantly inhibited the Chl a concentration, so that it was lower than that of the controls both under saline and non-saline conditions. The Chl a content was about 34%, 21%, and 46% lower than that of the control after treatment with GA, JA, and GA + JA mixture, respectively, under saline conditions. On the other hand, all hormonal treatments significantly increased the Chl b concentration, as compared to controls under both non-saline and saline conditions. In the best results recorded for JA treatment, it enhanced Chl b concentration by about 5-fold and 2-fold under non-saline and saline conditions, respectively, as compared with their corresponding controls. Seed priming with GA and the GA + JA mixture enhanced Chl b accumulation at comparable rates under non-saline and saline conditions. In the same context, all hormonal treatments significantly decreased the Chl a/b ratio, under non-saline and saline conditions, to lower levels than their corresponding controls. GA, JA, and the GA + JA mixture inhibited the Chl a/b ratio, under saline conditions, by about 59%, 63%, and 63% and, under non-saline conditions, by about 77.5%, 81%, and 77%, compared to their respective controls. Carotenoid concentration was significantly enhanced by hormonal treatments under saline and non-saline conditions. The highest values of Car were recorded for plants treated with JA; they were 24% and 58% higher than the controls under non-saline and saline conditions respectively. GA and the GA + JA mixture enhanced Car accumulation under saline conditions by about 18% and 14% and, under non-saline conditions, by about 22% and 31%, as compared to the respective controls (Table 2). The two-way ANOVA showed highly significant effects of salinity stress on Chl b concentration ($p < 0.001$) and Chl a/b ratio ($p < 0.01$) and a significant effect on Car concentration ($p < 0.05$), while no significant effect was observed for salinity stress on Chl a concentration. All hormonal treatments and the salinity × hormones interaction showed highly significant effects ($p < 0.001$) on Chl a, Chl b, and Car concentration and the Chl a/b ratio (Table 2).

Table 2. Effects of salinity stress and hormonal treatments on Chl a, Chl b, carotenoid contents, and Chl a/b ratio in summer squash shoot and root [a].

NaCl (mM)	Hormones	Chl a (mg/g DW)	Chl b (mg/g DW)	Chl a/b Ratio	Carotenoid (mg/g DW)
0	Non	11.60 ± 0.08 [a]	6.24 ± 0.11 [f]	1.9 ± 0.03 [b]	1.56 ± 0.07 [f]
	GA	5.07 ± 0.52 [ef]	30.27 ± 0.38 [c]	0.17 ± 0.02 [d]	1.92 ± 0.04 [d]
	JA	5.61 ± 0.16 [e]	39.79 ± 0.75 [a]	0.15 ± 0.01 [d]	2.48 ± 0.08 [a]
	GA + JA	5.74 ± 0.52 [e]	32.34 ± 0.88 [c]	0.17 ± 0.02 [d]	2.06 ± 0.05 [c]
50	Non	11.62 ± 0.03 [a]	16.96 ± 0.11 [e]	0.69 ± 0.01 [a]	1.80 ± 0.02 [e]
	GA	7.65 ± 0.06 [c]	29.37 ± 0.44 [c]	0.28 ± 0.02 [c]	2.14 ± 0.04 [c]
	JA	9.07 ± 0.58 [b]	35.82 ± 0.50 [b]	0.25 ± 0.02 [c]	2.25 ± 0.07 [b]
	GA + JA	6.24 ± 0.39 [d]	29.84 ± 0.92 [cd]	0.25 ± 0.02 [c]	2.07 ± 0.06 [c]
ANOVA					
Salinity (S)		ns	***	**	*
Hormones (H)		***	***	***	***
S × H		***	***	***	***

[a] Values represent the mean $\pm$ SD, n = 3. Different lowercase letters indicate significant differences within parameters ($p < 0.05$) as determined by Duncan's multiple range test. The stars indicate significant differences (* $p < 0.05$, ** $p < 0.01$, *** $p < 0.001$); ns, not significant.

2.3. Mineral Ions

The plants' mineral uptake was strongly influenced by both salinity and hormonal treatments (Table 3). Na^+ concertation significantly increased in plant shoots and roots in response to saline conditions, while all hormonal treatments significantly reduced Na^+ concentration in plant shoots and roots to lower levels than those of the corresponding stressed controls. Under saline conditions, GA, JA, and the GA + JA mixture reduced Na^+ concentration in shoots by about 21%, 61%, and 58%, and in roots by about 56%, 39%, and 54%, compared to their respective controls. Salinity stress significantly reduced K^+ concentration and the K^+/Na^+ ratio in plant shoots and roots. Under saline conditions, only GA treatment enhanced Na^+ concentration to a significantly higher level than that of the stressed control in shoot samples. In plant roots, seed priming with GA, JA, and the GA + JA mixture significantly enhanced K^+ concentration by about 36.1%, 89.6%, and 59.74%, as compared to the stressed control. Salinity stress significantly reduced the K^+/Na^+ ratio in plant roots and shoots. All hormones significantly enhanced the K^+/Na^+ ratio under saline conditions to higher levels than in stressed controls. GA, JA, and the GA + JA mixture increased the K^+/Na^+ ratio in summer squash shoots by about 47.15%, 122.76%, and 156.09%, respectively, as compared to the saline control. The alleviative effect of seed priming with the studied treatments on the K^+/Na^+ ratio was more pronounced in the root samples. GA, JA, and the GA + JA mixture increased the K^+/Na^+ ratio in plant roots by about 3-fold, as compared to the stressed control (Table 3).

Salinity stress noticeably reduced Mg^{++} concentration in both plant shoot and root. Under saline conditions, seed priming with JA and the GA + JA mixture significantly enhanced Mg+ concentration in summer squash shoots and roots to higher levels than those of the stressed controls. JA and the GA + JA mixture enhanced Mg^{++} concentration in shoots under saline conditions by about 14.41% and 12.40%, as compared to the stressed control. In root samples, seed priming with JA and the GA + JA mixture enhanced Mg^{++} concentration in shoots under saline conditions by about 80.24% and 55.45%, respectively, compared to the stressed control (Table 3).

The concentration of chloride ions markedly increased in both shoot and root of summer squash in response to salt treatment. Seed priming with GA significantly reduced Cl^- concentration of summer squash shoot under saline conditions by about 17.6%, as compared to the stressed control. All hormonal treatments significantly reduced Cl^- concentration of summer squash roots under saline conditions. GA, JA, and the GA + JA mixture decreased Cl^- concentration under saline conditions in roots by about 21.04%, 26.11%d and 13.35%, respectively, as compared to the stressed control. Under non-saline conditions, seed priming with JA significantly reduced Cl^- concentration in the shoot by about 31.33% and enhanced it in the root by about 21.22%, as compared to their corresponding unstressed controls (Table 3). The two-way ANOVA showed highly significant ($p < 0.001$; $p < 0.01$) effects of salinity stress on the concentration of mineral ions both in shoots and roots; hormonal treatments showed highly significant ($p < 0.01$) effects on Na^+ and K^+ concentrations and the Na^+/K^+ ratio in plant shoots and roots, and on the Cl^- concentration in roots, while no significant effects were observed for hormonal treatments and the salinity $\times$ hormones interaction on Mg^{++} and Cl^- in the plant shoots. The salinity $\times$ hormones interaction showed highly significant effects ($p < 0.001$) on Na^+ and the Na^+/K^+ ratio in plant shoots and on all studied mineral ions in roots, as well as significant effects ($p < 0.05$) on K^+ in shoots (Table 3).

Table 3. Effects of salinity stress and hormonal treatments on Na^+, K^+, Mg^{++}, and Cl^- concentration (g/kg DW) and Na^+/K^+ ratio in summer squash shoot and root [a].

NaCl (mM)	Hormones	Shoot					Root				
		Na^+	K^+	K^+/Na^+ Ratio	Mg^{++}	Cl^-	Na^+	K^+	K^+/Na^+ Ratio	Mg^{++}	Cl^-
0	Non	2.50 ± 0.10 [c]	16.33 ± 0.37 [a]	6.53 ± 0.15 [e]	8.17 ± 0.47 [b]	36.41 ± 2.38 [c]	12.25 ± 0.27 [f]	54.67 ± 7.60 [d]	4.46 ± 0.02 [b]	13.24 ± 2.30 [b]	118.40 ± 10.70 [e]
	GA	2.20 ± 0.11 [d]	11.14 ± 0.38 [d]	5.06 ± 0.49 [f]	8.95 ± 1.23 [a]	35.24 ± 2.90 [c]	12.61 ± 0.15 [e]	44.43 ± 3.64 [e]	3.52 ± 0.03 [d]	15.46 ± 3.88 [b]	115.10 ± 4.15 [e]
	JA	1.07 ± 0.04 [g]	11.37 ± 0.43 [d]	10.63 ± 0.12 [b]	7.80 ± 0.35 [c]	25.00 ± 1.85 [d]	10.37 ± 0.25 [g]	73.29 ± 6.40 [b]	7.07 ± 0.05 [a]	10.80 ± 1.30 [c]	143.50 ± 16.30 [c]
	GA + JA	1.04 ± 0.01 [g]	14.19 ± 0.27 [c]	13.64 ± 0.14 [a]	7.44 ± 0.53 [c]	33.24 ± 5.57 [c]	13.79 ± 0.56 [e]	53.67 ± 6.20 [d]	3.89 ± 0.01 [c]	11.90 ± 2.57 [c]	120.70 ± 8.32 [d]
50	Non	3.54 ± 0.11 [a]	13.09 ± 0.56 [c]	3.69 ± 0.02 [g]	6.65 ± 0.50 [d]	54.29 ± 5.31 [a]	54.25 ± 4.51 [a]	44.06 ± 3.10 [e]	0.81 ± 0.01 [f]	11.73 ± 1.52 [c]	251.10 ± 18.40 [a]
	GA	2.77 ± 0.08 [b]	15.05 ± 0.15 [b]	5.43 ± 0.06 [f]	6.20 ± 0.40 [d]	44.71 ± 2.57 [b]	23.76 ± 2.82 [d]	59.96 ± 6.90 [d]	2.52 ± 0.41 [e]	13.56 ± 1.93 [b]	198.30 ± 20.90 [b]
	JA	1.37 ± 0.03 [f]	11.26 ± 0.78 [d]	8.22 ± 0.02 [d]	7.61 ± 0.27 [c]	55.98 ± 3.16 [a]	32.68 ± 2.81 [b]	83.54 ± 7.27 [a]	2.56 ± 0.32 [e]	21.14 ± 1.69 [a]	185.30 ± 6.81 [b]
	GA + JA	1.45 ± 0.01 [e]	13.70 ± 0.10 [c]	9.45 ± 0.12 [c]	7.47 ± 0.67 [c]	55.68 ± 3.54 [a]	24.81 ± 3.27 [c]	70.38 ± 3.37 [c]	2.84 ± 0.31 [e]	18.23 ± 1.98 [a]	217.51 ± 11.63 [b]
ANOVA											
	Salinity (S)	**	**	***	**	***	***	***	***	**	***
	Hormones (H)	**	**	**	ns	ns	**	**	**	ns	**
	S × H	***	*	***	ns	ns	***	***	***	***	***

[a] Values represent the mean ± SE, $n = 3$. Different lowercase letters indicate significant differences within parameters ($p < 0.05$) as determined by Duncan's multiple range test. The stars indicate significant differences (* $p < 0.05$, ** $p < 0.01$, *** $p < 0.001$); ns, not significant.

2.4. Proline

The proline content significantly increased in plant shoots and roots due to salinity stress. Seed priming with JA and the GA + JA mixture significantly enhanced proline accumulation in the shoot system under saline and non-saline conditions to higher levels than in their corresponding controls (Figure 3A). JA and the GA + JA mixture showed significant increase in proline concentration under saline conditions in shoots, by about 42% and 43%, respectively, as compared to the stressed control. Under non-saline conditions, JA and the GA + JA mixture significantly enhanced the proline concentration in shoots by about 160% and 240%, respectively, as compared to the control (Figure 3A). In roots, GA and the GA + JA mixture significantly enhanced proline concentration under non-saline conditions by about 37% and 46%, respectively, as compared to the control. Seed priming with all investigated hormonal treatments did not enhance proline accumulation in plant roots under salt stress (Figure 3B). The two-way ANOVA indicated highly significant effects of salinity stress ($p < 0.001$) and hormonal treatments ($p < 0.01$) on proline concentration in plant shoots and roots. The interactive effect of the salinity $\times$ hormones interaction was highly significant ($p < 0.01$) on proline concentration in plant shoots and significant ($p < 0.05$) on proline concentration in roots (Table 1).

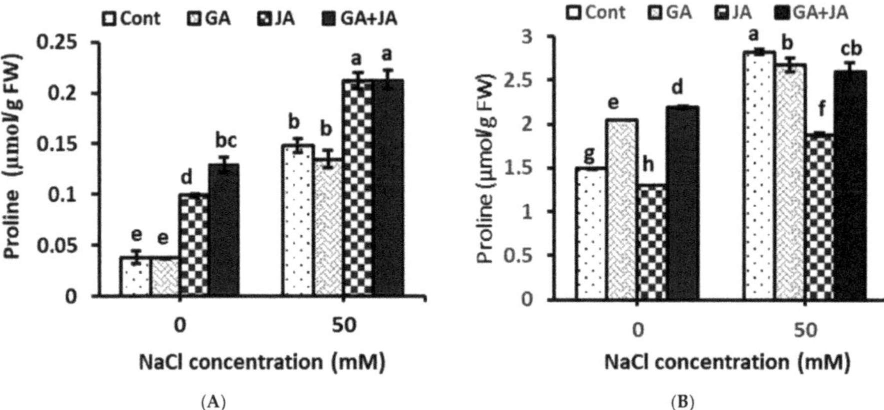

Figure 3. Proline concentration in (**A**) shoot and (**B**) root of non-salinized and salinized summer squash plant as affected by seed priming with 1.5 mM GA, 0.005 mM JA, and mixture of them. Data represent mean of three replicates ($n = 3$) with error bars indicating standard error of the mean. Bars carrying different lowercase letters are significantly different at $p < 0.05$ according to Duncan's multiple range test. p values for two-way ANOVA are reported in Table 1.

2.5. Thiobarbituric Acid Reactive Substances (TBARS)

The accumulation of lipid peroxidation products is the main sign of ROS acting on the bio-membrane. As shown in Figure 4, all hormonal treatments alleviated stress-induced lipid peroxidation, in terms of TBARS content, both in plant shoots and roots. Under saline conditions, GA, JA, and the GA + JA mixture decreased TBARS concentration in shoots by about 42%, 41%, and 22%, and in root by about 28%, 36%, and 26%, as compared to their corresponding controls. The two-way ANOVA indicated highly significant effects of salinity stress ($p < 0.001$) and significant effects of hormonal treatments and their interaction ($p < 0.05$) on TBARS content in plant shoots. Salinity stress, hormonal treatments, and their interaction showed highly significant effects ($p < 0.001$) on TBARS concentration in plant roots (Table 1).

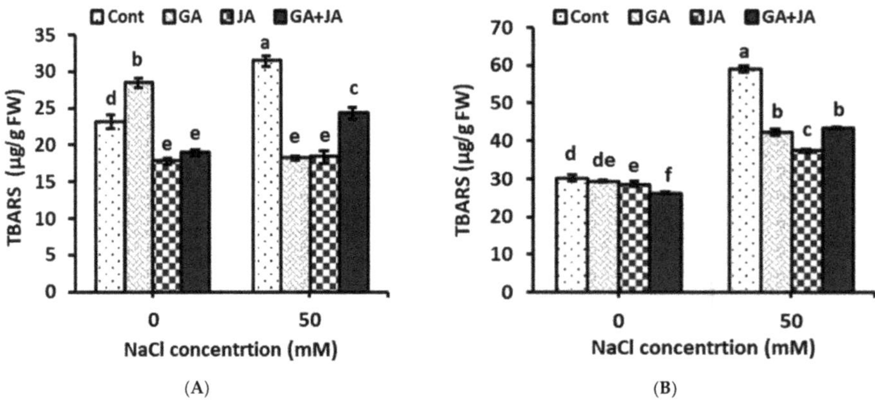

Figure 4. Concentrations of thiobarbituric acid reactive substances (TBARS) in (**A**) shoot and (**B**) root of non-salinized and salinized summer squash plants, as affected by seed priming with 1.5 mM GA, 0.005 mM JA, and mixture of them. Data represent mean of three replicates ($n = 3$) with error bars indicating standard error of the mean. Bars carrying different lowercase letters are significantly different at $p < 0.05$ according to Duncan's multiple range test. p values for two-way ANOVA are reported in Table 1.

2.6. Antioxidant Enzymes

The data represented in Figure 5 revealed that salt treatment significantly reduced CAT and APX activity in summer squash shoots and roots compared to unstressed controls. Only JA treatment significantly enhanced SOD activity in plant shoots and roots to levels higher than in the controls under saline conditions. Superoxide dismutase activity was enhanced due to JA priming under non-saline and saline conditions in plant shoot by about 77% and 209%, respectively, as compared to their corresponding untreated controls (Figure 5A). All investigated hormonal treatments failed to significantly enhance SOD activity in plant roots under non-saline and saline conditions. While JA significantly enhanced SOD activity under salinity stress by about 18% (Figure 5B).

All hormonal treatments showed significant increases in CAT activity in plant shoots and roots under saline conditions. Catalase activity increased by about 89%, 227%, and 147% in shoot after JGA, JA, and GA + JA, respectively, as compared to the untreated stressed control (Figure 5C). In plant roots, CAT activity under saline conditions enhanced by about 94%, 19%, and 77%, as compared with the stressed control after treatment with GA, JA, and GA + JA, respectively (Figure 5D).

All investigated hormonal treatments increased APX activity in shoots and roots significantly, as compared to the control, under saline conditions. In the best results recorded for JA treatment, the APX activity in shoots and roots was about 247% and 313% higher, respectively, than their corresponding controls. GA and the GA + JA mixture enhanced APX activity under saline conditions in shoots by about 151% and 105% and in root by about 182% and 146%, respectively, as compared to their corresponding controls (Figure 5E,F).

The two-way ANOVA indicated significant effects of salinity stress ($p < 0.05$) on SOD activity in plant roots only and highly significant effects ($p < 0.001$) on CAT and APX activity in both shoots and roots. Hormonal treatments showed highly significant effects ($p < 0.001$) on SOD activity in plant shoots and roots and significant effects ($p < 0.05$) on CAT activity in plant roots. The salinity × hormones interaction showed highly significant effects ($p < 0.001$) on CAT and APX activity in both plant shoots and roots (Table 1).

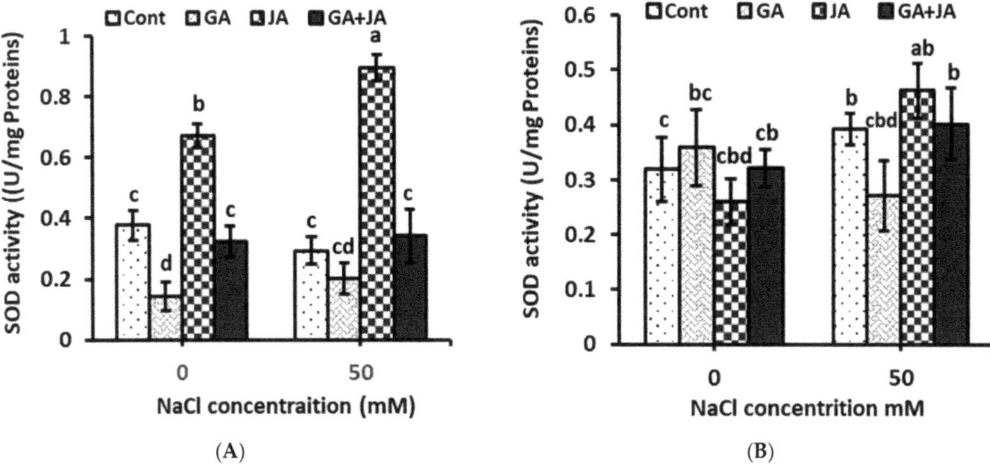

Figure 5. *Cont.*

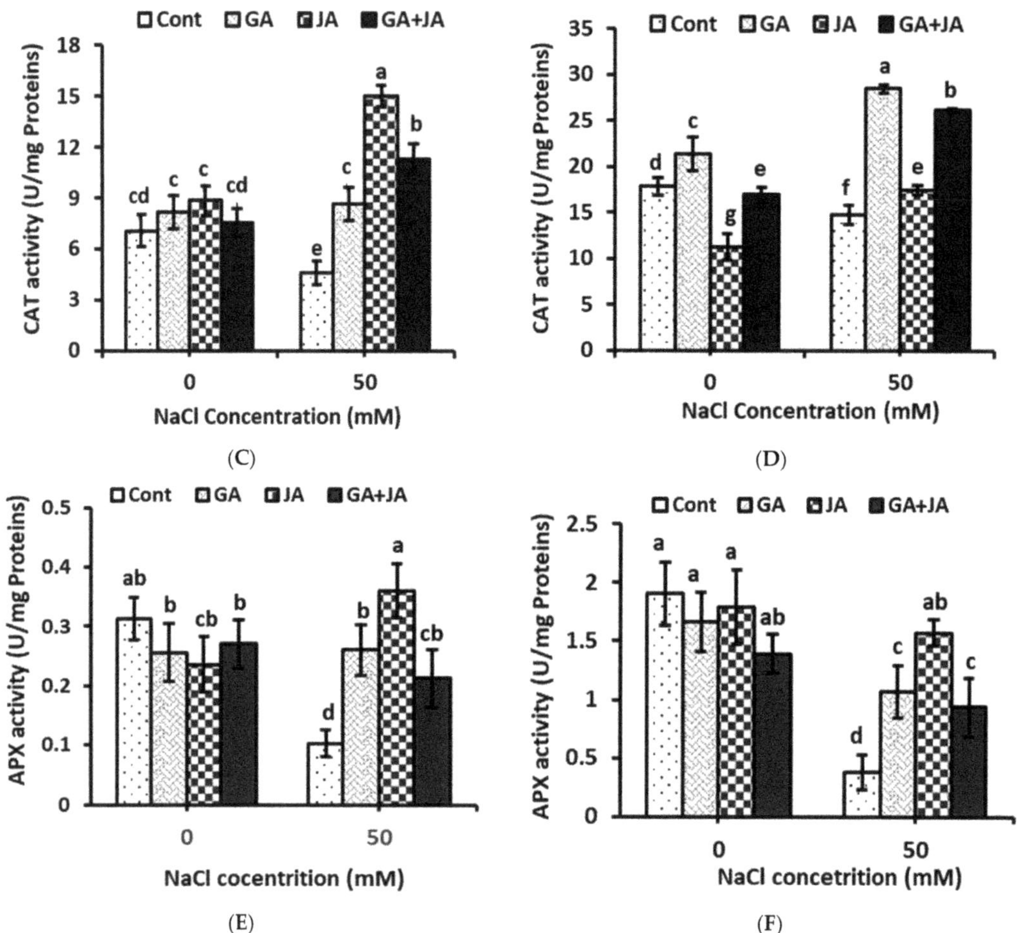

Figure 5. Antioxidant enzyme activity—(**A**) Superoxide dismutase (SOD) in shoot; (**B**) SOD in root; (**C**) Catalase (CAT) in shoot; (**D**) CAT in root; (**E**) Ascorbate peroxidase (APX) in shoot; and (**F**) APX in root—of non-salinized and salinized summer squash plants, as affected by seed priming with 1.5 mM GA, 0.005 mM JA, and mixture of them. Data represent mean of three replicates ($n = 3$) with error bars indicating standard error of the mean. Bars carrying different lowercase letters are significantly different at $p < 0.05$ according to Duncan's multiple range Test. p values for two-way ANOVA are reported in Table 1.

2.7. Nucleic Acids

Based on the results demonstrated in Figure 6, salinity stress significantly reduced RNA and DNA content in summer squash shoots and roots, while all hormonal treatments significantly increased RNA concertation in plant shoots and roots to higher levels than in untreated stressed plants. In shoot samples, GA, JA, and the GA + JA mixture increased RNA concentration by about 83%, 69%, and 38%, respectively, as compared to the control (Figure 6A). In plant roots, the best treatment under saline conditions was GA + JA, followed by GA alone (Figure 6B). GA, JA, and the GA + JA mixture increased RNA concentration by about 28%, 17%, and 55%, respectively, as compared to the control. In the same context, all hormonal treatments significantly increased DNA concentration in plant shoots and roots under saline condition only (Figure 6). GA, JA, and the GA + JA mixture caused significant and similar escalations in DNA content in plant shoots (around

25%) and roots (around 20%), as compared to the stressed control (Figure 6C,D). The two-way ANOVA indicated highly significant effects ($p < 0.001$) of salinity stress and the salinity × hormones interaction on RNA and DNA concentrations in plant shoots and roots, while hormonal treatments showed significant effects ($p < 0.05$) only on DNA concentrations both in plant shoots and roots (Table 1).

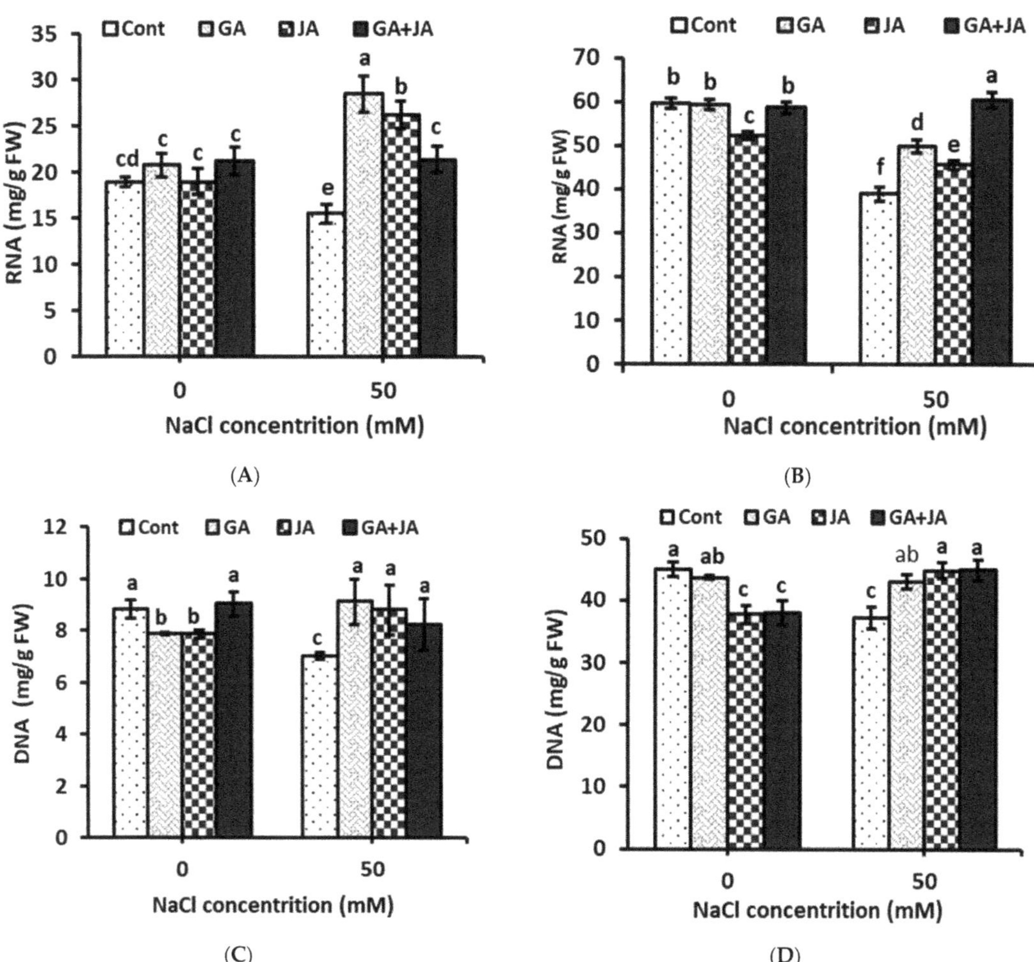

Figure 6. Concentration of nucleic acids—(**A**) RNA in shoot; (**B**) RNA in root; (**C**) DNA in shoot; and (**D**) DNA in root—of non-salinized and salinized summer squash plants, as affected by seed priming with 1.5 mM GA, 0.005 mM JA, and mixture of them. Data represent mean of three replicates ($n = 3$) with error bars indicating standard error of the mean. Bars carrying different lowercase letters are significantly different at $p < 0.05$ according to Duncan's multiple range Test. p values for two-way ANOVA are reported in Table 1.

3. Discussion

Improving the salt tolerance of crops is an essential target of plant breeders, in aiming to meet the future food demands of coming generations [23]. Pre-sowing priming with phytohormones is used to accelerate seed germination and homogenous seedling emergence, to enhance further plant growth, and to establish the stress resistance of seedlings and adult plants, which ultimately increases crop production [24]. Phytohormones interact with each other antagonistically as well as synergistically, making a super-complex network of closely intertwined pathways of biosynthesis, metabolism, transport, and signaling, thus causing responses to external stimuli [25]. Of these phytohormones, GAs and JAs have been documented as activators for plant growth under salinity stress, during which they can break seed dormancy, improve seed vitality, stimulate plant gene expression, and repair membrane damage [26,27]. The findings of this study demonstrate that salinity treatment significantly reduces the fresh and dry biomass of summer squash plants. This reduction in plant biomass under salinity stress can be attributed to reduced water absorption from the surrounding habitat as a result of physiological drought and the toxic effects of Na^+ ions [28].

The inhibitory effect of salt stress on summer squash biomass was alleviated partially or entirely by all investigated hormonal treatments. In accordance with our results, other studies reported that treatment of plants with GA and JA is efficient in relieving salinity stress effects at various stages of plant growth by enhancing plant height, root length, root diameter, and shoot fresh weight [27,29]. Exogenous JA treatment under saline conditions may affect plant hormonal balance, e.g., ABA, which presents important evidence for understanding protection mechanisms against salt stress [30]. In the same context, the protecting influence of JA might be due to its ability to avoid the decline in cytokinin levels under salinity stress that results from the retardation of both cytokinin oxidase gene expression and activity [31]. On the other hand, GA application has been shown to improve plant growth by altering the ratio between endogenous ABA and SA, reducing the quantity of polyamines, which are involved in the regulation of aging [7].

Although Chl a content did not change significantly due to salinity stress in this study, Chl b content was significantly enhanced. Therefore, the Chl a/b ratio decreased in salinized plants. A few conflicting reports pinpointed that the amounts of Chl a and Chl b under saline conditions were more than those of the control [32]. All hormonal treatments in this study cause significant induction in Chl b and Car content, compared to the controls, under both saline and normal conditions. In accordance with our results, Misratia et al. [33] demonstrated that GA plays an important role in improving plant salt tolerance by enhancing chlorophyll biosynthesis. JA is reported also to counteract the negative effects of salinity on Chl b and Car content [34]. Nevertheless, our results showed that hormonal treatments significantly reduced Chl a content and increased Chl b content, which reduced the Chl a/b ratio. Chlorophyll b was shown to be the main constituent of the photosystems [35]. It is worth mentioning that the enhancement in Chl b content in response to hormonal treatments both under saline and non-saline conditions could be, mostly, a clear transformation of Chl a into Chl b or, at least, an improved de novo Chl b biosynthesis (Table 2). This conversion was enhanced by K fertilization under salinity stress in bitter almond trees [36]. The results of the current study also recorded the alleviating effects of GA and JA on K^+ uptake by summer squash plants (Table 3). Several studies showed that GA and JA implement an effect on plant metabolism by regulating nutrition utilizations, mainly by meditating carbon metabolism; photosynthetic pigments were accordingly accumulated, and the content of Chl b then increased [10,12,31,34]. The transformation of Chl a into Chl b seems to be part of the overall Chl a/b interconversion cycle, which is supposed to play a major role in the formation and reorganization of the photosynthetic apparatus, and which aids plants to cope with different adverse stress conditions [37]. In their study, Yan et al. [38] proved that the decrease in the Chl a/b ratio is associated with a rise in maize crop production.

Sodium chloride stress has been shown to correlate with the disorder in Na^+ ions homeostasis and essential minerals [39]. It has been shown that salinity stress retards plant growth owing to the excessive accumulation of Na^+ and Cl^- ions in plant tissues [40]. In the current study, salt stress markedly increased Na^+ and Cl^- concentrations and reduced Mg^{++} and K^+ concentrations in summer squash shoots and roots. Related to this, Cakmak [41] demonstrated that enhancing Na^+ content in plant leaves results in K^+ deficiency owing to the antagonistic effects of Na^+ and K^+ ions. Seed priming with the studied hormonal treatments reduced Na^+ and Cl^- ion accumulation to lower levels than those of the stressed control in summer squash shoots and roots. On the other hand, Mg^{++} and K^+ concentrations in plant shoots and roots were significantly enhanced in response to seed priming with all hormonal treatments under saline conditions. In accordance with our results, one study showed that a remarkable suppression in Na^+ and Cl^- accumulation was concomitant with enhancements in K^+, Ca^{++} and Mg^{++} levels in stressed plants due to seed priming in GA3 [42]. The impact of GA3 on the mechanism of ion uptake might be allied with its influence on membrane stability, since the rate of ion uptake via the cell membrane was increased and, consequently, the translocation of ions from root to shoot was enhanced [43]. Moreover, Kang et al. [44] indicated that exogenous JA decreases Na^+ accumulation and increases K^+ and Mg^{++} content in salt-stressed rice plants. Balkaya et al. [45] showed that salt tolerance correlated with the ability of the plant species to accumulate higher levels of K^+ ions. A plant's tolerance to salinity may be more related to the K^+/Na^+ ratio in the cell than the absolute Na^+ concentration [46]. In our study, all hormonal treatments significantly enhanced the K^+/Na^+ ratio under saline conditions to higher levels than in the stressed controls, which helps in increasing plant growth under saline conditions as shown from the results of the growth parameters in this study. More K^+ can be taken through active transport by increasing the osmotic potential so that more water can enter the plant cell [47]. Potassium ion content in the cell is important for conservation of osmotic equilibrium, during which it activates a range of enzymes that are responsible for stomatal movement in response to changes in bulk leaf water status under salt stress [47]. Several studies have shown that plant treatments with phytohormones increase mineral uptake by enhancing root mass (as indicated from the results in Figure 1), root volume, root hair, and lateral root formation and by stimulating high levels of root activity by increasing the roots' active absorption area [48–50].

Proline content in summer squash shoots and roots showed a significant increase after salinity stress. Similar observations have been reported in salt-stressed crops in several studies [51–53]. Proline is an important parameter to evaluate plant stress tolerance capability [54]. It is a highly water-soluble amino acid that protects cell membranes from the toxic effects of an excess of inorganic ions; in addition to its role as an osmolyte, it also helps the cells to alleviate oxidative stress in salt-affected plants [55]. Seed priming with JA and the JA + GA mixture significantly enhanced proline content in summer squash shoots, while all hormonal treatments significantly reduced proline content in root samples. The most pronounced reduction was recorded for JA treatment. Some conflicting results are available in the literature regarding the effect of JA application on proline content under water stress. Huguet-Robert et al. [56] observed that MeJA restricted proline accumulation in canola leaf discs subjected to osmotic stress. However, Maslenkova et al. [57] reported that proline was accumulated in barley plants treated with JA. In accordance with our findings, Dheeba et al. [58] detected that exogenous GA decreases the level of proline content in salt-stressed plants. The reduction in proline content after hormonal treatments indicates that these phytohormones reduce the stress caused by salinity

Thiobarbituric acid reactive substances (TBARS) are the decomposition products of the polyunsaturated fatty acids of cell membranes. Therefore, TBARS accumulation under salt stress has been used as an indicator of lipid peroxidation, which may indicate salt-induced oxidative stress [59]. In the present study, TBARS content in summer squash shoots and roots was increased markedly due to salinity stress. All hormonal treatments alleviate stress-induced lipid peroxidation in summer squash plants as inferred from the reduction in TBARS concentration in plant shoots and roots. The exogenous application of GA and JA reduced TBARS accumulation under various stress factors [60,61].

One of the most important mechanisms involved in the salt tolerance response is the harmonized up-regulation of the antioxidative system, since salt tolerance is associated with elevated activity of definite antioxidant enzymes [62]. Plants have inclusive antioxidative machinery, which plays a vital role in ROS scavenging, whereas CAT and SOD alleviate the destructive effects of oxidative stress [63]. In this study, salinity stress increased SOD activity in summer squash roots. Similar enhancement in the activity of SOD enzymes has been recorded in many plants subjected to salt stress [64]. Seed priming with JA significantly enhances SOD activity in summer squash shoots and roots under saline conditions. The increased SOD activity due to JA treatment in salt-stressed plants could be connected to its important role in plant survival. When SOD activity was elevated, the scavenging of superoxide radicals was performed properly, which protected the cell membrane against oxidative stress damage; consequently, tolerance to oxidative stress increased [65]. On the other hand, the results of this investigation showed that the activity of SOD in salinized plants significantly decreased in response to seed priming with GA. Likewise, it has been shown that the exogenous application of GA inhibits SOD activity in *Vigna radiata* plants under salt stress [43].

The down-regulation of CAT and APX activity is considered to be a general response to several stress factors [66]. Under environmental stresses, the decline in CAT activity is allegedly owing to the inhibition of enzyme synthesis or to the alteration in the assembly of enzyme subunits. In the present study, CAT and APX activity in the shoots and roots increased in hormone-treated plants to higher levels than in untreated controls under saline conditions. In support of our results, Qiu et al. [34] reported that JA significantly enhanced the CAT and APX activity under saline conditions. Overproduction of CAT and APX can be an adaptive mechanism of plants to stressful ecosystems, and JA contributes to this. The mitigation influence of the GA application on antioxidant enzyme activity has also been recorded in salt-stressed mung bean plants by Chakrabarti and Mukherji [43]. It has been noted that exogenous GA is able to overcome the influence of different salinity levels on CAT, APX, and SOD and restores their activity to values around that of the control [67].

It has been postulated that ROS, which accumulates as a result of salt stress, can damage nucleic acids [68]. The results of the current study show significant reductions in RNA and DNA contents in summer squash shoots and roots due to salinity stress. The findings of our study are in accordance with those recorded by Yupsanis et al. [69] who obtained significant decreased in RNA and DNA contents in alfalfa and lentil plants under salinity stress. In his study using five species of Chenopodiaceae, Abo-Kassem [70] concluded that the reduction in nucleic acid content, along with the enhancement in RNase activity, might be related to the increased levels of salinity which cause inhibition in the biosynthesis of nucleic acids and/or stimulation in their degradation. In the present study, seed priming with the investigated hormonal treatments increased RNA and DNA contents in shoots and roots of summer squash under saline conditions to comparable values. In support of our results, Ismail [71] recorded that salt stress reduces RNA and DNA content in sorghum plants, and pre-sowing priming of barley grains with GA enhances DNA content in salt-stressed plants. Comparable findings are recorded by Fujii et al. [72], who revealed the role of the phytohormones, especially GA, in regulating gene expression and mRNA induction by high salinity levels and the possible correlation between the endogenous GA content and the achievement of stress protection. Taken together, under salt stress conditions, the pre-sowing priming of summer squash seeds with GA or JA stabilized ionic homeostasis increased the content of chlorophyll b, carotenoid, proline, nucleic acids, and antioxidant enzyme activity and decreased the level of membrane lipid peroxidation. All these subsidized the plants and enhanced their salt tolerance. However, the GA and JA combination showed antagonistic effects for the regulation of plant growth and for tolerance responses, while JA treatment alleviated salinity stress induced in summer squash more efficiently than GA treatment or the GA + JA combination. The crosstalk between GA and JA signaling underlying this antagonistic effect needs further investigation.

4. Materials and Methods

4.1. Plant Cultivation and Hormonal Treatments

This study was conducted at King Abdulaziz University Experimental Station, Saudi Arabia during the summer season, 2018. The plants were grown in the glasshouse under natural day/night conditions with average maximum/minimum temperature of 41/25 °C and daylength of 13 h. The seeds of the summer squash (*Cucurbita pepo* L., cv. Sucheimie No. 2) were bought from Holler Co., Jeddah, Saudi Arabia. Gibberellic acid and jasmonic acid, were purchased from Sigma–Aldrich Chemie GmbH, Mittelfranken, Germany. The seeds were dispersed in plastic pots (30 cm diameter and 30 cm depth) at 1 cm depth; pot bottoms were sealed to avoid salt leaching, and each pot contained 5 kg of mixed soil, consisting of sand and pitmoss (2:1). Before sowing, summer squash seeds were primed in 1.5 mM GA, 0.005 mM JA, or a mixture of them both for 1 h at room temperature under continuous shaking. Seeds primed in distilled water were used as a control. The concentrations of GA and JA were chosen depending on a preliminary experiment, conducted using Petri dishes over 10 days with four concentrations for GA (0.1, 0.5, 1, 1.5 mM) and four for JA (0.001, 0.005, 0.01 and 0.1 mM). The best results for seed germination and seedling growth were attained when using 1.5 mM GA and 0.005 mM JA under saline conditions. In the same experiment, four concentrations of NaCl (50, 100, 150, and 200 mM) were studied and 50 mM was chosen as the effective NaCl concentration to cause moderate inhibition (around 50%) in summer squash seed germination and seedling growth (EC50 level) (see Supplementary Figures S1–S3 for more details). After pre-sowing treatments, the seeds were surface dried on filter paper and then used for cultivation. The plants were irrigated with tap water regularly every two days. Pots were arranged in a randomized complete design with three replicates. Each replicate consisted of 10 uniform plants per pot. After two weeks (3 true leaves stage), each treatment was divided into 2 sets; one of them was irrigated with 50 mM NaCl (EC: 5 dS/m) dissolved in tap water to induce salinity stress, whereas the other was irrigated with tap water. A final volume of 1500 mL of the saline solution was added to the soil, giving a final concentration of 10 mM NaCl/100 g soil. To avoid osmotic shock, 750 mL of salty solution was added at the first emergence of the third true leaf; then the other 750 mL was added 3 days later. Afterwards, irrigation was applied up to approximately 90% of the pots' water-holding capacity, using tap water every 2 days to compensate for the loss of water due to evapotranspiration. After three weeks, plant samples were collected. For all assays, the collected shoot and root samples were frozen immediately in liquid nitrogen and kept at −80 °C.

4.2. Plant Biomass and Water Content

When the plants were harvested (5 weeks after planting), they were carefully separated into shoots and roots, washed using tap water and blotted with filter paper to remove excess water. Summer squash growth under different treatments was determined by measuring the fresh and dry weights in grams (g) of the shoots and roots. Dry weights were determined after drying at 70 °C to constant weight. Shoot and root water content (WC), as a percentage of fresh weight, was estimated according to Sumithra et al. [73] using the formula:

$$WC\ (\%) = -\ [(FW - DW) \times 100]/FW$$

4.3. Chlorophyll and Carotenoid Determination

Chlorophyll a, Chl b, and carotenoid were measured using UV-VIS spectroscopy according to Su et al. [74]. A total of 0.05 g of leaf tissue was suspended in 5 mL of 95% ethyl alcohol in a test tube at 60 °C, until it was colorless. Then the total volume was refilled to 5 mL with 95% ethyl alcohol. The green solution was placed in a cuvette against 1 mL of 95% ethyl alcohol as a blank. The absorbance readings were measured using spectrophotometry with a Lamda 25 UV-Vis spectrophotometer at wavelengths of 664, 649, and 470 nm.

4.4. Mineral Ion Determination

A total of 0.5 g of dried shoots and roots was placed in a 250 mL round bottom flask, 2 mL of concentrated nitric acid was added to the flask, and the flask was placed on a hot plate; the mixer was boiled on medium heat for 10–15 min or until the mixture was completely oxidized; afterwards, the mixture was left to cool at room temperature. A total of 1 mL of 70% perchloric acid was added to the mixture and boiled until white fumes appeared; the mixture cooled down at room temperature. A total of 5 mL of distilled water was added to the mixture, which was boiled until the white fumes stopped appearing. Finally, the solution was filtered by Whatman 1 filter paper, and the volume was refilled to 25 mL with distilled water. Then concentrations of the mineral ions were estimated via inductively coupled plasma emission optical spectrometry [75].

4.5. Proline Determination

Proline was measured in plant shoots and roots following the method of Bates et al. [76]. A total of 0.6 g of plant tissue was homogenized using liquid N_2 in 1.5 mL 3% (w/v) sulfosalicylic acid and then centrifuged for 10 min at 10,000 rpm. One milliliter of the supernatant was mixed with one milliliter of ninhydrin reagent (250 mg ninhydrin, 20 mL glacial acetic acid and 30 mL of 6 M phosphoric acid) and boiled in a water bath for 1 h. The resulted color was extracted in 2 mL toluene and estimated calorimetrically at 520 nm.

4.6. Lipid Peroxidation Products Determination

Lipid peroxidation products were estimated by the formation of thiobarbituric acid reactive substances (TBARS) as described by Heath and Packer [77]. The crude extract was mixed with the same volume of a 0.5% (w/v) thiobarbituric acid, which contained 20% (w/v) trichloroacetic acid. Then the mixture was heated for 30 min at 95 °C, quickly cooled in an ice-bath, and centrifuged for 10 min at 3000 rpm. The absorbance of the supernatant was assayed spectrophotometrically at 532 and 600 nm TBARS concentration was calculated using the molar extinction coefficient (155 mM^{-1} cm^{-1}).

4.7. Antioxidant Enzymes Activity Determination

Antioxidant enzymes extraction: Enzyme extraction was assayed according to Cakmak and Marschner's method [78]. A total of 0.5 g of shoot and root tissue was ground to a fine powder in liquid (N_2). After that, it was homogenized in 5 mL of 100 mM potassium phosphate buffer (pH 7.8), which contained 0.1 mM ethylenediamine tetraacetic acid (EDTA) and 0.1 g polyvinylpyrrolidone. The mixture was centrifuged under cooling (4 °C) for 10 min. at 18,000 rpm and the supernatants were collected and used for the analyses of enzyme activity.

Superoxide dismutase (SOD, EC 1.15.1.1) activity was assayed by following the autoxidation of epinephrine according to Misra and Fridovich [79]. Enzyme activity was assayed in a final volume of 2 mL of the reaction medium containing 25 mM of sodium carbonate buffer (pH 10.2), 200 µL 0.5 mM EDTA, and 100 µL enzyme extract. The reaction was initiated by adding 100 µL of 15 mM epinephrine (dissolved in 10 mM HCl, pH 2.4). Enzyme activity was determined by the increased absorption at 480 nm and was calculated using the molar extinction coefficient (ε = 43.6 mM^{-1} cm^{-1}).

Catalase (CAT, EC 1.11.1.6) activity was assayed spectrophotometrically by monitoring the change in A240 due to the decreased absorption of H_2O_2 [80]. The reaction medium contained 50 mM potassium phosphate buffer (pH 7), and 500 µL of enzyme extract in a 3 mL final volume. The reaction was started by adding 100 µL of 10 mM H_2O_2. The enzyme activity was calculated using the extinction coefficient (ε = 39.4 mM^{-1} cm^{-1}).

Ascorbate peroxidase (APX, EC 1.11.1.11) activity was determined according to Zhang and Kirkham's method [80]. The rate of hydrogen peroxide-dependent oxidation of ascorbic acid was determined in a reaction mixture, which contained 50 mM potassium phosphate buffer (pH 7), 5 mM H_2O_2, 0.1 mM Na_2-ETDA, 0.5 mM ascorbic acid, and 50 µL enzyme extract. Ascorbic acid oxidation rate was estimated from the reduction in absorbance at 290 nm. The ascorbate peroxidase activity was calculated using the molar extinction coefficient (ε = 2.8 mM^{-1} cm^{-1}).

4.8. Nuclein Acids Determination

The determination of RNA and DNA was carried out according to the method of Schmidt and Thannhuser [81] and its modification as described by Morse and Carter [82]. A known weight of plant materials was extracted with 5% TCA, and then it was washed three times with 5 mL methanol chloroform in the ratio of 1:2; the delipidated material was dissolved in 2 mL of 1N KOH at 37 °C for 16–20 h and precipitated with 0.4 mL of 6N HCl; then it was centrifuged. The precipitate contained the DNA fraction, while the supernatant contained RNA. TCA was added to the supernatant to give the final concentration of 5% TCA. It was then centrifuged, and the supernatant constituted the RNA fraction. The precipitate was hydrolyzed in 5 mL of 5% TCA at 90 °C for 30 min, cooled, and then centrifuged, and the supernatant constituted the DNA fraction. Estimated quantitative determination of RNA and DNA, as described by Abd El-Wahab [83] and Burton [84], was carried out.

4.9. Statistical Analysis

All data were subjected to two-way analysis of variance (ANOVA) using the statistical software SPSS version 21.0. It was performed to examine the effects of salinity stress, hormonal treatments, and their interactions upon all investigated traits. Significant differences between mean values ($p < 0.05$) were confirmed using Duncan's multiple range test. All values were expressed with their standard error (SE) as a mean value of three replicates.

5. Conclusions

This study compares the effects of exogenous GA and JA on summer squash growth and metabolic responses to figure out which phytohormone is more efficient in alleviating salinity stress in this plant. We also hypothesized that, with respect to the great abilities of GA and JA under stress, their combination would show synergistic effects in alleviating salinity drawbacks on summer squash growth by enhancing antioxidant enzymes better than GA or JA separately. The important findings of the present paper showed that GA, JA, and their mixture enhanced salt tolerance of summer squash. The first piece of evidence is that all hormonal treatments significantly enhance Chl b, carotenoid, K, and Mg contents and decreased Na and Cl contents under saline conditions. Seed priming with all hormonal treatment significantly reduced proline content in summer squash roots. All hormonal treatments alleviated stress-induced lipid peroxidation both in shoots and roots by increasing MDA content. The activity of antioxidant enzymes including SOD, CAT, and APX was enhanced due to GA and JA seed priming under non-saline and saline conditions. All investigated hormonal treatments significantly increased RNA and DNA concentration in plant shoots and roots to higher levels than in untreated stressed plants under saline conditions. Finally, the results of this research did not support our initial hypothesis: JA treatment alleviated salinity stress induced in summer squash more efficiently than GA treatment or the GA + JA combination.

Supplementary Materials: The following are available online at https://www.mdpi.com/article/10.3390/plants10122768/s1, Figure S1. Effect of different salinity levels on seed germination and seedling growth of summer squash seedlings grown for 10 days under salt stress. All values are the mean of three replicates ± SE. Values carrying different litters are significantly different at $p < 0.05$ (A) Germination percentage (B) Plumule and radical length (C) Fresh and dry weight, Figure S2. Effect of different gibberellic acid concentrations on seedling growth of summer squash seedlings grown for 10 days under salt stress. All values are the mean of three replicates ± SE. Values carrying different litters are significantly different at $p < 0.05$ (A) Plumule length (B) Radical length (C) Fresh weight (D) Dry weigh, Figure S3. Effect of different jasmonic acid concentrations on seedling growth of summer squash seedlings grown for 10 days under salt stress. All values are the mean of three replicates ± SE. Values carrying different litters are significantly different at $p < 0.05$ (A) Plumule length (B) Radical length (C) Fresh weight (D) Dry weigh.

Author Contributions: Conceptualization, S.O.B. and M.E.-Z.; methodology, M.M.A.-h. and M.E.-Z.; software, M.M.A.-h.; validation, M.M.A.-h., S.O.B. and M.E.-Z.; formal analysis, M.M.A.-h.; investigation, M.E.-Z.; resources, S.O.B.; data curation, M.M.A.-h.; writing—original draft preparation, M.E.-Z.; writing—review and editing, M.M.A.-h., S.O.B. and M.E.-Z.; visualization, M.M.A.-h.; supervision, S.O.B. and M.E.-Z.; project administration, S.O.B. and M.E.-Z. All authors have read and agreed to the published version of the manuscript.

Funding: This research received no external funding.

Institutional Review Board Statement: Not applicable.

Informed Consent Statement: Not applicable.

Data Availability Statement: All data are presented within the article.

Conflicts of Interest: The authors declare no conflict of interest.

References

1. Evelin, H.; Devi, T.S.; Gupta, S.; Kapoor, R. Mitigation of salinity stress in plants by arbuscular mycorrhizal symbiosis: Current understanding and new challenges. *Front. Plant Sci.* **2019**, *10*, 470. [CrossRef] [PubMed]
2. Castillo-Campohermoso, M.A.; Broetto, F.; Rodríguez-Hernández, A.M.; Soriano-Melgar, L.D.A.A.; Mounzer, O.; Sánchez-Blanco, M.J. Plant-available water, stem diameter variations, chlorophyll fluorescence, and ion content in *Pistacia lentiscus* under salinity stress. *Terra Latinoam.* **2020**, *38*, 103–111. [CrossRef]
3. Abdel-Hamid, A.M. Physiological and molecular markers for salt tolerance in four barley cultivars. *Eur. Sci. J.* **2014**, *10*, 252–272.
4. Sahin, O.; Karlik, E.; Meric, S.; Ari, S.; Gozukirmizi, N. Genome organization changes in GM and non-GM soybean [*Glycine max* (L.) Merr.] under salinity stress by retro-transposition events. *Genet. Resour. Crop Evol.* **2020**, *67*, 1551–1566. [CrossRef]
5. Augé, R.M.; Toler, H.D.; Saxton, A.M. Arbuscular mycorrhizal symbiosis and osmotic adjustment in response to NaCl stress: A meta-analysis. *Front. Plant Sci.* **2014**, *5*, 562. [PubMed]
6. Wang, Y.; Mostafa, S.; Zeng, W.; Jin, B. Function and mechanism of jasmonic acid in plant responses to abiotic and biotic stresses. *Int. J. Mol. Sci.* **2021**, *22*, 8568. [CrossRef]
7. Kosakivska, I.V.; Vedenicheva, N.P.; Babenko, L.M.; Voytenko, L.V.; Romanenko, K.O.; Vasyuk, V.A. Exogenous phytohormones in the regulation of growth and development of cereals under abiotic stresses. *Mol. Biol. Rep.* **2021**. [CrossRef]
8. Teklić, T.; Parađiković, N.; Špoljarević, M.; Zeljković, S.; Lončarić, Z.; Lisjak, M. Linking abiotic stress, plant metabolites, biostimulants and functional food. *Ann. Appl. Biol.* **2021**, *178*, 169–191. [CrossRef]
9. Khan, N.; Bano, A.; Ali, S.; Babar, M.A. Crosstalk amongst phytohormones from planta and PGPR under biotic and abiotic stresses. *Plant Growth Regul.* **2020**, *90*, 189–203. [CrossRef]
10. Miceli, A.; Moncada, A.; Sabatino, L.; Vetrano, F. Effect of gibberellic acid on growth, yield, and quality of leaf lettuce and rocket grown in a floating system. *Agronomy* **2019**, *9*, 382. [CrossRef]
11. Alonso-Ramírez, A.; Rodríguez, D.; Reyes, D.; Jiménez, J.A.; Nicolás, G.; López-Climent, M.; Nicolás, C. Evidence for a role of gibberellins in salicylic acid-modulated early plant responses to abiotic stress in Arabidopsis seeds. *Plant Physiol.* **2009**, *150*, 1335–1344. [CrossRef]
12. Rady, M.M.; Boriek, S.H.; El-Mageed, A.; Taia, A.; Seif El-Yazal, M.A.; Ali, E.F.; Abdelkhalik, A. Exogenous gibberellic acid or dilute bee honey boosts drought stress tolerance in *Vicia faba* by rebalancing osmoprotectants, antioxidants, nutrients, and phytohormones. *Plants* **2021**, *10*, 748. [CrossRef] [PubMed]
13. Emamverdian, A.; Ding, Y.; Xie, Y. The role of new members of phytohormones in plant amelioration under abiotic stress with an emphasis on heavy metals. *Pol. J. Environ.* **2020**, *29*, 1009–1020. [CrossRef]
14. Schenk, P.M.; Kazan, K.; Wilson, I.; Anderson, J.P.; Richmond, T.; Somerville, S.C.; Manners, J.M. Coordinated plant defense responses in Arabidopsis revealed by microarray analysis. *Proc. Natl. Acad. Sci. USA* **2000**, *97*, 11655–11660. [CrossRef]
15. Walling, L.L. The myriad plant responses to herbivores. *J. Plant Growth Regul.* **2000**, *19*, 195–216. [CrossRef]
16. Browse, J. Jasmonate passes muster: A receptor and targets for the defense hormone. *Annu. Rev. Plant Biol.* **2009**, *60*, 183–205. [CrossRef] [PubMed]
17. Achard, P.; Gong, F.; Cheminant, S.; Alioua, M.; Hedden, P.; Genschik, P. The cold-inducible CBF1 factor–dependent signaling pathway modulates the accumulation of the growth-repressing DELLA proteins via its effect on gibberellin metabolism. *Plant Cell* **2008**, *20*, 2117–2129. [CrossRef]
18. Wingler, A.; Tijero, V.; Müller, M.; Yuan, B.; Munné-Bosch, S. Interactions between sucrose and jasmonate signalling in the response to cold stress. *BMC Plant Biol.* **2020**, *20*, 176. [CrossRef]
19. Soto, A.; Ruiz, K.B.; Ziosi, V.; Costa, G.; Torrigiani, P. Ethylene and auxin biosynthesis and signaling are impaired by methyl jasmonate leading to a transient slowing down of ripening in peach fruit. *J. Plant Physiol.* **2012**, *169*, 1858–1865. [CrossRef]
20. Paris, H.S. Summer squash. In *Handbook of Plant Breeding, Vegetables I*; Prohens, J., Nuez, F., Eds.; Springer: New York, NY, USA, 2008; pp. 351–379.

21. Adepoju, G.K.A.; Adepanjo, A.A. Effect of consumption of *Cucurbita pepo* seeds on hematological and biochemical parameters. *Afr. J. Pharm. Pharmacol.* **2011**, *5*, 18–22. [CrossRef]

22. Huang, B.; NeSmith, D.S.; Bridges, D.C.; Johnson, J.W. Responses of squash to salinity, waterlogging, and subsequent drainage: II. Root and shoot growth. *J. Plant Nutr.* **1995**, *18*, 141–152. [CrossRef]

23. Mantri, N.; Patade, V.; Penna, S.; Ford, R.; Pang, E. Abiotic stress responses in plants: Present and future. In *Abiotic Stress Responses in Plants*; Springer: New York, NY, USA, 2012; pp. 1–19.

24. Rhaman, M.S.; Imran, S.; Rauf, F.; Baskin, C.C. Seed priming with phytohormones: An effective approach for the mitigation of abiotic stress. *Plants* **2021**, *10*, 37. [CrossRef] [PubMed]

25. Liu, J.; Moore, S.; Chen, C.; Lindsey, K. Crosstalk complexities between auxin, cytokinin and ethylene in Arabidopsis root development: From experiments to systems modeling and back again. *Mol. Plant* **2017**, *10*, 1480–1496. [CrossRef] [PubMed]

26. Muniandi, S.K.M.; Hossain, M.A.; Abdullah, M.P.; Ab Shukor, N.A. Gibberellic acid (GA3) affects growth and development of some selected kenaf (*Hibiscus cannabinus* L.) cultivars. *Ind. Crops Prod.* **2018**, *118*, 180–187. [CrossRef]

27. Kaur, H.; Sirhindi, G.; Sharma, P. Effect of jasmonic acid on some biochemical and physiological parameters in salt-stressed Brassica napus seedlings. *Plant Physiol. Biochem.* **2017**, *9*, 36–42.

28. Saeedipour, S.; Alipoor, R.; Farhoudi, R. Feasibility of Seed Priming for Improving Salt Tolerance of Summer Squash (*Cucurbita pepo*) Seedling. *Int. J. Appl. Agric. Res.* **2010**, *5*, 205–213.

29. Ashraf, M.; Karim, F.; Rasul, E. Interactive effects of gibberellic acid (GA 3) and salt stress on growth, ion accumulation and photosynthetic capacity of two spring wheat (*Triticum aestivum* L.) cultivars differing in salt tolerance. *Plant Growth Regul.* **2002**, *36*, 49–59. [CrossRef]

30. Seo, H.S.; Kim, S.K.; Jang, S.W.; Choo, Y.S.; Sohn, E.Y.; Lee, I.J. Effect of jasmonic acid on endogenous gibberellins and abscisic acid in rice under NaCl stress. *Biol. Plant.* **2005**, *49*, 447–450. [CrossRef]

31. Avalbaev, A.; Yuldashev, R.; Fedorova, K.; Somov, K.; Vysotskaya, L.; Allagulova, C.; Shakirova, F. Exogenous methyl jasmonate regulates cytokinin content by modulating cytokinin oxidase activity in wheat seedlings under salinity. *J. Plant Physiol.* **2016**, *191*, 101–110. [CrossRef] [PubMed]

32. Djanaguiraman, M.; Sheeba, J.A.; Shanker, A.K.; Devi, D.D.; Bangarusamy, U. Rice can acclimate to lethal level of salinity by pretreatment with sublethal level of salinity through osmotic adjustment. *Plant Soil* **2006**, *284*, 363–373. [CrossRef]

33. Misratia, K.M.; Ismail, M.R.; Hakim, M.A.; Musa, M.H.; Puteh, A. Effect of salinity and alleviating role of gibberellic acid (GA3) for improving the morphological, physiological and yield traits of rice varieties. *Aust. J. Crop Sci.* **2013**, *7*, 1682–1692.

34. Qiu, Z.; Guo, J.; Zhu, A.; Zhang, L.; Zhang, M. Exogenous jasmonic acid can enhance tolerance of wheat seedlings to salt stress. *Ecotoxicol. Environ. Saf.* **2014**, *104*, 202–208. [CrossRef] [PubMed]

35. Mauchamp, A.; Méthy, M. Submergence-induced damage of photosynthetic apparatus in *Phragmites australis*. *Environ. Exp. Bot.* **2004**, *51*, 227–235. [CrossRef]

36. Zrig, A.; Tounekti, T.; Ben-Mohamed, H.; AbdElgawad, H.; Vadel, A.M.; Valero, D.; Khemira, H. Differential response of two almond rootstocks to chloride salt mixtures in the growing medium. *Russ. J. Plant Physiol.* **2016**, *63*, 143–151. [CrossRef]

37. Terletskaya, N.; Zobova, N.; Stupko, V.; Shuyskaya, E. Growth and photosynthetic reactions of different species of wheat seedlings under drought and salt stress. *Period. Biol.* **2017**, *119*, 37–45. [CrossRef]

38. Yan, Y.; Hou, P.; Duan, F.; Niu, L.; Dai, T.; Wang, K.; Zhou, W. Improving photosynthesis to increase grain yield potential: An analysis of maize hybrids released in different years in China. *Photosynth. Res.* **2021**, *150*, 295–311. [CrossRef]

39. Keutgen, A.J.; Pawelzik, E. Impacts of NaCl stress on plant growth and mineral nutrient assimilation in two cultivars of strawberry. *Environ. Exp. Bot.* **2009**, *65*, 170–176. [CrossRef]

40. Isayenkov, S.V.; Maathuis, F.J. Plant salinity stress: Many unanswered questions remain. *Front. Plant Sci.* **2019**, *10*, 80. [CrossRef] [PubMed]

41. Cakmak, I. The role of potassium in alleviating detrimental effects of abiotic stresses in plants. *J. Plant. Nutr. Soil Sci.* **2005**, *168*, 521–530. [CrossRef]

42. Mohammed, A.H.M.A. Physiological aspects of mungbean plant (*Vigna radiata* L. Wilczek) in response to salt stress and gibberellic acid treatment. *Res. J. Biol. Sci.* **2007**, *3*, 200–213.

43. Chakrabarti, N.; Mukherji, S. Effect of phytohormone pretreatment on nitrogen metabolism in *Vigna radiata* under salt stress. *Biol. Plant.* **2003**, *46*, 63–66. [CrossRef]

44. Kang, D.J.; Seo, Y.J.; Lee, J.D.; Ishii, R.; Kim, K.U.; Shin, D.H.; Lee, I.J. Jasmonic acid differentially affects growth, ion uptake and abscisic acid concentration in salt-tolerant and salt-sensitive rice cultivars. *J. Agron. Crop Sci.* **2005**, *191*, 273–282. [CrossRef]

45. Balkaya, A. Effects of salt stress on vegetative growth parameters and ion accumulations in cucurbit rootstock genotypes. *Ekin J. Crop Breed. Genet.* **2016**, *2*, 11–24.

46. Qian, Y.L.; Fu, J.M. Response of creeping bentgrass to salinity and mowing management: Carbohydrate availability and ion accumulation. *HortScience* **2005**, *40*, 2170–2174. [CrossRef]

47. Abdelgawad, Z.A.; Khalafaallah, A.A.; Abdallah, M.M. Impact of methyl jasmonate on antioxidant activity and some biochemical aspects of maize plant grown under water stress condition. *J. Agric. Sci.* **2014**, *5*, 1077. [CrossRef]

48. Onanuga, A.O.; Jiang, P.; Adl, S. Effect of phytohormones, phosphorus and potassium on cotton varieties (*Gossypium hirsutum*) root growth and root activity grown in hydroponic nutrient solution. *J. Agric. Sci.* **2012**, *4*, 93–110. [CrossRef]

49. Benková, E.; Hejátko, J. Hormone interactions at the root apical meristem. *Plant Mol. Biol.* **2009**, *69*, 383–396. [CrossRef] [PubMed]

50. Ivanchenko, M.G.; Muday, G.K.; Dubrovsky, J.G. Ethylene–auxin interactions regulate lateral root initiation and emergence in *Arabidopsis thaliana*. *Plant J.* **2008**, *55*, 335–347. [CrossRef]
51. Khan, M.A.; Gul, B.; Weber, D.J. Improving seed germination of *Salicornia rubra* (Chenopodiaceae) under saline conditions using germination regulating chemicals. *West. N. Am. Nat.* **2002**, *62*, 101–105.
52. de Castro, R.D.; van Lammeren, A.A.; Groot, S.P.; Bino, R.J.; Hilhorst, H.W. Cell division and subsequent radicle protrusion in tomato seeds are inhibited by osmotic stress but DNA synthesis and formation of microtubular cytoskeleton are not. *Plant Physiol.* **2000**, *122*, 327–336. [CrossRef] [PubMed]
53. Rodriguez, P.; Alarcon, J.J. Effect of salinity on growth, shoot water relation and root hydraulic conductivity in tomato plants. *J. Agric. Sci.* **1997**, *128*, 439–444. [CrossRef]
54. Heidari, M. Effects of salinity stress on growth, chlorophyll content and osmotic components of two basil (*Ocimum basilicum* L.) genotypes. *Afr. J. Biotechnol.* **2012**, *11*, 379–384.
55. Singh, S.P.; Singh, B.B.; Singh, M. Effect of kinetin on chlorophyll, nitrogen and proline in mungbean *(Vigna radiata)* under saline condition. *Indian J. Plant Physiol.* **1991**, *37*, 37–39.
56. Huguet-Robert, V.; Sulpice, R.; Lefort, C.; Maerskalck, V.; Emery, N.; Larher, F.R. The suppression of osmoinduced proline response of *Brassica napus* L. var *oleifera* leaf discs by polyunsaturated fatty acids and methyl-jasmonate. *Plant Sci.* **2003**, *164*, 119–127. [CrossRef]
57. Maslenkova, L.T.; Miteva, T.S.; Popova, L.P. Changes in the polypeptide patterns of barley seedlings exposed to jasmonic acid and salinity. *Plant Physiol.* **1992**, *98*, 700–707. [CrossRef] [PubMed]
58. Dheeba, B.; Selvakumar, S.; Kannan, M.; Kannan, K. Effect of gibberellic acid on black gram (*Vigna mungo*) irrigated with different levels of saline water. *Res. J. Pharm. Biol. Chem. Sci.* **2015**, *6*, 709–720.
59. Meloni, D.A.; Oliva, M.A.; Martinez, C.A.; Cambraia, J. Photosynthesis and activity of superoxide dismutase, peroxidase and glutathione reductase in cotton under salt stress. *Environ. Exp. Bot.* **2003**, *49*, 69–76. [CrossRef]
60. Saeidi-Sar, S.; Abbaspour, H.; Afshari, H.; Yaghoobi, S.R. Effects of ascorbic acid and gibberellin A 3 on alleviation of salt stress in common bean (*Phaseolus vulgaris* L.) seedlings. *Acta Physiol. Plant.* **2013**, *35*, 667–677. [CrossRef]
61. Bandurska, H.; Stroiński, A.; Kubiś, J. The effect of jasmonic acid on the accumulation of ABA, proline and spermidine and its influence on membrane injury under water deficit in two barley genotypes. *Acta Physiol. Plant.* **2003**, *25*, 279–285. [CrossRef]
62. Ahanger, M.A.; Mir, R.A.; Alyemeni, M.N.; Ahmad, P. Combined effects of brassinosteroid and kinetin mitigates salinity stress in tomato through the modulation of antioxidant and osmolyte metabolism. *Plant Physiol. Biochem.* **2020**, *147*, 31–42. [CrossRef]
63. Shah, T.; Latif, S.; Saeed, F.; Ali, I.; Ullah, S.; Alsahli, A.A.; Ahmad, P. Seed priming with titanium dioxide nanoparticles enhances seed vigor, leaf water status, and antioxidant enzyme activities in maize (*Zea mays* L.) under salinity stress. *J. King Saud. Univ. Sci.* **2021**, *33*, 101207. [CrossRef]
64. Hasanuzzaman, M.; Bhuyan, M.H.M.; Zulfiqar, F.; Raza, A.; Mohsin, S.M.; Mahmud, J.A.; Fotopoulos, V. Reactive oxygen species and antioxidant defense in plants under abiotic stress: Revisiting the crucial role of a universal defense regulator. *Antioxidants* **2020**, *9*, 681. [CrossRef]
65. Esfandiari, E.; Shakiba, M.R.; Mahboob, S.A.; Alyari, H.; Toorchi, M. Water stress, antioxidant enzyme activity and lipid peroxidation in wheat seedling. *J. Food Agric. Environ.* **2007**, *5*, 149.
66. Gunes, A.; Pilbeam, D.J.; Inal, A.; Coban, S. Influence of silicon on sunflower cultivars under drought stress, I: Growth, antioxidant mechanisms, and lipid peroxidation. *Commun. Soil Sci. Plant Anal.* **2008**, *39*, 1885–1903. [CrossRef]
67. Rouhi, H.R.; Aboutalebian, M.A.; Moosavi, S.A.; Karimi, F.A.; Karimi, F.; Saman, M.; Samadi, M. Change in several antioxidant enzymes activity of berseem clover (*Trifolium alexandrinum* L.) by priming. *Int. J. Agric. Sci.* **2012**, *2*, 237–243.
68. Shalata, A.; Neumann, P.M. Exogenous ascorbic acid (vitamin C) increases resistance to salt stress and reduces lipid peroxidation. *J. Exp. Bot.* **2001**, *52*, 2207–2211. [CrossRef] [PubMed]
69. Yupsanis, T.; Kefalas, P.S.; Eleftheriou, P.; Kotinis, K. RNase and DNase activities in the alfalfa and lentil grown in iso-osmotic solutions of NaCl and mannitol. *J. Plant Physiol.* **2001**, *158*, 921–927. [CrossRef]
70. Abo-kassem, E.E. Effects of salinity: Calcium interaction on growth and nucleic acid metabolism in five species of Chenopodiaceae. *Turk. J. Bot.* **2007**, *31*, 125–134.
71. Ismail, A.M. Physiological studies on the influence of cytokinin or GA3 in the alleviation of salt stress in sorghum plants. *Acta Agron. Hung.* **2003**, *51*, 371–380. [CrossRef]
72. Fujii, H.; Shimada, T.; Sugiyama, A.; Endo, T.; Nishikawa, F.; Nakano, M.; Omura, M. Profiling gibberellin (GA3)-responsive genes in mature mandarin fruit using a citrus 22K oligoarray. *Sci. Hortic.* **2008**, *116*, 291–298. [CrossRef]
73. Sumithra, K.; Jutur, P.P.; Carmel, D.B.; Reddy, A.R. Salinity-induced changes in two cultivars of *Vigna radiata*: Responses of antioxidative and proline metabolism. *Plant Growth Regul.* **2006**, *50*, 11–22. [CrossRef]
74. Su, X.; Howell, A.B.; D'Souza, D.H. Antiviral effects of cranberry juice and cranberry proanthocyanidins on foodborne viral surrogates–a time dependence study in vitro. *Food Microbiol.* **2010**, *27*, 985–991. [CrossRef]
75. Hseu, Z.Y. Evaluating heavy metal contents in nine composts using four digestion methods. *Bioresour. Technol.* **2004**, *95*, 53–59. [CrossRef] [PubMed]
76. Bates, L.S.; Waldren, R.P.; Teare, I.D. Rapid determination of free proline for water-stress studies. *Plant Soil* **1973**, *39*, 205–207. [CrossRef]

77. Heath, R.L.; Packer, L. Photoperoxidation in isolated chloroplasts: I. Kinetics and stoichiometry of fatty acid peroxidation. *Arch. Biochem. Biophys.* **1968**, *125*, 189–198. [CrossRef]
78. Cakmak, I.; Marschner, H. Magnesium deficiency and high light intensity enhance activities of superoxide dismutase, ascorbate peroxidase, and glutathione reductase in bean leaves. *Plant Physiol.* **1992**, *98*, 1222–1227. [CrossRef] [PubMed]
79. Misra, H.P.; Fridovich, I. The role of superoxide anion in the autoxidation of epinephrine and a simple assay for superoxide dismutase. *J. Biol. Chem.* **1972**, *247*, 3170–3175. [CrossRef]
80. Zhang, J.; Kirkham, M.B. Antioxidant responses to drought in sunflower and sorghum seedlings. *New Phytol.* **1996**, *132*, 361–373. [CrossRef]
81. Schmidt, G.; Thannhauser, S.J. A method for the determination of desoxyribonucleic acid, ribonucleic acid, and phosphoproteins in animal tissues. *J. Biol. Chem.* **1945**, *161*, 83–89. [CrossRef]
82. Morse, M.L.; Carter, C.E. The synthesis of nucleic acids in cultures of *Escherichia coli*, strains B and B/r. *J. Bacteriol.* **1949**, *58*, 317–326. [CrossRef]
83. Abd El-Wahab, N.H. Hormonal and Metabolic Effects of Coumarin on Germination of Sorghum. Master's Thesis, Faculty of Science, Ain Shams University, Cairo, Egypt, 1982.
84. Burton, K. A study of the conditions and mechanism of the diphenylamine reaction for the colorimetric estimation of deoxyribonucleic acid. *Biochem. J.* **1956**, *62*, 315. [CrossRef] [PubMed]

 *plants* 

Article

Salicylic Acid Stimulates Antioxidant Defense and Osmolyte Metabolism to Alleviate Oxidative Stress in Watermelons under Excess Boron

Mohamed Moustafa-Farag [1,2,3], Heba I. Mohamed [4], Ahmed Mahmoud [1,2], Amr Elkelish [5], Amarendra N. Misra [6,7], Kateta Malangisha Guy [1], Muhammad Kamran [3], Shaoying Ai [3] and Mingfang Zhang [1,*]

[1] Lab of Germplasm Improvement and Molecular Breeding, Agriculture and Biotechnology College, Zhejiang University, Hangzhou 310029, China; m_m_kamel2005@gdaas.cn (M.M.-F.); 11716103@zju.edu.cn (A.M.); malangisha@zju.edu.cn (K.M.G.)

[2] Horticulture Research Institute, Agriculture Research Center, Giza 12619, Egypt

[3] Institute of Agricultural Resources and Environment, Guangdong Academy of Agricultural Sciences, Guangzhou 510640, China; drkamran2017@nwsuaf.edu.cn (M.K.); aishaoying@gdaas.cn (S.A.)

[4] Biological and Geological Sciences Department, Faculty of Education, Ain Shams University, Cairo 11566, Egypt; hebaebrahem@edu.asu.edu.eg

[5] Botany Department, Faculty of Science, Suez Canal University, Ismailia 41522, Egypt; amr.elkelish@science.suez.edu.eg

[6] Department for Life Sciences, Central University of Jharkhand, Brambe, Ranchi-835205, India; anm@cuj.ac.in

[7] Department of Biosciences & Biotechnology, Khallikote University, GMax Building, Konishi, Berhampur 761008, India

* Correspondence: mfzhang@zju.edu.cn

Received: 18 May 2020; Accepted: 2 June 2020; Published: 8 June 2020

Abstract: Boron (B) is a microelement required in vascular plants at a high concentration that produces excess boron and toxicity in many crops. B stress occurs widely and limits plant growth and crop productivity worldwide. Salicylic acid (SA) is an essential hormone in plants and is a phenolic compound. The goal of this work is to explore the role of SA in the alleviation of excess B (10 mg L^{-1}) in watermelon plants at a morphological and biochemical level. Excess boron altered the nutrient concentrations and caused a significant reduction in morphological criteria; chlorophyll a, b, and carotenoids; net photosynthetic rate; and the stomatal conductance and transpiration rate of watermelon seedlings, while intercellular carbon dioxide (CO_2) was significantly increased compared to the control plants (0.5 mg L^{-1} B). Furthermore, excess boron accelerated the generation of reactive oxygen species (ROS), such as hydrogen peroxide (H_2O_2) and induced cellular oxidative injury. The application of exogenous SA significantly increased chlorophyll and carotenoid contents in plants exposed to excess B (10 mg L^{-1}), in line with the role of SA in alleviating chlorosis caused by B stress. Exogenously applied SA promoted photosynthesis and, consequently, biomass production in watermelon seedlings treated with a high level of B (10 mg L^{-1}) by reducing B accumulation, lipid peroxidation, and the generation of H_2O_2, while significantly increasing levels of the most reactive ROS, OH^-. SA also activated antioxidant enzymes, such as superoxide dismutase (SOD), peroxidase (POD), and ascorbate peroxidase (APX) and protected the seedlings from an ROS induced cellular burst. In conclusion, SA can be used to alleviate the adverse effects of excess boron.

Keywords: salicylic acid; chlorophyll fluorescence; excess boron; lipid peroxidation; enzymatic antioxidant; glutathione; proline; stomatal conductance

1. Introduction

Boron (B) is an essential plant micronutrient and has a functional role in the creation and function of cell walls [1]. Boron stress causes drastic effects worldwide. Toxic concentrations of B occur in soils that are irrigated with water contaminated with B or an excessive use of B-rich fertilizer, sewage sludge, as well as natural deposits discovered throughout the globe in arid and semi-arid areas [2]. Boron stress induces certain biochemical and morphological failures, such as reduced shoot and root development [3,4]; photosynthesis inhibition; reduced stomatal conductance [5]; the generation of reactive oxygen species (ROS) causing oxidative stress to lipids, proteins, and nucleic acids [6]; reduced root proton extrusion [7], reduced root cell division [8]; lignin and suberin accumulation in roots [9]; increased permeability of the membrane; lipid peroxidation; and altered antioxidant enzyme activities [10]. Scientists have been interested in developing potential strategies to enhance the seeds germination and the plants growth to achieve higher crop production [11,12]. Using phytohormones such as salicylic acid (SA) can enhance plant tolerance and counteract the toxic impacts of heavy metals on the germination of seedlings and the growth and development of plants. SA is a natural plant hormone that acts as a signal molecule to regulate plant growth [11], seedling germination, glycolysis, the flowering process, fruit yield [11], the uptake and transport of ions [12], stomatal conductance, the transpiration and photosynthetic rate [13], and the regulation of antioxidant enzyme synthesis during both biotic and abiotic stress [14,15]. Additionally, SA is a crucial signal biomolecule mediated systemic resistance to pathogen attacks by plants [16]. SA has also been shown to protect winter wheat plants [17] against cold stress and heat stress [18,19] and to modulate plant responses to salt and osmotic stresses [15,20], ozone or UV light [21], drought [22], and heavy metals, such as Cd [23–27], Mn [28], Hg [29], B [30], and Pb [31].

Although SA is widely used to protect economic crops against abiotic stress, only a few reports investigated its protective impact under excess boron stress [32,33]. SA protects plants under excess boron associated salinity, such as spinach [34]. The protective role of SA is expressed mainly by regulating ROS and antioxidants, inducing gene expression, and absorbing and distributing elements [23,35,36]. The most effective effect of SA is to improve the function of antioxidant enzymes in vivo [37,38]. Although the role of SA in the antioxidant system is very well studied under different biotic and abiotic stresses [14,39], its role under B stress remains unclear. In addition, the relationship between SA and boron toxicity has had contrary results in different studies. For instance, in mung beans, SA inhibits lipid peroxidation and reduced boron toxicity disorders through a reduction in lipoxygenase activity [40]. Watermelon belongs to the Cucurbitaceous family and is considered one of the world's top 20 cultivated crops with elevated economic value. The principal countries producing watermelon, such as China and Turkey [41,42], are affected by B stress, which has drastic effects on watermelon growth and yield. Nevertheless, there is insufficient knowledge on the mechanisms of boron toxicity-induced watermelons injury [4], highlighting the importance of alleviating B stress in watermelon crops.

Therefore, the aim of the present study is to (1) study the different responses of excess boron on watermelon and (2) evaluate the changes induced by SA in the antioxidant system and mineral uptake during excess boron stress. This research will help us to understand the mechanisms associated with excess boron in watermelon, help recommend the correct treatment of SA to alleviate B stress and aid the scientific community in solving the B stress problem.

2. Results

2.1. Plant Growth

The results show that the shoot and root dry weight of watermelon plants is significantly reduced by excess boron compared to control plants. In addition, the root projected area, surface area, diameter, volume, and number of tips was significantly decreased under excess boron except root volume, which experienced insignificant effects (Table 1). On the other hand, SA treatment (0.3 mM) increased watermelon growth and development under excess boron (Table 1).

Table 1. Effect of 0.3 mM salicylic acid (SA) on watermelon root system development under excess boron at 35 days old.

Treatment	Shoot Dry Weight	Root Dry Weight	Project Area (cm^2)	Surface Area (cm^2)	Avg Diameter (mm)	Root Volume (cm^3)	Number of Tips
Control	7.0 ± 0.42 [a]	0.17 ± 0.010 [b]	42.85 ± 2.57 [b]	90 ± 5.4 [c]	0.64 ± 0.04 [d]	1.48 ± 0.09 [b]	3173 ± 190 [b]
Excess B	1.54 ± 0.09 [d]	0.11 ± 0.007 [e]	20.25 ± 1.21 [e]	59 ± 3.6 [e]	0.72 ± 0.04 [c]	1.23 ± 0.07 [c]	1161 ± 69 [e]
Control + SA	3.78 ± 0.29 [b]	0.29 ± 0.022 [a]	51.16 ± 3.99 [a]	160 ± 12.5 [a]	0.80 ± 0.06 [b]	3.20 ± 0.25 [a]	4162 ± 324 [a]
Excess B + SA	2.29 ± 0.22 [c]	0.14 ± 0.014 [d]	24.03 ± 2.40 [d]	73 ± 7.4 [d]	0.49 ± 0.05 [e]	1.54 ± 0.15 [b]	1921 ± 192 [d]

Columns stand for mean ± SD. ANOVA were analyzed variations between the four treatments. Various letters indicate a significant difference from $p < 0.05$.

2.2. Boron Uptake and Translocation

The accumulation of boron in the roots and leaves of watermelon was increased significantly by approximately 5.5- and 10-fold, respectively, in response to excess boron stress compared to the untreated plants (0.5 mg L^{-1}) (Table 2). Watermelon plants translocated more B from the root to leaf tissues, as shown by the high translocation factor value (TF, the ratio of total B in the leaf and root tissues). The translocation factor showed a significant increase in excess boron treated plants. Treatment with SA caused a significant decrease in B contents in the roots and leaves of watermelon seedlings under excess boron stress compared to non-SA-treated plants.

Table 2. Effect of 0.3 mM SA on boron (B) uptake and translocation on watermelon leaves and roots under excess boron at 35 days old.

Treatment	B Level in Roots (mg kg^{-1})	B Level in Leaves (mg kg^{-1})	Translocation Factor
Control	73.9 ± 3.05 [c,d]	150.9 ± 20 [c]	2.04 ± 0.35 [b]
Excess B	408.0 ± 40 [a]	1592.3 ± 400 [a]	3.92 ± 1.36 [a]
Control + SA	80.3 ± 6 [c]	90.2 ± 6 [c]	1.12 ± 0.07 [c]
Excess B + SA	276.3 ± 40 [b]	1146.3 ± 200 [b]	4.18 ± 1.33 [a]

Columns stand for mean ± SD. ANOVA were analyzed variations between the four treatments. Various letters indicate a significant difference from $p < 0.05$.

2.3. Mineral Uptake

Without the SA treatment, a non-significant effect was found for the different macro-nutrients (Ca^{2+}, Mg^{2+} and K^+) in the shoots and root tissues under excess boron stress, but higher amounts of Na^+ was recorded in the shoots of watermelon and lower amounts in roots under excess boron (Table 3). In the presence of SA, Ca^{2+} content was significantly reduced in shoots and a marked increase in roots compared to their respective non-SA plants under excess boron stress. Furthermore, treatment with SA caused a significant increase in the K^+ content of watermelon shoots but decreased in roots under excess boron stress. SA-treated plants under excess boron stress had more Mg^{2+} in their root tissues, although the concentration in shoots remained unaffected. In addition, SA significantly decreased the shoot Na^+ concentration in the boron-stressed plants and control plants. On the other side, the Na^+ concentration increased dramatically in roots under excess boron and control conditions compared to non-SA plants (Table 3).

Table 3. Effect of 0.3 mM SA on nutrient uptake and translocation in watermelon shoot and root tissues under excess boron at 35 days old.

Treatment	Macro-Nutrient (mg kg^{-1} Shoot DW)			
	Ca^{2+}	K$^+$	Mg^{2+}	Na$^+$
Control	60.8 ± 5.03 [a]	87.5 ± 4 [b,c]	16.1 ± 2 [a]	15.1 ± 2 [c]
Excess B	56.9 ± 4 [a,b]	79.8 ± 4 [c]	14.0 ± 2 [a]	19.0 ± 2 [b]
Control + SA	54.6 ± 4 [b]	103.7 ± 20 [a]	14.1 ± 2 [a]	1.2 ± 0.2 [d]
Excess B + SA	49.8 ± 4 [c]	96.7 ± 12 [a,b]	15.4 ± 4 [a]	1.7 ± 0.2 [d]

Treatment	Macro-Nutrient (mg kg^{-1} Root DW)			
	Ca^{2+}	K$^+$	Mg^{2+}	Na$^+$
Control	12.4 ± 4 [c]	107.5 ± 20 [a]	3.4 ± 0.4 [c]	4.2 ± 0.4 [b]
Excess B	14.5 ± 4 [c]	112.8 ± 20 [a]	3.5 ± 0.4 [c]	3.4 ± 0.4 [e]
Control + SA	24.8 ± 2 [b]	68.2 ± 4 [b]	6.0 ± 0.2 [a]	8.9 ± 0.2 [a]
Excess B + SA	22.7 ± 2 [b]	34.8 ± 4 [c]	4.9 ± 0.2 [b]	4.2 ± 0.2 [b]

Columns stand for mean ± SD. ANOVA were analyzed variations between the four treatments. Various letters indicate a significant difference from $p < 0.05$.

2.4. Chlorophyll and Carotenoid Content

Excess boron stress caused a significant decrease in chlorophyll a, b, and carotenoid content in watermelon leaves compared to the untreated plants (Figure 1). The application of SA precipitated significant increases in chl a, chl b, and carotenoid content in stressed plants. These results confirm the role of SA in alleviating chlorosis symptoms caused by excess boron stress.

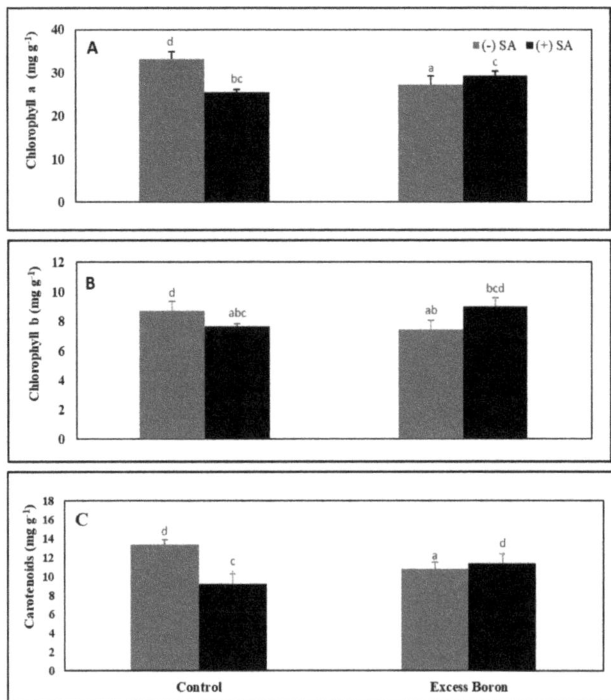

Figure 1. Effects of 0.3 mM SA on photosynthetic pigments (chlorophyll a (**A**), chlorophyll b (**B**) and carotenoids (**C**) in leaves under excess boron at 35 days old. The bars stand for mean ± SD. ANOVA were analyzed variations between the four treatments. The various letters indicate a significant difference from $p < 0.05$.

2.5. Leaf Gas Exchange

Excess boron exhibited significantly decreased PN, Gs, and Tr in watermelon seedlings and increased intercellular carbon dioxide (CO_2) compared to the unstressed plants (Figure 2). In addition, treatment with SA induced a significant reduction in PN, Gs, Tr but a significant rise in intercellular carbon dioxide (CO_2) under excess boron and in the control (Figure 2).

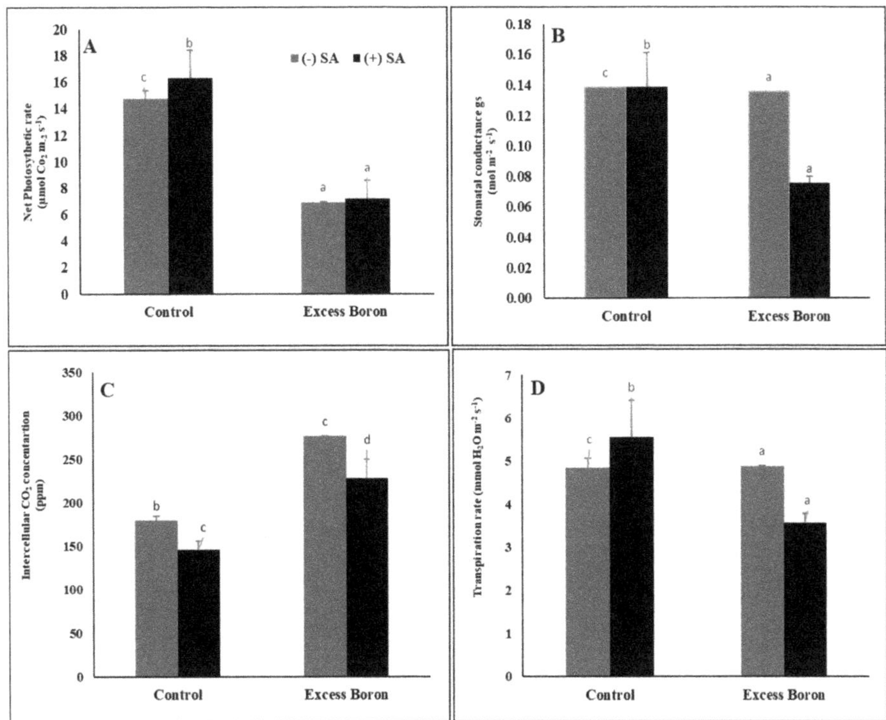

Figure 2. Effects of 0.3 mM SA on leaf gas exchange under excess boron at 35 days old. (**A**) Net photosynthesis rate, (**B**) stomatal conductance, (**C**) intercellular CO_2 concentration, and (**D**) transpiration rate. Bars stand for mean ± SD. ANOVA were analyzed variations between the four treatments. The various letters indicate a significant difference from $p < 0.05$.

2.6. Chlorophyll Fluorescence

Under excess boron stress, there are variable responses in the chlorophyll fluorescence of watermelon leaves (Figure 3). Excess boron induced a significant reduction in Fm but exerted a non-significant effect in F0 and PSII (Fv/Fm) compared to the control plants (0.5 B). Treatment with SA was significantly decreased in the PSII (Fv/Fm) of the plants under excess boron compared to the other treatments without SA (Figure 3), while F0 and Fm remained unaffected.

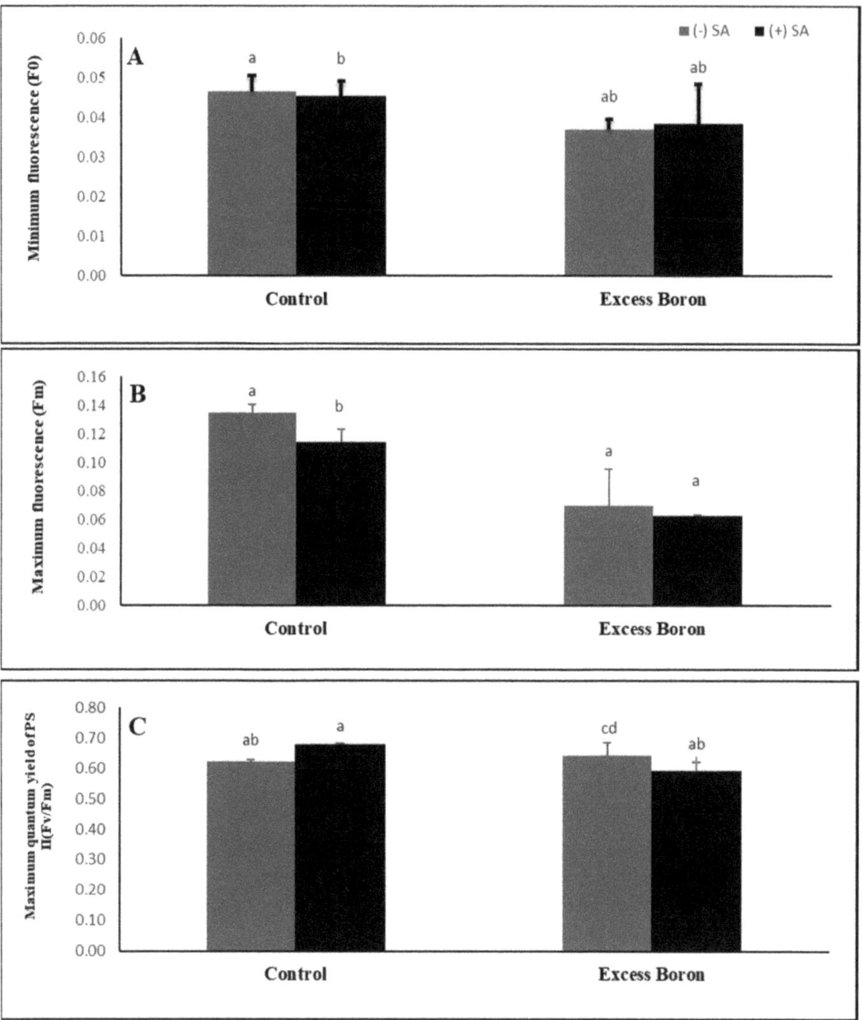

Figure 3. Effects of 0.3 mM SA on chlorophyll fluorescence under excess boron at 35 days old. (**A**) minimum fluorescence, (**B**) maximum fluorescence, and (**C**) maximum quantum yield. The bars stand for mean ± SD. ANOVA were analyzed variations between the four treatments. The various letters indicate a significant difference from $p < 0.05$.

2.7. MDA Content and the Endogenous ROS Generation Rate

A higher amount of the malondialdehyde (MDA) was recorded in boron-stressed plants as compared with the unstressed plants (Figure 4). Treatment with SA (0.3 mM) caused a significant decreased in the MDA content of watermelon leaves under control conditions and excess boron stress (Figure 4).

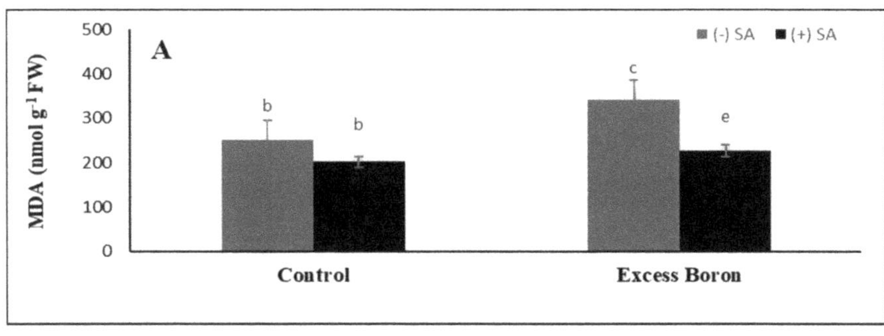

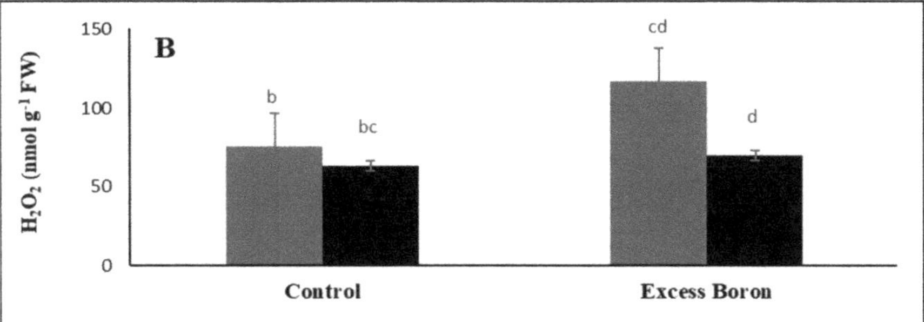

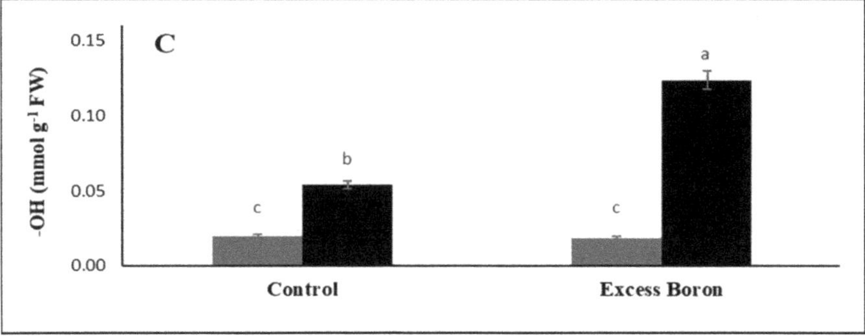

Figure 4. Effects of 0.3 mM SA on oxidative stress parameters under excess boron at 35 days old. (**A**) lipid peroxidation, (**B**) hydrogen peroxide, and (**C**) extra-cellular hydroxyl radicals. The bars stand for mean ± SD. ANOVA were analyzed variations between the four treatments. The various letters indicate a significant difference from $p < 0.05$.

The generation of ROS such as H_2O_2 and OH^- formation was determined in B treated watermelon leaves (Figure 3). The production of ROS molecules in watermelon leaves has increased significantly by boron excess compared to the unstressed plants. Moreover, SA significantly reduced the generation of H_2O_2 in non-B-stressed plants and stressed plants. For instance, SA more effectively reduced H_2O_2 generation in excess boron plants (Figure 4).

2.8. Antioxidant Enzyme Activities

The antioxidant enzymes activities (superoxide dismutase (SOD), ascorbate peroxidase (APX), peroxidase (POD), catalase (CAT), and glutathione reductase (GR) were significantly up-regulated in response to excess boron (Figure 5). The most pronounced accumulation (in POD, CAT, and APX) was detected in watermelon leaves under excess boron. SA treatments further improved the activities of POD, CAT, APX, and reduced glutathione (GSH) in the excess boron-stressed plants (Figure 5). In addition, SA significantly activated SOD, POD, CAT, and GSH in the control plants compared to their respective non-SA leaves (Figure 5).

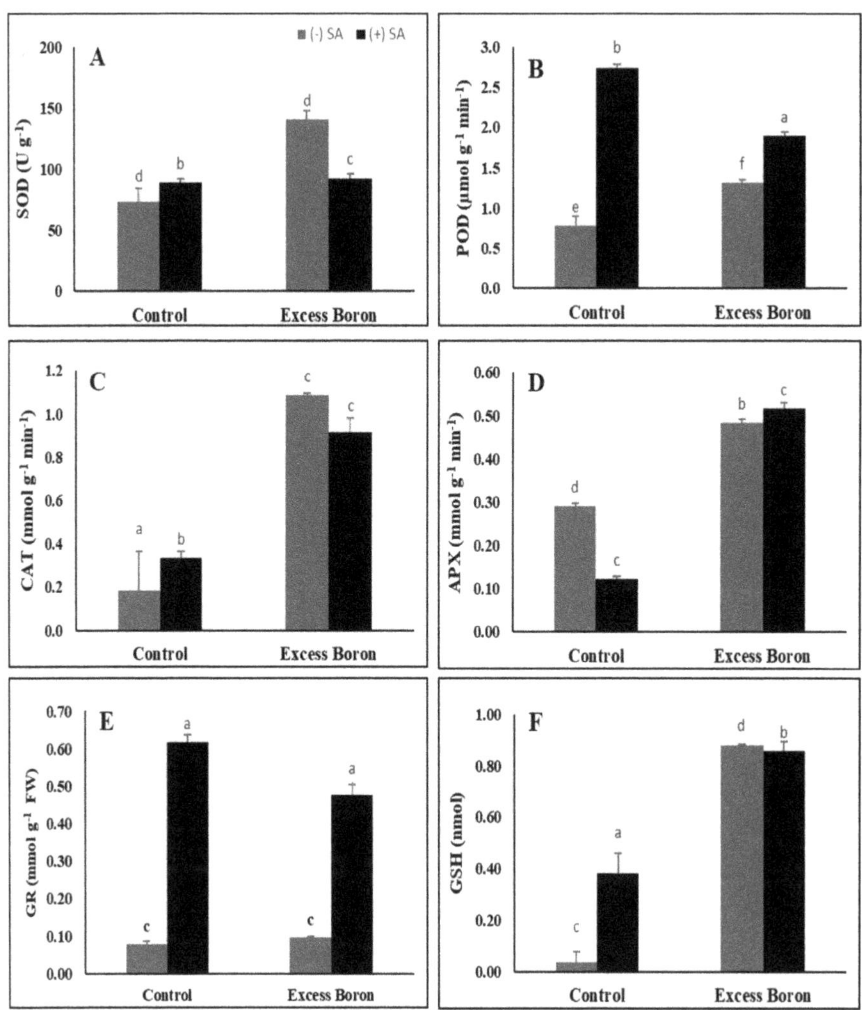

Figure 5. Effects of 0.3 mM SA on antioxidant enzyme activities under excess boron at 35 days old. (**A**) superoxide dismutase, (**B**) peroxidase, (**C**) catalase, (**D**) ascorbic peroxidase, (**E**) glutathione reductase, and (**F**) reduced glutathione. The bars stand for mean ± SD. ANOVA were analyzed variations between the four treatments. The various letters indicate a significant difference from $p < 0.05$.

2.9. Total Soluble Protein (TSP) and Proline Content

TSP in the leaves of watermelon plants showed non-significant change under excess boron compared to the untreated plants (Figure 6). The application of SA exerted non-significant effects on the TSP in watermelon leaves under excess boron. On the other hand, the amount of proline in watermelon leaves was unaffected under excess boron, while the application of SA caused a significant inhibition of proline content in plants grown under excess boron (Figure 6).

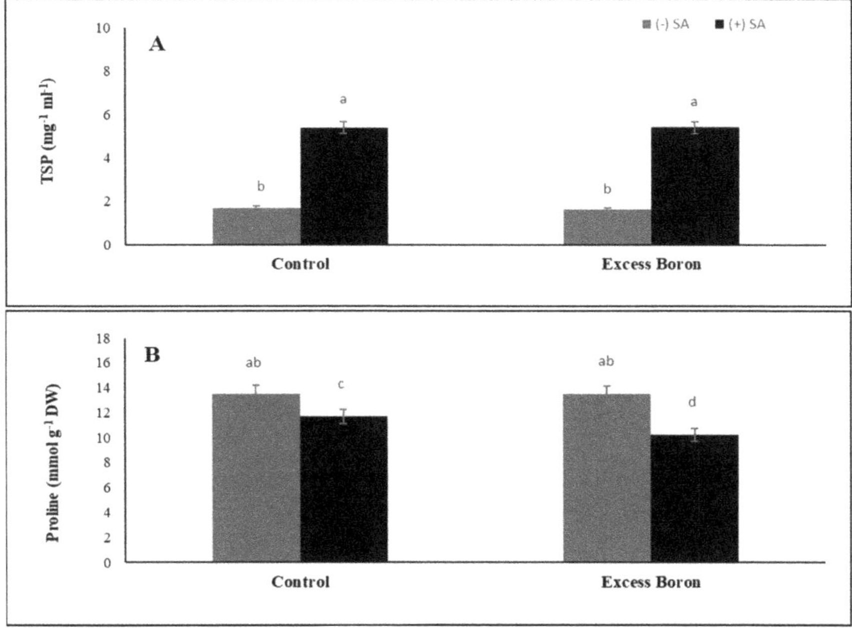

Figure 6. Effects of 0.3 mM SA on total soluble protein (**A**) and proline (**B**) content under excess boron at 35 days old. The bars stand for mean ± SD. ANOVA were analyzed variations between the four treatments. The various letters indicate a significant difference from $p < 0.05$.

3. Discussion

Boron is an important mineral nutrient needed for most plant species to progress and grow adequately. Excess boron can lead to lower growth, alter photosynthetic levels and cause morphological changes in the leaves [4]. In this experiment, we observed excess boron-induced chlorophyll and carotenoid deficiency in watermelon leaves. Analogous results were recorded by Aftab et al. [43]. The chlorophyll deficiency may be due to the production of degradation enzymes such as δ-aminolevulinic acid and protochlorophyllide under B stress [4]. Moreover, excess boron caused inhibition in plant growth (Table 1). The inhibition of growth in watermelon might result from the excess-boron-induced modification of essential metabolic processes, such as photosynthesis, and the nutrient uptake [4]. Furthermore, the inhibition of plant growth via excess boron may be due to the root cell division modulation and modification of the gene expression patterns associated with abscisic acid (ABA), or cell wall modifications, mitosis, water transport, and cell elongation [44,45].

The drop in photosynthetic pigments may be attributed to the accumulation of MDA and H_2O_2, resulting in the oxidation of chlorophyll and chloroplast membranes, which could be aggravated by excess B levels [4]. Treatment with SA protects photosynthetic pigments under excess boron. Some reports also indicate that SA is an effective photosynthesis regulator because it has a positive impact on the structure of the leaves and chloroplasts [46], as well as chlorophyll and carotenoid contents [47]. Alternatively, SA acts in the biogenesis for chloroplasts, protecting them against ROS and increasing chlorophyll stability [46]. In the present study, we observed an oxidative stress-induced reduction in the PN, Gs, and Tr of watermelon under excess boron stress (Figure 3). Similar results were reported by Lovatt et al. [5], who found that B reduced both Gs and the Pn transpiration rate.

Consequently, because of CO_2 limits, PSII electrons are available from thylakoid and stromal fatty acids [48]. An SA supply significantly enhanced the Pn, Gs, and Tr of BD watermelon leaves compared to plants without an SA supply, which may prevent the oxidation of auxin, whose elevated contents increased Pn in the leaf [49]. Treatment with SA significantly decreased the PSII (Fv/Fm) of the plants under excess boron compared to their respective treatments without SA, suggesting that the exclusion of SA induced changes in chlorophyll fluorescence, from watermelon tolerance to B stress. This inhibition in the fluorescence of chlorophyll under excess boron may be due to oxidation of chlorophyll and chloroplast membranes, which may be exacerbated by excess B, as recorded in hot pepper [50].

SA participates in the regulation of multiple biochemical processes under excess boron stress. Our results showing excess boron are contrary to those in previous studies, which showed that SA decreased boron toxicity stress [32,33,40]. These differences may be due to the different excess boron and SA concentrations and genetic resources used in our study. Watermelon plants translocated relatively more B from the root to leaf tissues, as evidenced by the high value of the translocation factor (Table 2). Similarly, El-Feky et al. [32] found that the accumulation of B was significantly higher in barley shoots than in roots under boron toxicity [51]. SA not only reduced the stress conditions of BT by reducing the accumulation of B within the plant organs, but also by reducing the distribution of root to leaf B (Table 2). These results are in agreement with the findings of El-Feky et al. [32] and El-Feky et al. [32,52] on barley tissue. SA may work through certain specific processes, such as reducing the absorption or triggering the efflux from the roots, i.e., processes leading to lower cytoplasmic B content [30].

Unbalance in nutrients may result from excessive boron stress based on nutrient availability, absorption, transport and competition (Table 3). One mechanism for alleviating excess boron stress may be the shift of the mineral absorption after SA treatments. We postulate that SA lowered the damaging impacts of excess boron and enhanced the stability of the membrane and, therefore, the tolerance of the plant. In SA plants, the possible mechanism of B tolerance is the detoxification of excess B through cell exclusion and/or vacuolar compartmentalization. Furthermore, apoplast-formed B complexes could play a significant role in the excess boron tolerance of the plant. In this regard, Reid et al. [53] suggested that the transfer of B from the symplast into the apoplast or from the cytoplasm into the vacuole may facilitate plant tolerance to boron toxicity [54]. In plants, there are several reports that show excess boron stress-induced changes in the activity of both antioxidant enzymes and soluble antioxidant levels [3], accompanied by an enhancement of lipid peroxidation [55]. Excess boron can inhibit the transport of electrons and may result in the production and accumulation of ROS, such as H_2O_2 and OH^-, in watermelon (Figure 4).

Moreover, SA application successfully inhibited the accumulation of MDA and the generation of H_2O_2 under excess boron stress compared to their respective SA alone. A high exogenous B supply could facilitate the transportation of boric acid to the cell, which could be partially converted to borate due to the cytosol's elevated inner pH [56], thus potentially releasing free ROS, such as O_2, OH, and H_2O_2 [8]. Similar findings were noted by [57], who clarified that MDA and H_2O_2 concentrations in safflower crops subjected to excess Zn are decreased significantly after supplementation with SA. It has been observed that SA relieves heavy metal-induced injury through metal chelation, scavenges lipid peroxyl radicals, and prevents lipid peroxidation through the activation of the antioxidant system and is, therefore, capable of protecting membrane integrity [58–63].

The protective effect of SA towards B stress strongly correlates with severalfold increases in levels of the hydroxyl radical (OH.), the most reactive ROS (Figure 4). SA can significantly elevate OH. levels even in unstressed (normal B) plants but OH. levels increased in SA-treated plants with B stress. It is known that OH. is mainly generated from H_2O_2 and Fe^{2+} via the Fenton reaction. The presence of free Fe is key to this process, e.g., in plants, where excess Fe is stored by ferritin proteins [64]. Therefore, it could be worth monitoring changes in Fe levels in SA-treated, B-stressed plants. Moreover, it is worth testing the effect of scavenging OH. and/or chelating Fe on the protective action of SA.

The activities of antioxidant defense enzymes, such as SOD, APX, POD, GR, and GSH, were significantly increased in response to B stress (Figure 5). These antioxidants play a critical role in the alleviation of abiotic stresses. For instance, SOD catalyzes the dismutation of superoxide to

H_2O_2 and O_2 as a significant scavenger. However, H_2O_2 is also poisonous to cells, and must be further drained by CAT or POD or both into water and O_2 [65,66]. CAT and POD, when coordinated with SOD, seem to play an important protective role in scavenging ROS [67]. Our results are in agreement with those of Garcia et al. [68], who found that boron toxicity induced SOD activity in tobacco leaves. SA treatments further enhanced the activities of POD, CAT, and GR that were activated under excess boron conditions (Figure 5), suggesting that POD, CAT, and GR play a key role in the alleviation of both aspects of B stress. Similar results found that the application of SA activated antioxidant enzymes under zinc [57] and nickel stress [69].

Glutathione (GSH), a multi-functional plant metabolite, plays a significant role in oxidative stress in cell defense and safety. GSH protects thiol protein groups during stress from oxidation [70]. GSH content was significantly increased in watermelon plants under excess boron and SA treatments in stressed plants (Figure 5). Moreover, Cervilla et al. [2] found that GSH was significantly increased in only one tomato species under B stress. Yadav [71], likewise, suggested that the availability of GSH is related to the formation of phytochelatins, which are used to alleviate the toxicity of heavy metals. SA also function as an antioxidant [72], and the prevention of oxidative cell damage during stress has been suggested to be one of the stress tolerance mechanisms [73]. This level of safety was due to the increased production of antioxidants [18].

Proline content in watermelon plants showed a decrease under excess boron and control conditions after SA treatment (Figure 6). These findings are consistent with those of Namdjoyan et al. [57], who found that the addition of SA and sodium nitroprusside in Zn-stressed safflower crops caused a sharp decrease in proline content. SA's function in proline degradation may be due to enhanced proline dehydrogenase activity [69].

The total soluble protein (TSP) in the leaves of watermelon plants experienced non-significant effects under excess boron and the application of SA but increased for the SA-treated plants compared to the non-stressed and stressed plants (Figure 6) [74,75]. SA induced a considerable increase in the content of protein fractions in sunflower plants under Cu stress [76] and chamomile plants under Cd and Ni stresses [77].

4. Materials and Methods

4.1. Plant Materials, Growth Conditions, and Treatments

Watermelon seeds are sterilized and then soaked in distilled water for one day at room temperature. Then, the seeds are germinated for 5 days in the dark on filter papers. The seedlings were transplanted into 5 L plastic boxes containing full Hoagland, while excess boron stress was applied by adding boric acid (H_3BO_3) into the solution at 10 mg L^{-1}. The concentrations of B were chosen based on preliminary studies using different concentrations of B to choose the suitable concentrations for excess boron and according to the literature [3,4]. SA (0.3 mM) supplementations were added to the Hoagland solution directly after transplanting into the plastic boxes. In our experiments, we used 2 concentrations of B and 2 concentrations of SA (4 treatments). Each treatment used 30 plants divided into three boxes (10 plants for each box). The plants grew for 30 days in growth chambers under 20–25/15–20 °C (day/night) and 60% humidity. The solutions were changed twice a week, and the pH of the solution was preserved between at 6.0 and 6.3.

Leaves from the third middle part were used for photosynthetic gas exchange and biochemical aspects.

4.2. Morphological Measurements

Following 1 month of the treatments, 10 seedlings were harvested from each treatment. The plants were dried out at 80 °C for 3 days. The root system, including surface areas, root volume, root diameter, and the number of root tips, was analyzed with WinRhizo Pro (S) v. 2009a software, in accordance with Yu et al. [78].

4.3. Determination of B and Uptake of Other Nutrients

The dried shoot and root samples were digested in a mixture of nitric acid and hydrogen peroxide at 130 °C. Mineral concentrations of B, Ca^{2+}, Mg^{2+}, K^+, and Na^+ were assayed by coupled plasma mass spectrometry (ICP-MS, Agilent 7500a, Agilent Technologies, Santa Clara, CA, USA) [79].

4.4. Malondialdehyde (MDA) and Reactive Oxygen Species (ROS)

Lipid peroxidation was measured by malondialdehyde (MDA), according to Zhou et al. [80].

Hydrogen peroxide was measured via the method described by Yordanov et al. [81] with a slight modification. The absorbance was read at 630 nm after 15 min.

Extra-cellular hydroxyl radicals (OHs) from the leaf samples were estimated following the protocol of Halliwell et al. [82]. The reaction mixtures contained, in a final volume of 1.0 mL, the following reagents at the following concentrations: deoxyribose (variable concentration), KH_2PO_4-KOH buffer, pH 7.4 (20 mM), $FeCl_3$ (100 µM), EDTA (104 µM), H_2O_2 (1 mM), and ascorbate (100 µM). Solutions of $FeCl_3$ and ascorbate were produced immediately before use in deaerated water. Reaction mixtures were incubated at 37 °C for 1 h, and the color was developed at 532 nm using a spectrophotometer.

4.5. Chlorophyll and Carotenoid Determination

Fresh leaf tissues were ground in the presence of liquid nitrogen. Methanol (95%) was added to the samples, and the mixture was centrifuged for 5 min at 4500 rpm at 20 °C. The absorbance of the extracted solution was recorded at wavelengths of 470, 649 and 665 nm to estimate the chlorophyll-a, chlorophyll-b [83], and carotenoids [84] using a spectrophotometer (Shimadzu UV-1700, Tokyo, Japan).

4.6. Photosynthetic Gas Exchange Parameters

The photosynthetic rate (Pn), stomatal conductance (Gs), transpiration rate (Tr), and intercellular CO_2 were recorded using a portable LI-COR 6400 photosynthesis system (Lincoln, NE, USA). The intrinsic water use efficiency (WUE) was calculated from the ratio of the photosynthesis rate and stomatal conductance [84].

4.7. Chlorophyll Fluorescence

The maximal quantum yield of PSII (Fv/Fm), initial fluorescence (F0), and maximum fluorescence (Fm) were determined by an imaging pulse amplitude modulation (PAM) device (IMAG-MAXI; Heinz Walz, Effeltrich, Germany).

4.8. Antioxidant Enzymes

Fresh leaf samples (0.5 g) (4 replicates in total) were homogenized in cold phosphate buffer (pH 7.8). The homogenate was centrifuged at $13,000 \times g$ for 20 min at 4 °C, and the supernatant was used to determine the enzymatic activities.

Superoxide dismutase activity (SOD: EC 1.15.1.1) was determined following the method of Zhang et al. [85]. One unit of SOD was defined as the amount contained in the volume of extract that caused a 50% inhibition of the SOD-inhibitable fraction of the Nitro blue tetrazolium (NBT) reduction.

Peroxidase activity (POD: EC 1.11.1.7) was assayed according to Velikova et al. [86]. The absorbance was recorded at 470 nm.

Catalase activity (CAT: EC 1.11.1.6) was measured according to Aebi [87]. The consumption of H_2O_2 was monitored using a spectrophotometer at 240 nm ($e = 39.4 \text{ mM}^{-1} \cdot \text{cm}^{-1}$) for 3 min.

Ascorbate peroxidase activity (APX: EC 1.11.1.11) was estimated by the method of Nakano et al. [88]. The absorbance was read at 265 nm ($e = 13.7 \text{ mM}^{-1} \cdot \text{cm}^{-1}$).

Glutathione reductase (GR: EC 1.6.4.2) activity was assayed according to Jiang et al. [89] after the oxidation of NADPH at 340 nm (with an extinction coefficient of $6.2 \text{ mM} \cdot \text{cm}^{-1}$) for 1 min. Reduced glutathione (GSH) was determined according to Law et al. [90]. An increase in the absorbance at 412 nm was measured after the addition of 5,5-dithiobis (2-nitrobenzoic acid) (DTNB).

4.9. Total Soluble Protein and Proline

Total soluble protein (TSP) was determined according to the Bradford [91] method using bovine serum albumin as a standard. For proline determination, sulphosalicylic acid methods are being applied and the absorbance was recorded at 520 nm using a spectrophotometer [92].

4.10. Statistical Analysis

A completely randomized block design was used in our study, and the biostatistics were analyzed using CoStat v. 6.4 software. The results are denoted as the mean ± SD.

5. Conclusions

The present work demonstrates that SA can alleviate excess boron. The mechanisms by which SA alleviates excess boron mainly involve the following activities: (1) promoting chlorophyll content, (2) modulating the balance of mineral elements, (3) protecting against oxidative stress by decreasing MDA and ROS production, and (4) acting directly as an antioxidant to scavenge reactive oxygen species and/or indirectly modulating redox balance through the activation of the antioxidant responses. Our results could facilitate an integrated and optimized management strategy to alleviate excess boron in watermelon, but further study is needed.

Author Contributions: Conceptulazation, M.M.-F., A.M., M.C., S.A.; formal analysis, M.M.-F., H.I.M., A.M., A.E., M.C., S.A.; data analysis, M.M.-F., M.Z., H.I.M., A.E.; K.M.G., M.K., S.A., H.I.M., A.E.; resources, H.I.M., A.N.M., K.M.G., M.Z.; supervision, A.E., K.M.G., M.Z.; visualization, H.I.M., A.M., A.E., A.N.M., K.M.G., M.C., S.A., M.Z.; funding, M.M.-F., A.N.M., K.M.G., M.C., S.A.; writing the manuscript, M.M.-F., H.I.M., A.E.; editing the revised version, A.M., A.N.M., M.Z. All authors have read and agreed to the published version of the manuscript.

Funding: This work was supported by the earmarked fund for Modern Agro-Industry Technology Research System (CARS-26-17) of China, the National Key Technology R&D Program of China (2011BAD12B04), National Natural Science Foundation of China (31372077), Science and Technology Program of Zhejiang Province (2011C12001) and key science and technology breeding program of agricultural new variety of Zhejiang Province (2012C129031-2-11).

Acknowledgments: We would like to thank my group team in Guangdong Academy of Agricultural Sciences; Linfeng Li, Yichun Li, Mingdeng Tang and Yanhong Wang for their support and cooperation during the writing of this article. Moreover, we would like to thank Muhammad Dawood, Basharat Ali and Tariq Rafique for their support during biochemical analyses of the manuscript.

Conflicts of Interest: The authors state that no conflict of interest exists.

References

1. Tanaka, M.; Fujiwara, T. Physiological roles and transport mechanisms of boron: Perspectives from plants. *Pflügers Arch. Eur. J. Physiol.* **2008**, *456*, 671–677. [CrossRef] [PubMed]
2. Cervilla, L.M.; Blasco, B.; RÍOs, J.J.; Romero, L.; Ruiz, J.M. Oxidative stress and antioxidants in tomato (solanum lycopersicum) plants subjected to boron toxicity. *Ann. Bot.* **2007**, *100*, 747–756. [CrossRef]
3. Moustafa-Farag, M.; Bingsheng, F.; Malangisha Guy, K.; Hu, Z.; Yang, J.; Zhang, M. Activated antioxidant enzymes-reduced malondialdehyde concentration, and improved mineral uptake-promoted watermelon seedlings growth under boron deficiency. *J. Plant Nutr.* **2016**, *39*, 1989–2001. [CrossRef]
4. Moustafa-Farag, M.; Najeeb, U.; Yang, J.; Hu, Z.; Fang, Z.M. Nitric oxide protects carbon assimilation process of watermelon from boron-induced oxidative injury. *Plant Physiol. Biochem.* **2017**, *111*, 166–173. [CrossRef] [PubMed]
5. Lovatt, C.J.; Bates, L.M. Early effects of excess boron on photosynthesis and growth of Cucurbita pepo. *J. Exp. Bot.* **1984**, *35*, 297–305. [CrossRef]
6. Ardic, M.; Sekmen, A.H.; Turkan, I.; Tokur, S.; Ozdemir, F. The effects of boron toxicity on root antioxidant systems of two chickpea (*Cicer arietinum* L.) cultivars. *Plant Soil* **2009**, *314*, 99–108. [CrossRef]

7. Roldán, M.; Belver, A.; Rodríguez-Rosales, P.; Ferrol, N.; Donaire, J.P. *In vivo* and *in vitro* effects of boron on the plasma membrane proton pump of sunflower roots. *Physiol. Plant.* **1992**, *84*, 49–54. [CrossRef]

8. Choi, E.Y.; Kolesik, P.; McNeill, A.N.N.; Collins, H.; Zhang, Q.; Huynh, B.L.; Graham, R.; Stangoulis, J. The mechanism of boron tolerance for maintenance of root growth in barley (*Hordeum vulgare* L.). *Plant Cell Environ.* **2007**, *30*, 984–993. [CrossRef]

9. Ghanati, F.; Morita, A.; Yokota, H. Induction of suberin and increase of lignin content by excess boron in tobacco cells. *Soil Sci. Plant Nutr.* **2002**, *48*, 357–364. [CrossRef]

10. Senousy, H.H.; Abd Ellatif, S.; Ali, S. Assessment of the antioxidant and anticancer potential of different isolated strains of cyanobacteria and microalgae from soil and agriculture drain water. *Environ. Sci. Pollut. Res.* **2020**, *27*, 18463–18474. [CrossRef]

11. Vwioko, E.; Adinkwu, O.; El-Esawi, M.A. Comparative Physiological, Biochemical, and Genetic Responses to Prolonged Waterlogging Stress in Okra and Maize Given Exogenous Ethylene Priming. *Front. Physiol.* **2017**, *8*, 632. [CrossRef]

12. Mohamed, H.I.; Elsherbiny, E.A.; Abdelhamid, M.T. Physiological and biochemical responses of *Vicia faba* plants to foliar application with zinc and iron. *Gesunde Pflanzen* **2016**, *68*, 201–212. [CrossRef]

13. El-Esawi, M.A.; Alaraidh, I.A.; Alsahli, A.A.; Alzahrani, S.M.; Ali, H.M.; Alayafi, A.A.; Ahmad, M. Serratia liquefaciens KM4 Improves Salt Stress Tolerance in Maize by Regulating Redox Potential, Ion Homeostasis, Leaf Gas Exchange and Stress-Related Gene Expression. *Int. J. Mol. Sci.* **2018**, *19*, 3310. [CrossRef]

14. Elkelish, A.A.; Soliman, M.H.; Alhaithloul, H.A.; El-Esawi, M.A. Selenium protects wheat seedlings against salt stress-mediated oxidative damage by up-regulating antioxidants and osmolytes metabolism. *Plant Physiol. Biochem.* **2019**, *137*, 144–153. [CrossRef]

15. Abdelaal, K.A.; EL-Maghraby, L.M.; Elansary, H.; Hafez, Y.M.; Ibrahim, E.I.; El-Banna, M.; El-Esawi, M.; Elkelish, A. Treatment of sweet pepper with stress tolerance-inducing compounds alleviates salinity stress oxidative damage by mediating the physio-biochemical activities and antioxidant systems. *Agronomy* **2020**, *10*, 26. [CrossRef]

16. Loake, G.; Grant, M. Salicylic acid in plant defence—The players and protagonists. *Curr. Opin. Plant Biol.* **2007**, *10*, 466–472. [CrossRef]

17. Taşgín, E.; Atící, Ö.; Nalbantoğlu, B. Effects of salicylic acid and cold on freezing tolerance in winter wheat leaves. *Plant Growth Regul.* **2003**, *41*, 231–236. [CrossRef]

18. El-Esawi, M.A.; Alayafi, A.A. Overexpression of Rice Rab7 Gene Improves Drought and Heat Tolerance and Increases Grain Yield in Rice (Oryza sativa L.). *Genes* **2019**, *10*, 56. [CrossRef]

19. Elkelish, A.; Qari, S.H.; Mazrou, Y.S.A.; Abdelaal, K.A.A.; Hafez, Y.M.; Abu-Elsaoud, A.M.; Batiha, G.E.-S.; El-Esawi, M.A.; El Nahhas, N. Exogenous Ascorbic Acid Induced Chilling Tolerance in Tomato Plants Through Modulating Metabolism, Osmolytes, Antioxidants, and Transcriptional Regulation of Catalase and Heat Shock Proteins. *Plants* **2020**, *9*, 431. [CrossRef]

20. El-Esawi, M.A.; Al-Ghamdi, A.A.; Ali, H.M.; Alayafi, A.A. Azospirillum lipoferum FK1 confers improved salt tolerance in chickpea (Cicer arietinum L.) by modulating osmolytes, antioxidant machinery and stress-related genes expression. *Environ. Exp. Bot.* **2019**, *159*, 55–65. [CrossRef]

21. Sharma, Y.K.; León, J.; Raskin, I.; Davis, K.R. Ozone-Induced Responses in Arabidopsis thaliana: The Role of Salicylic Acid in the Accumulation of Defense-Related Transcripts and Induced Resistance. *Proc. Natl. Acad. Sci. USA* **1996**, *93*, 5099–5104. [CrossRef] [PubMed]

22. Elkeilsh, A.; Awad, Y.M.; Soliman, M.H.; Abu-Elsaoud, A.; Abdelhamid, M.T.; El-Metwally, I.M. Exogenous application of β-sitosterol mediated growth and yield improvement in water-stressed wheat (Triticum aestivum) involves up-regulated antioxidant system. *J. Plant Res.* **2019**, *132*, 881–901. [CrossRef] [PubMed]

23. Metwally, A.; Finkemeier, I.; Georgi, M.; Dietz, K.-J. Salicylic acid alleviates the cadmium toxicity in barley seedlings. *Plant Physiol.* **2003**, *132*, 272–281. [CrossRef]

24. Palma, J.M.; Sandalio, L.M.; Javier Corpas, F.; Romero-Puertas, M.C.; McCarthy, I.; del Río, L.A. *Plant Proteases, Protein Degradation, and Oxidative Stress: Role of Peroxisomes*; Elsevier Masson SAS: Paris, France, 2002; Volume 40, pp. 521–530.

25. Drazic, G.; Mihailovic, N. Modification of cadmium toxicity in soybean seedlings by salicylic acid. *Plant Sci.* **2005**, *168*, 511–517. [CrossRef]

26. Li, Q.; Wang, Y.; Wang, G.; Yang, D.; Guan, C.; Ji, J. Foliar application of salicylic acid alleviate the cadmium toxicity by modulation the reactive oxygen species in potato. *Ecotoxicol. Environ. Saf.* **2019**, *172*, 317–325. [CrossRef] [PubMed]

27. Moradkhani, S.; Nejad, R.A.K.; Dilmaghani, K.; Chaparzadeh, N. Salicylic acid decreases Cd toxicity in sunflower plants. *Ann. Biol. Res.* **2013**, *4*, 135–141.

28. Shi, Q.; Zhu, Z. Effects of exogenous salicylic acid on manganese toxicity, element contents and antioxidative system in cucumber. *Environ. Exp. Bot.* **2008**, *63*, 317–326. [CrossRef]

29. Zhou, Z.S.; Guo, K.; Elbaz, A.A.; Yang, Z.M. Salicylic acid alleviates mercury toxicity by preventing oxidative stress in roots of Medicago sativa. *Environ. Exp. Bot.* **2009**, *65*, 27–34. [CrossRef]

30. Metwally, A.M.; Radi, A.A.; El-Shazoly, R.M.; Hamada, A.M. The role of calcium, silicon and salicylic acid treatment in protection of canola plants against boron toxicity stress. *J. Plant Res.* **2018**, *131*, 1015–1028. [CrossRef]

31. Saleem, M.H.; Ali, S.; Rehman, M.; Rana, M.S.; Rizwan, M.; Kamran, M.; Imran, M.; Riaz, M.; Soliman, M.H.; Elkelish, A.; et al. Influence of phosphorus on copper phytoextraction via modulating cellular organelles in two jute (Corchorus capsularis L.) varieties grown in a copper mining soil of Hubei Province, China. *Chemosphere* **2020**, *248*, 126032. [CrossRef]

32. El-Feky, S.S.; El-Shintinawy, F.A.; Shaker, E.M.; El-Din, H.A.S. Effect of elevated boron concentrations on the growth and yield of barley (hordeum vulgare l.) and alleviation of its toxicity using different plant growth modulators. *Aust. J. Crop Sci.* **2012**, *6*, 1687–1695.

33. El-Shazoly, R.M.; Metwally, A.A.; Hamada, A.M. Salicylic acid or thiamin increases tolerance to boron toxicity stress in wheat. *J. Plant Nutr.* **2019**, *42*, 702–722. [CrossRef]

34. Eraslan, F.; Inal, A.; Pilbeam, D.J.; Gunes, A. Interactive effects of salicylic acid and silicon on oxidative damage and antioxidant activity in spinach (*Spinacia oleracea* L. cv. Matador) grown under boron toxicity and salinity. *Plant Growth Regul.* **2008**, *55*, 207–219. [CrossRef]

35. Chen, W.; Singh, K.B. The auxin, hydrogen peroxide and salicylic acid induced expression of the Arabidopsis GST6 promoter is mediated in part by an ocs element. *Plant J.* **1999**, *19*, 667–677. [CrossRef] [PubMed]

36. Wang, Y.-S.; Wang, J.; Yang, Z.-M.; Wang, Q.-Y.; Lü, B.; Li, S.-Q.; Lu, Y.-P.; Wang, S.-H.; Sun, X. Salicylic acid modulates aluminum-induced oxidative stress in roots of Cassia tora. *Acta Bot. Sin.* **2004**, *46*, 819–828.

37. He, Y.; Liu, Y.; Cao, W.; Huai, M.; Xu, B.; Huang, B. Effects of Salicylic Acid on Heat Tolerance Associated with Antioxidant Metabolism in Kentucky Bluegrass. *Crop Sci.* **2005**, *45*, 988–995. [CrossRef]

38. Wang, L.-J.; Li, S.-H. Thermotolerance and Related Antioxidant Enzyme Activities Induced by Heat Acclimation and Salicylic Acid in Grape (*Vitis vinifera* L.) Leaves. *Plant Growth Regul.* **2006**, *48*, 137–144. [CrossRef]

39. Gholamnezhad, J.; Sanjarian, F.; Goltapeh, E.M.; Safaie, N.; Razavi, K. Effect of Salicylic Acid on Enzyme Activity in Wheat in Immediate Early Time after Infection with Mycosphaerella Graminicola. *Sci. Agric. Bohem.* **2016**, *47*, 1–8. [CrossRef]

40. Muhammad, A.J.; Shaheed, A.I. Effects of salicylic acid and silicon on oxidative damage and antioxidant activity in mung bean cuttings under boron toxicity. *J. Univ. Babylon* **2012**, *22*, 335–348.

41. Ozturk, M.; Sakcali, S.; Gucel, S.; Tombuloglu, H. Boron and Plants. In *Plant Adaptation and Phytoremediation*; Ashraf, M., Ozturk, M., Ahmad, M.S.A., Eds.; Springer: Dordrecht, The Netherlands, 2010; pp. 275–311. [CrossRef]

42. Liu, Z.; Zhu, Q.Q.; Tong, L.H. Boron-deficient soils and their distribution in China. *Acta Pedol. Sin.* **1980**, *17*, 228–239.

43. Aftab, T.; Khan, M.M.A.; Naeem, M.; Idrees, M.; Moinuddin; Teixeira da Silva, J.A.; Ram, M. Exogenous nitric oxide donor protects Artemisia annua from oxidative stress generated by boron and aluminium toxicity. *Ecotoxicol. Environ. Saf.* **2012**, *80*, 60–68. [CrossRef] [PubMed]

44. Aquea, F.; Federici, F.; Moscoso, C.; Vega, A.; Jullian, P.; Haseloff, J.I.M.; Arce-Johnson, P. A molecular framework for the inhibition of Arabidopsis root growth in response to boron toxicity. *Plant Cell Environ.* **2012**, *35*, 719–734. [CrossRef] [PubMed]

45. Josten, P.; Kutschera, U. The Micronutrient Boron Causes the Development of Adventitious Roots in Sunflower Cuttings. *Ann. Bot.* **1999**, *84*, 337–342. [CrossRef]

46. Uzunova, A.N.; Popova, L.P. Effect of Salicylic Acid on Leaf Anatomy and Chloroplast Ultrastructure of Barley Plants. *Photosynthetica* **2000**, *38*, 243–250. [CrossRef]

47. Fariduddin, Q.; Hayat, S.; Ahmad, A. Salicylic Acid Influences Net Photosynthetic Rate, Carboxylation Efficiency, Nitrate Reductase Activity, and Seed Yield in Brassica juncea. *Photosynthetica* **2003**, *41*, 281–284. [CrossRef]

48. Moustafa-Farag, M.; Almoneafy, A.; Mahmoud, A.; Elkelish, A.; Arnao, M.B.; Li, L.; Ai, S. Melatonin and Its Protective Role against Biotic Stress Impacts on Plants. *Biomolecules* **2020**, *10*, 54. [CrossRef]

49. Arteca, R.N. *Plant Growth Substances: Principles and Applications*; Chapman & Hall: New York, NY, USA, 1996.

50. Supanjani, S.; Lee, K.D. Hot pepper response to interactive effects of salinity and boron. *Plant Soil Environ.* **2006**, *52*, 227–233. [CrossRef]

51. Reid, R. Physiology and Metabalism of Boron in Plants. In *Advances in Plant and Animal Boron Nutrition*; Springer: Dordrecht, The Netherlands, 2007; pp. 83–90.

52. El-Feky, S.S.; El-Feky, S.S.; El-Shintinawy, F.A.; Shaker, E.M. Role of CaCl2 and salicylic acid on metabolic activities and productivity of boron stressed barley (*Hordeum vulgare* L.). *Int. J. Curr. Microbiol Appl. Sci.* **2014**, *3*, 368–380.

53. Reid, R.; Fitzpatrick, K. Influence of Leaf Tolerance Mechanisms and Rain on Boron Toxicity in Barley and Wheat. *Plant Physiol.* **2009**, *151*, 413–420. [CrossRef]

54. Pang, Y.; Li, L.; Ren, F.; Lu, P.; Wei, P.; Cai, J.; Xin, L.; Zhang, J.; Chen, J.; Wang, X. Overexpression of the tonoplast aquaporin AtTIP5;1 conferred tolerance to boron toxicity in Arabidopsis. *J. Genet. Genom.* **2010**, *37*, 389–397. [CrossRef]

55. Karabal, E.; Yücel, M.; Öktem, H.A. Antioxidant responses of tolerant and sensitive barley cultivars to boron toxicity. *Plant Sci.* **2003**, *164*, 925–933. [CrossRef]

56. Wimmer, M.A.; Mühling, K.H.; LÄuchli, A.; Brown, P.H.; Goldbach, H.E. Boron Toxicity: The Importance of Soluble Boron. In *Boron in Plant and Animal Nutrition*; Goldbach, H.E., Brown, P.H., Rerkasem, B., Thellier, M., Wimmer, M.A., Bell, R.W., Eds.; Springer: Boston, MA, USA, 2002; pp. 241–253. [CrossRef]

57. Namdjoyan, S.; Kermanian, H.; Soorki, A.A.; Modarres Tabatabaei, S.; Elyasi, N. Effects of exogenous salicylic acid and sodium nitroprusside on α-tocopherol and phytochelatin biosynthesis in zinc-stressed safflower plants. *Turk. J. Bot.* **2018**, *42*, 271–279. [CrossRef]

58. Soliman; Alhaithloul; Hakeem; Alharbi; El-Esawi; Elkelish Exogenous Nitric Oxide Mitigates Nickel-Induced Oxidative Damage in Eggplant by Upregulating Antioxidants, Osmolyte Metabolism, and Glyoxalase Systems. *Plants* **2019**, *8*, 562. [CrossRef] [PubMed]

59. Szalai, G.; Krantev, A.; Yordanova, R.; Popova, L.P.; Janda, T. Influence of salicylic acid on phytochelatin synthesis in Zea mays during Cd stress. *Turk. J. Bot.* **2013**, *37*, 708–714. [CrossRef]

60. Fodor, J.; Gullner, G.; Adam, A.L.; Barna, B.; Kömives, T.; Kiràly, Z. Local and systemic responses of antioxidants to tobacco mosaic virus infection and to salicylic acid in tobacco. *Plant Physiol* **1997**, *114*, 1443–1451. [CrossRef] [PubMed]

61. Király, Z.; Barna, B.; Kecskés, A.; Fodor, J. Down-regulation of antioxidative capacity in a transgenic tobacco which fails to develop acquired resistance to necrotization caused by tobacco mosaic virus. *Free. Radic. Res.* **2002**, *36*, 981–991. [CrossRef] [PubMed]

62. Janda, T.; Lejmel, M.A.; Molnár, A.B.; Majláth, I.; Pál, M.; Nguyen, Q.T.; Nguyen, N.T.; Le, V.N.; Szalai, G. Interaction between elevated temperature and different types of Na-salicylate treatment in Brachypodium dystachion. *PLoS ONE* **2020**, *15*, e0227608. [CrossRef] [PubMed]

63. Janda, T.; Szalai, G.; Tari, I.; Paldi, E. Hydroponic treatment with salicylic acid decreases the effect of chilling injury in maize (*Zea mays* L.) plants. *Planta* **1999**, *208*, 175–180. [CrossRef]

64. Deák, M.; Horváth, G.V.; Davletova, S.; Török, K.; Sass, L.; Vass, I.; Barna, B.; Király, Z.; Dudits, D. Plants ectopically expressing the ironbinding protein, ferritin, are tolerant to oxidative damage and pathogens. *Nat. Biotechnol.* **1999**, *17*, 192–196. [CrossRef]

65. Mohamed, H.I. Molecular and biochemical studies on the effect of gamma rays on lead toxicity in cowpea (Vigna sinensis) plants. *Biol Trace Elem Res* **2011**, *144*, 1205–1218. [CrossRef]

66. Horváth, E.; Szalai, G.; Janda, T. Induction of abiotic stress tolerance by salicylic acid signaling. *J. Plant Growth Regul.* **2007**, *26*, 290–300. [CrossRef]

67. Jaleel, C.A.; Manivannan, P.; Wahid, A.; Farooq, M.; Al-Juburi, H.J.; Somasundaram, R.; Panneerselvam, R. Drought stress in plants: A review on morphological characteristics and pigments composition. *Int. J. Agric. Biol.* **2009**, *11*, 100–105.

68. Garcia, P.C.; Rivero, R.M.; López-Lefebre, L.R.; Sánchez, E.; Ruiz, J.M.; Romero, L. Response of oxidative metabolism to the application of carbendazim plus boron in tobacco. *Aust. J. Plant Physiol.* **2001**, *28*, 801–806. [CrossRef]

69. Kazemi, N.; Khavari-Nejad, R.A.; Fahimi, H.; Saadatmand, S.; Nejad-Sattari, T. Effects of exogenous salicylic acid and nitric oxide on lipid peroxidation and antioxidant enzyme activities in leaves of *Brassica napus* L. under nickel stress. *Sci. Hortic.* **2010**, *126*, 402–407. [CrossRef]

70. Noctor, G.; Gomez, L.; Vanacker, H.; Foyer, C.H. Interactions between biosynthesis, compartmentation and transport in the control of glutathione homeostasis and signalling. *J. Exp. Bot.* **2002**, *53*, 1283–1304. [CrossRef]

71. Yadav, S.K. Heavy metals toxicity in plants: An overview on the role of glutathione and phytochelatins in heavy metal stress tolerance of plants. *South Afr. J. Bot.* **2010**, *76*, 167–179. [CrossRef]

72. Cheng, I.F.; Zhao, C.P.; Amolins, A.; Galazka, M.; Doneski, L. A hypothesis for the *in vivo* antioxidant action of salicyclic acid. *Biometals* **1996**, *9*, 285–290. [CrossRef]

73. El-Beltagi, H.S.; Mohamed, H.I. Reactive oxygen species, lipid peroxidation and antioxidative defense mechanism. *Notulae Botanicae Horti Agrobotanici ClujNapoca.* **2013**, *41*, 44–57. [CrossRef]

74. Akladious, S.A.; Mohamed, H.I. Physiological role of exogenous nitric oxide in improving performance, yield and some biochemical aspects of sunflower plant under zinc stress. *Acta Biol. Hung.* **2017**, *68*, 101–114. [CrossRef]

75. Mohamed, H.I.; Latif, H.H.; Hanafy, R.S. Influence of Nitric Oxide Application on Some Biochemical Aspects, Endogenous Hormones, Minerals and Phenolic Compounds of Vicia faba Plant Grown under Arsenic Stress. *Gesunde Pflanzen* **2016**, *68*, 99–107. [CrossRef]

76. El-Tayeb, M.A.; El-Enany, A.E.; Ahmed, N.L. Salicylic acid-induced adaptive response to copper stress in sunflower (*Helianthus annuus* L.). *Plant Growth Regul.* **2006**, *50*, 191–199. [CrossRef]

77. Kováčik, J.; Grúz, J.; Hedbávný, J.; Klejdus, B.; Strnad, M. Cadmium and Nickel Uptake Are Differentially Modulated by Salicylic Acid in Matricaria chamomilla Plants. *J. Agric. Food Chem.* **2009**, *57*, 9848–9855. [CrossRef]

78. Yu, M.; Shen, R.; Xiao, H.; Xu, M.; Wang, H.; Wang, H.; Zeng, Q.; Bian, J. Boron alleviates aluminum toxicity in pea (*Pisum sativum*). *Plant Soil* **2009**, *314*, 87–98. [CrossRef]

79. Wu, C.-y.; Lu, L.-l.; Yang, X.-e.; Feng, Y.; Wei, Y.-y.; Hao, H.-l.; Stoffella, P.J.; He, Z.-l. Uptake, Translocation, and Remobilization of Zinc Absorbed at Different Growth Stages by Rice Genotypes of Different Zn Densities. *J. Agric. Food Chem.* **2010**, *58*, 6767–6773. [CrossRef] [PubMed]

80. Zhou, W.; Leul, M. Uniconazole-induced alleviation of freezing injury in relation to changes in hormonal balance, enzyme activities and lipid peroxidation in winter rape. *Plant Growth Regul.* **1998**, *26*, 41–47. [CrossRef]

81. Yordanov, I.; Velikova, V.; Tsonev, T. Plant Responses to Drought, Acclimation, and Stress Tolerance. *Photosynthetica* **2000**, *38*, 171–186. [CrossRef]

82. Halliwell, B.; Gutteridge, J.M.C.; Aruoma, O.I. The deoxyribose method: A simple "test-tube" assay for determination of rate constants for reactions of hydroxyl radicals. *Anal. Biochem.* **1987**, *165*, 215–219. [CrossRef]

83. Vernon, L.P.; Seely, R.G. *The Chlorophylls*; Academic Press: New York, NY, USA, 1966.

84. Lichtenthaler, H.K. Chlorophylls and carotenoids: Pigments of photosynthetic biomembranes. *Methods Enzymol.* **1987**, *148*, 350–382.

85. Zhang, W.F.; Zhang, F.; Raziuddin, R.; Gong, H.J.; Yang, Z.M.; Lu, L.; Ye, Q.F.; Zhou, W.J. Effects of 5-aminolevulinic acid on oilseed rape seedling growth under herbicide toxicity stress. *J. Plant Growth Regul.* **2008**, *27*, 159–169. [CrossRef]

86. Velikova, V.; Yordanov, I.; Edreva, A. Oxidative stress and some antioxidant systems in acid rain-treated bean plants: Protective role of exogenous polyamines. *Plant Sci.* **2000**, *151*, 59–66. [CrossRef]

87. Aebi, H. Catalase in vitro. *Methods Enzymol.* **1984**, *105*, 121–126. [PubMed]

88. Nakano, Y.; Asada, K. Hydrogen peroxide is scavenged by ascorbate-specific peroxidase in spinach chloroplasts. *Plant Cell Physiol.* **1981**, *22*, 867–880. [CrossRef]

89. Jiang, M.; Zhang, J. Water stress-induced abscisic acid accumulation triggers the increased generation of reactive oxygen species and up-regulates the activities of antioxidant enzymes in maize leaves. *J. Exp. Bot.* **2002**, *53*, 2401–2410. [CrossRef] [PubMed]

90. Law, M.Y.; Charles, S.A.; Halliwell, B. Glutathione and ascorbic acid in spinach (*Spinacia oleracea*) chloroplasts. The effect of hydrogen peroxide and of Paraquat. *Biochem. J.* **1983**, *210*, 899–903. [CrossRef]

91. Bradford, M.M. A rapid and sensitive method for the quantitation of microgram quantities of protein utilizing the principle of protein-dye binding. *Anal. Biochem.* **1976**, *72*, 248. [CrossRef]

92. Bates, L.S.; Waldren, R.P.; Teare, I.D. Rapid determination of free proline for water-stress studies. *Plant Soil* **1973**, *39*, 205–207. [CrossRef]

plants

MDPI

Article

Exogenous Application of Methyl Jasmonate and Salicylic Acid Mitigates Drought-Induced Oxidative Damages in French Bean (*Phaseolus vulgaris* L.)

Mohammed Mohi-Ud-Din [1,*], Dipa Talukder [1], Motiar Rohman [2], Jalal Uddin Ahmed [1], S. V. Krishna Jagadish [3], Tofazzal Islam [4,*] and Mirza Hasanuzzaman [5,*]

[1] Department of Crop Botany, Bangabandhu Sheikh Mujibur Rahman Agricultural University, Gazipur 1706, Bangladesh; medipaone@gmail.com (D.T.); jahmed06@bsmrau.edu.bd (J.U.A.)
[2] Plant Breeding Division, Bangladesh Agricultural Research Institute, Gazipur 1701, Bangladesh; mrahman@bari.gov.bd
[3] Department of Agronomy, Kansas State University, Manhattan, KS 66506, USA; kjagadish@ksu.edu
[4] Institute of Biotechnology and Genetic Engineering (IBGE), Bangabandhu Sheikh Mujibur Rahman Agricultural University, Gazipur 1706, Bangladesh
[5] Department of Agronomy, Faculty of Agriculture, Sher-e-Bangla Agricultural University, Dhaka 1207, Bangladesh
* Correspondence: mmu074@bsmrau.edu.bd (M.M.-U.-D.); tofazzalislam@bsmrau.edu.bd (T.I.); mhzsauag@yahoo.com (M.H.)

Citation: Mohi-Ud-Din, M.; Talukder, D.; Rohman, M.; Ahmed, J.U.; Jagadish, S.V.K.; Islam, T.; Hasanuzzaman, M. Exogenous Application of Methyl Jasmonate and Salicylic Acid Mitigates Drought-Induced Oxidative Damages in French Bean (*Phaseolus vulgaris* L.). *Plants* 2021, 10, 2066. https://doi.org/10.3390/plants10102066

Academic Editor: Fermin Morales

Received: 27 August 2021
Accepted: 27 September 2021
Published: 30 September 2021

Publisher's Note: MDPI stays neutral with regard to jurisdictional claims in published maps and institutional affiliations.

Abstract: Drought stress impairs the normal growth and development of plants through various mechanisms including the induction of cellular oxidative stresses. The aim of this study was to evaluate the effect of the exogenous application of methyl jasmonate (MeJA) and salicylic acid (SA) on the growth, physiology, and antioxidant defense system of drought-stressed French bean plants. Application of MeJA (20 µM) or SA (2 mM) alone caused modest reductions in the harmful effects of drought. However, combined application substantially enhanced drought tolerance by improving the physiological activities and antioxidant defense system. The drought-induced generation of $O_2^{\bullet-}$ and H_2O_2, the MDA content, and the LOX activity were significantly lower in leaves when seeds or leaves were pre-treated with a combination of MeJA (10 µM) and SA (1 mM) than with either hormone alone. The combined application of MeJA and SA to drought-stressed plants also significantly increased the activities of the major antioxidant enzymes superoxide dismutase, catalase, peroxidase, glutathione peroxidase, and glutathione-S-transferase as well as the enzymes of the ascorbate–glutathione cycle. Taken together, our results suggest that seed or foliar application of a combination of MeJA and SA restore growth and normal physiological processes by triggering the antioxidant defense system in drought-stressed plants.

Keywords: abiotic stress; antioxidant defense; phytohormones; photosynthesis; pulse crop; water deficit

1. Introduction

French bean (*Phaseolus vulgaris* L.) is one of the world's most widely cultivated bean species [1]. For example, in Central and South America it accounts for 90% of total bean production [2]. It is a dual-purpose crop that is grown as pulse (grain) and consumed in the immature stage as a tender vegetable [3]. However, bean cultivation in approximately 60% of the regions around the world is affected by drought stress during different periods of crop growth [1,4]. Drought stress decreases plant growth and development by changing plant morphology and triggering variations in a suite of key physiological and biochemical processes [5]. Drought stress is known to increase leaf proline content and decrease leaf chlorophyll content, relative water content, stomatal conductance, cell membrane stability, and maximal efficiency of photosystem II by interrupting the electron transport chain and chloroplast integrity [6]. Drought stress during the bean flowering period decreases the number of pods per plant and the number of seeds per pod [7], thereby reducing productivity. Frequent, longer, and more severe droughts, accompanied by erratic rainfall, are expected in the 21st century across many regions of the world [8,9], including Bangladesh [10], due to the fact of climate change. This predicted increase in episodes of drought stress will constrain crop cultivation and productivity in the future.

Drought-stressed plants experience oxidative damage due to the generation of reactive oxygen species (ROS; singlet oxygen [$^1O_2$], superoxide radical [$O_2^{\bullet-}$], hydrogen peroxide [H_2O_2], and hydroxyl radical [$^\bullet OH$]). The ROS-induced damage to biomolecules is one of the major factors that limit plant growth under drought stress [11]. Increased ROS accumulation severely damages cell membrane integrity by accelerating lipid peroxidation [12,13], protein degradation, and nucleic acid damage [14]. Plants can mitigate the toxicity of ROS by activating a dynamic antioxidative defense system comprised of both non-enzymatic and enzymatic constituents [15,16]. The enzymatic components of this antioxidant defense mechanism contain superoxide dismutase (SOD), four enzymes of the ascorbate–glutathione cycle (i.e., ascorbate peroxidase (APX), monodehydroascorbate reductase (MDHAR), dehydroascorbate reductase (DHAR), and glutathione reductase (GR)), catalase (CAT), glutathione peroxidase (GPX), and glutathione s-transferase (GST) [15,17]. The non-enzymatic antioxidants are ascorbate (AsA), glutathione (GSH), tocopherol, flavanones, carotenoids, and anthocyanins, among others [16].

Exogenous application of cellular protectants, such as plant hormones, signaling molecules, and trace elements, is a popular approach in research aimed at enhancing abiotic stress tolerance [18–20]. For instance, methyl jasmonate (MeJA) and salicylic acid (SA) are the regulatory phytohormones that play a pivotal role in plant signaling responses to environmental cues through the orchestration of a myriad of defensive mechanisms [20–23]. The exogenous application of these two hormones together to mustard [18,24], grasses [25], jatropha [22], *Verbascum* [26], and maize [27] has been shown to protect plants against abiotic stresses, including drought, by regulating important physiological processes ranging from photosynthesis to nitrogen and proline metabolism and by activating the antioxidant defense system. Previous studies on wheat [28], maize [27], sweet potato [29], and Eureka lemon [30] have suggested that treatment with 5–20 µM MeJA and/or 0.5–2 mM SA effectively enhances the defense signaling routes to mitigate damage due to the abiotic stresses.

A number of reports have confirmed the protective role of MeJA and SA under abiotic stress conditions; however, few investigations have examined the combined application of these plant signaling molecules for mitigation of oxidative damage in plants under stressful conditions. Moreover, to the best of our knowledge, no studies have aimed to elucidate the impact of exogenous MeJA and SA, independently or in combination, on the amelioration of drought-induced oxidative stress in French bean. Recently, Tayyab et al. [27] demonstrated that the combined application of MeJA and SA significantly improved drought tolerance in maize seedlings. Therefore, the goal of this study was to examine how drought-stressed French bean plants respond to exogenous MeJA and SA, applied both separately and in combination, as seed and foliar pre-treatments. The findings from this study provide new mechanistic insights into the coordinated actions of MeJA and SA on the antioxidant defense system in French beans to enhance tolerance against drought-induced oxidative stress. The specific objectives of the study were (i) to determine the effects of exogenous application of MeJA and SA, separately and in combination, on the morpho-physiological traits of French beans and (ii) to evaluate the role of exogenous MeJA and SA and their combined application on the antioxidant defense system of French bean plants exposed to drought stress conditions.

2. Materials and Methods

2.1. Plant Material, Experimental Conditions, and Treatments

A popular, high yielding, and widely cultivated French bean variety, BARI Jharsheem-1, was used as the experimental material. Seeds were collected from the Vegetable Division, Horticulture Research Center, Bangladesh Agricultural Research Institute, Gazipur, Bangladesh. The experiment was carried out in a dry and cool environment (Table S1).

Methyl jasmonate (CAS #39924-52-2) and salicylic acid (CAS #69-72-7) were purchased from Sigma–Aldrich Chemie GmbH, Germany. Seeds were surface-sterilized with 1% (*v/v*) sodium hypochlorite for 10 min and rinsed thoroughly 3 times with sterile distilled water and then soaked in solutions of 20 μM MeJA and 2 mM SA for 18 h prior to sowing, while seeds for the combined application were soaked in an equal volume mixture of 10 μM MeJA and 1 mM SA [27]. Seeds for the control and drought groups were soaked in sterile water. Seeds were then sown in pots (18 cm in diameter × 23 cm in height) containing 13 kg of silt loam soil (sand 26%, silt 50%, and clay 24%), maintained at a full pot capacity of 30.6% volumetric soil water content. The soil was fertilized properly as recommended by Ahmmed et al. [31]. Ten healthy seeds were sown per pot, maintaining uniform spacing in each pot. Seven days after germination, seedlings were thinned to retain five uniform and healthy seedlings in each pot. Fifteen days after germination, the seedlings, which were previously soaked in the hormones, were again treated with MeJA (20 μM), SA (2 mM), and combined (10 μM MeJA + 1 mM SA) solutions as foliar spray. Another set of plants were sprayed with distilled water. Drought stress was imposed a day after the foliar spray by stopping irrigation until the end of the experiment, while the control pots were well watered. The experiment was a completely randomized design (CRD) with five treatments (control, drought (D), drought-stressed plants treated with methyl jasmonate (D + MeJA), salicylic acid (D + SA), and their combination (D + MeJA + SA)) with four replications.

2.2. Pot Capacity and Water Content of the Experimental Soil

The pot capacity of the soil used was measured prior to the start of the experiment gravimetrically using the procedure of Ogbaga et al. [32]. Briefly, pots of completely water-saturated soil were weighed and then dried to a constant weight at 105 °C. The differences between the weights of the water-saturated and oven-dried soils were used to determine the amount of water required to bring the pots to pot capacity, and volumetric water content (%) was determined accordingly. In addition, the water content (%) of the pot soils (at a 15 cm depth) was monitored daily using a digital soil moisture meter (PMS-714, Lutron Electronic Enterprise Co., Ltd., Taiwan, China).

2.3. Growth Parameters

At harvest (25 days after foliar application), root length (RL), shoot length (SL), root dry weight (RDW), shoot dry weight (SDW), leaf dry weight (LDW), and total dry weight (TDW) were calculated from 5 plants in each replication. Root and shoot length were determined from the root–shoot junction to the tip of the longest root and shoot, respectively, using a meter scale. Dry weights were weighed after drying the plant samples at 80 °C until a stable weight was reached.

2.4. Physiological Parameters

2.4.1. Leaf Chlorophyll and Pigment Content

At harvest, three 3rd fully expanded leaves from each replicate were used to estimate leaf SPAD value (chlorophyll index) and pigment contents. SPAD value was recorded just before the final harvest with a Chlorophyll Meter (Model: SPAD-502, Minolta Co., Ltd., Tokyo, Japan). Then, the leaves were harvested into Ziplock polybags and brought to the laboratory for pigment extraction. Leaf pigments were extracted in 80% (*v/v*) acetone and absorbance of the supernatant was determined with a UV-visible spectrophotometer (GENESYS 10S UV-VIS, Thermo Fisher Scientific, Waltham, MA, USA) at 663, 645, and 470 nm for Chl a, Chl b, and carotenoid content, respectively, and calculated according to Arnon [33].

2.4.2. Canopy Temperature Depression

Canopy temperature was measured on alternate days by a hand-held infrared thermometer (Model- MT4, HTC Instruments, Taipei, Taiwan, China; distance–spot ratio, 12:1). An angle of approximately 30° to the horizontal line and a distance of 30 cm from the 3rd fully opened leaf surface was maintained during measurement of the canopy temperature. Canopy temperature depression (CTD) was determined using the procedure of Fischer et al. [34] as ambient temperature minus leaf temperature. CTD was measured from five leaves in each replicate.

2.4.3. Leaf Relative Water Content

Leaf relative water content (LRWC) was estimated following the procedure of Meher et al. [35]. Briefly, 0.5 g leaf sample was immersed in 100 mL of distilled water for 4 h. The weights of turgid leaf samples were then measured and then oven-dried at 80 °C for 48 h. The dry weights of the samples were recorded until a constant weight was achieved. The procedure was repeated thrice for each replicate.

$$\text{LRWC (\%)} = [(\text{Fresh weight} - \text{Dry weight})/(\text{Turgid weight} - \text{Dry weight})] \times 100$$

2.4.4. Cell Membrane Stability

Cell membrane stability (CMS) was determined following the procedure of Sairam et al. [36]. Briefly, in two sets, 30 leaf discs (0.7 cm in diameter) were taken from three completely opened third leaves and put in test tubes containing 10 mL deionized water. One set of 15 leaf discs was incubated at 40 °C for 30 min and the second set at 100 °C in a boiling water bath for 15 min, and then their electrical conductivities, C_1 and C_2, respectively, were read with a conductivity meter. The process was repeated thrice for each replicate. CMS was calculated following the equation:

$$\text{CMS (\%)} = [1 - (C_1/C_2)] \times 100$$

2.5. Biochemical Observations

2.5.1. Proline Content

Leaf proline content was measured spectrophotometrically by an acid-ninhydrin method using the procedure outlined by Bates et al. [37]. The proline content was calculated using a standard curve and reported as μmol g^{-1} fresh weight.

2.5.2. Oxidative Stress Indicators

Generation of superoxide radicals ($O_2^{\bullet-}$) were measured following the method of Elstner and Heupel [38] with some modifications. Briefly, 0.3 g fresh leaf tissue was homogenized in 3 mL of 65 mM K–P buffer (pH 7.8) and centrifuged at 5000× *g* for 10 min. With 750 μL supernatant, 675 μL K–P buffer and 70 μL of 10 mM hydroxylamine hydrochloride were added, vortexed, and incubated at 25 °C for 20 min. To the mixture, 375 μL of 17 mM sulfanilamide, 37.5 μL α-naphthylamine, and 337.5 μL K–P buffer were added and vortexed. Then, 2.25 mL diethyl ether was added to the mixture, vortexed again, and incubated for 10 min. The absorbance of the upper clear fraction was recorded at 530 nm. The $O_2^{\bullet-}$ generation was calculated by comparing a standard curve of $NaNO_2^{-}$.

Fresh leaf tissues (0.5 g) were homogenized in 3 mL of 5% (*w/v*) trichloroacetic acid (TCA). After centrifugation at 11,500× *g* for 10 min, the supernatant was used to determine H_2O_2 and malondialdehyde (MDA). The H_2O_2 content was determined spectrophotometrically according to the procedure of Yang et al. [39] with some modifications. Briefly, 400 μL supernatant was added to 400 mL of 10 mM potassium phosphate buffer (pH 7.0) and 800 mL of 1 M potassium iodide (KI). The reaction was allowed to proceed in the dark for 1 h before measuring the absorbance at 390 nm. The H_2O_2 concentration was calculated using the extinction coefficient of 0.28 μM^{-1} cm^{-1}. The methods of Heath and Packer [40] and Mohi-Ud-Din et al. [41]

were followed for MDA determination. The MDA content was calculated using an extinction coefficient of 155 mM^{-1} cm^{-1} and represented as nmol g^{-1} FW).

2.5.3. Extraction and Quantitation of Soluble Protein

Fresh leaf tissue (1:2) (*w/v*) was extracted in 0.5 M potassium–phosphate (K–P) buffer (pH 7.0) in ice-cold mortar. The extraction buffer included 1 mM ascorbic acid, 1 M KCl, β-mercaptoethanol, and glycerol. The homogenate was centrifuged at 11,500× *g* for 15 min, and the supernatant was used for enzymatic activities assays. Bradford's [42] rapid quantification method was used to determine the protein content of each enzyme solution.

2.5.4. Assays of Enzymatic Activities

The lipoxygenase (LOX, EC: 1.13.11.12) activity was assayed spectrophotometrically at 234 nm as per Doderer et al. [43] using linoleic acid as a substrate. The activity was calculated using an extinction coefficient of 25,000 M^{-1} cm^{-1}. The superoxide dismutase (SOD, EC: 1.15.1.1) activity was assayed based on the inhibition method of Spitz and Oberley [44]. The catalase (CAT, EC: 1.11.1.6) activity was measured at 240 nm according to the method of Noctor et al. [45] and calculated using an extinction coefficient of 39.4 M^{-1} cm^{-1}.

The guaiacol peroxidase (POD, EC: 1.11.1.7) activity was quantified as per the description of Castillo et al. [46] at 470 nm after 1 min and calculated considering an extinction coefficient of 26.6 mM^{-1} cm^{-1}. The activity of glutathione peroxidase (GPX, EC: 1.11.1.9) was measured following Elia et al. [47] at 340 nm for 1 min. An extinction coefficient of 6.62 mM^{-1} cm^{-1} was used to compute the activity. The glutathione *S*-transferase (GST, EC: 2.5.1.18) activity was measured spectrophotometrically by following the method of Hossain et al. [48] with model substrate 1-chloro-2,4-dinitrobenzene (CDNB) at 340 nm. The activity was calculated using an extinction coefficient of 9.6 mM^{-1} cm^{-1}.

The activities of glutathione reductase (GR, EC: 1.6.4.2), ascorbate peroxidase (APX, EC: 1.11.1.11), monodehydroascorbate reductase (MDHAR, EC: 1.6.5.4), and dehydroascorbate reductase (DHAR, EC: 1.8.5.1) were measured according to Noctor et al. [45] at 340, 290, 340, and 265 nm, respectively. For all enzymes, absorbance changes were observed for 1 min, and extinction coefficients of 6.2 mM^{-1} cm^{-1}, 2.8 mM^{-1} cm^{-1}, 6.2 mM^{-1} cm^{-1}, and 14 mM^{-1} cm^{-1} were used for the quantification of GR, APX, MDHAR, and DHAR, respectively.

2.6. Statistical Analysis

The data were analyzed using Statistix 10 (https://www.statistix.com/) (accessed on 26 May 2021). The least significant difference (LSD) test was used to compare the means at the $p \leq 0.05$ significance level. R v.4.1.0 for Windows was used to create a heatmap and to perform principal component analysis (PCA) (http://CRAN.R-project.org/) (accessed on 26 May 2021). Trait mean values were normalized and the library pheatmap was adapted to generate heatmap and hierarchical clusters (distance = Euclidean and method = ward.D2) [49]. PCA was performed with the packages ggplot2, factoextra, and FactoMineR [50,51].

3. Results

3.1. Soil Water Content

By the 25th day of treatment, visible effects, including reduced plant height and appearance, increased droopiness, chlorosis, and wilting of leaves, were observed in the French bean plants. Plants were harvested on that day to record observations and conduct chemical analyses (Figure 1). The silt loam used in this study gave a pot capacity of 30.6% volumetric soil water content, determined gravimetrically prior to the start of the experiment (Figure 2). The soil water content was considerably lower for the drought-stressed pots than in the control throughout the experiment (Figure 2). At plant harvest, the soil water contents recorded from the control, drought (D), D + MeJA, D + SA, and D + MeJA + SA treatments were equivalent to 59%, 24%, 26%, 25%, and 25% of pot capacity, respectively (Figure 2), confirming that the symptoms observed in the drought-stressed French bean plants were indicative of water deficit.

Figure 1. Phenotypic appearance of French bean plants: control, drought (D), and drought-stressed plants treated with methyl jasmonate (D + MeJA), salicylic acid (D + SA), and their combination (D + MeJA + SA) at the time of harvest. Control—plants grown under non-stress, well-irrigated conditions; drought—plants grown with a steady decline in moisture availability; D + MeJA—drought-stressed plants pre-treated (seed and foliar) with 20 μM methyl jasmonate; D + SA— drought-stressed plants pre-treated (seed and foliar) with 2 mM salicylic acid; D + MeJA + SA—drought-stressed plants pre-treated (seed and foliar) with a combination of 10 μM methyl jasmonate and 1 mM salicylic acid.

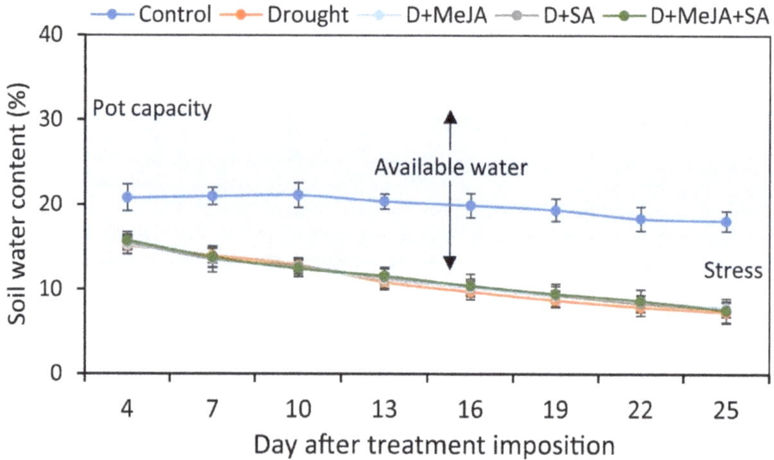

Figure 2. Soil water status of the control, drought (D), and drought-stressed plants treated with methyl jasmonate (D + MeJA), salicylic acid (D + SA), and their combination (D + MeJA + SA) throughout the experimental period. Soil water content was recorded daily for every pot. Each data point indicates the 3-day running average values across four replicates. Vertical bars represent +/− values of the mean. The experimental soil was silt loam (clay:silt:sand = 24:50:26) with a full pot water capacity of 30.6% volumetric soil water content. (Adapted from Pardossi et al. [52]; Weng and Luo [53]).

3.2. MeJA and SA Enhanced Plant Growth

Drought stress decreased the root and shoot lengths of the French bean plants by 5% and 42%, respectively, and the decrease in shoot length was statistically significant (Table 1, Figure S1A). The shoot and root lengths were longer in the drought-stressed plants given hormone treatments than in the drought-stressed plants without hormone treatments (Table 1). Pretreatment of drought-stressed plants with MeJA or SA alone did not result in statistically significant increases in shoot length compared to drought, but the combined (MeJA + SA) application significantly increased the shoot lengths by 20% (Figure S1B). Application of MeJA and SA also improved the plant phenotypic appearance under drought conditions (Figure 2).

Table 1. Effect of methyl jasmonate (MeJA), salicylic acid (SA), and their combined application on the growth parameters of French beans under drought stress.

Treatment	Root Length (cm)	Shoot Length (cm)	Root Dry Weight (g)	Shoot Dry Weight (g)	Leaf Dry Weight (g)	Total Dry Weight (g)
Control	58.63 ± 3.45 [a]	36.49 ± 0.87 [a]	2.88 ± 0.26 [a]	9.04 ± 0.74 [a]	9.93 ± 0.77 [a]	21.86 ± 1.62 [a]
Drought (D)	55.69 ± 1.23 [a]	21.27 ± 0.65 [c]	1.38 ± 0.12 [b]	2.77 ± 0.21 [c]	2.61 ± 0.28 [d]	6.76 ± 0.35 [c]
D + MeJA	57.59 ± 2.13 [a]	22.10 ± 0.49 [c]	1.53 ± 0.17 [b]	3.55 ± 0.19 [c]	3.91 ± 0.37 [bc]	8.99 ± 0.64 [c]
D + SA	58.91 ± 2.55 [a]	22.46 ± 0.63 [c]	1.54 ± 0.18 [b]	3.69 ± 0.49 [c]	3.62 ± 0.13 [cd]	8.85 ± 0.65 [c]
D + MeJA + SA	59.81 ± 1.07 [a]	25.63 ± 0.55 [b]	1.86 ± 0.19 [b]	5.09 ± 0.28 [b]	4.89 ± 0.12 [b]	11.84 ± 0.23 [b]

Values represent the mean ± SE. Values in a column with distinct letter(s) were significantly different at $p \leq 0.05$. Control—plants grown under non-stress, well-irrigated conditions; drought—plants grown with a steady decline in moisture availability; D + MeJA—drought-stressed plants pre-treated (seed and foliar) with 20 µM methyl jasmonate; D + SA—drought-stressed plants pre-treated (seed and foliar) with 2 mM salicylic acid; D + MeJA + SA—drought-stressed plants pre-treated (seed and foliar) with a combination of 10 µM methyl jasmonate and 1 mM salicylic acid.

Drought stress resulted in a substantial decrease in dry weights. Compared to the unstressed control plants, the root, shoot, leaf, and total dry weights decreased by 52%, 69%, 74%, and 69%, respectively, due to the drought stress (Table 1 and Figure S1A). Exogenous use of MeJA and SA, both alone and in combination, increased the dry weights under the drought-stressed condition, but the increases induced by the combined treatment, except for the root dry weight, were statistically significant. Increases of approximately 35%, 84%, 87%, and 75% in root, shoot, leaf, and total dry weight, respectively, were recorded in the combined treatment compared to drought stress alone (Figure S1B).

3.3. Impact of MeJA and SA on Photosynthetic Pigments

The non-destructive chlorophyll index (SPAD) value significantly decreased (20%) under drought stress compared to unstressed control plants (Table 2, Figure S1A). Single and combined applications of MeJA and SA increased the SPAD value of drought-stressed plants compared to untreated drought-stressed plants, but the increase due to the combined treatment (22%) was highly statistically significant (Figure S1B).

Table 2. Effect of methyl jasmonate (MeJA), salicylic acid (SA), and their combination on SPAD value and leaf pigment contents of French bean plants under drought stress.

Treatments	SPAD	Leaf Pigments (mg g^{-1} Fresh Weight)			
		Chl a	Chl b	Total Chl	Carotenoids
Control	49.85 ± 0.49 [a]	1.39 ± 0.03 [a]	0.49 ± 0.02 [a]	1.97 ± 0.06 [a]	0.50 ± 0.03 [a]
Drought (D)	39.65 ± 1.18 [c]	0.65 ± 0.10 [c]	0.24 ± 0.05 [c]	0.98 ± 0.07 [c]	0.26 ± 0.01 [c]
D + MeJA	43.65 ± 0.93 [b]	0.87 ± 0.11 [bc]	0.29 ± 0.04 [bc]	1.25 ± 0.17 [bc]	0.34 ± 0.05 [bc]
D + SA	44.35 ± 1.52 [b]	0.91 ± 0.16 [bc]	0.30 ± 0.07 [bc]	1.30 ± 0.25 [bc]	0.33 ± 0.06 [bc]
D + MeJA + SA	48.18 ± 0.94 [a]	1.12 ± 0.06 [b]	0.39 ± 0.02 [ab]	1.61 ± 0.08 [ab]	0.42 ± 0.03 [ab]

Values represent the mean ± SE. Values in a column with distinct letter(s) were significantly different at $p \leq 0.05$. Additional details are listed in Table 1.

Drought stress significantly decreased the leaf levels of chlorophyll a, chlorophyll b, total chlorophyll, and carotenoids by 53%, 51%, 50%, and 48%, respectively, compared to the control (Table 2 and Figure S1A). The exogenous applications of MeJA and SA, separately or in combination, lessened the negative effects of drought stress. However, the increases in pigment levels following application of MeJA or SA alone were not statistically significant. On the contrary, the combined application of MeJA and SA significantly increased chlorophyll a, chlorophyll b, total chlorophyll, and carotenoid contents by 72%, 63%, 64%, and 62%, respectively, compared to the untreated drought-stressed plants (Figure S1B).

3.4. MeJA and SA Improved Physiological Traits

A profound and significant decrease in the mean canopy temperature depression (CTD) was recorded due to the drought stress. The mean CTD decreased by 67% in the drought-stressed plants compared to the control (Figures 3B and S1A), whereas the drought-stressed plants treated with MeJA, SA, or MeJA + SA showed significant increases in mean CTD (73%, 83%, and 127%, respectively) compared to the untreated drought-stressed plants (Figure S1B). Plants treated with the combination of MeJA and SA maintained a greater CTD throughout the course of the experiment compared to plants treated with either MeJA or SA alone (Figure 3A,B).

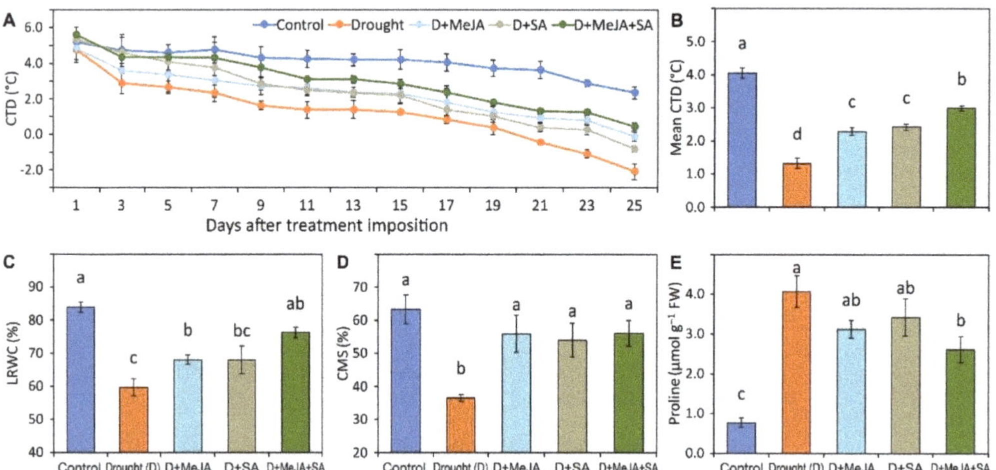

Figure 3. (**A**) Canopy temperature depression (CTD) throughout the experimental period; (**B**) mean CTD; (**C**) leaf relative water content (LRWC); (**D**) cell membrane stability; (**E**) proline content of French bean plants grown under control, drought (D), and drought-stressed plants treated with methyl jasmonate (D + MeJA), salicylic acid (D + SA), and their combination (D + MeJA + SA). Vertical bars represent +/− SE values. Different letter(s) denote a significant difference at $p \leq 0.05$. FW—Fresh weight. Additional details are shown in Figure 1.

The leaf relative water content (LRWC) significantly decreased (29% over the control) by drought stress (Figures 3C and S1A). The drought-stressed plants treated with MeJA and SA and their combination showed increases of 14%, 14%, and 28%, respectively, compared to untreated drought-stressed plants; however, the increase was significant only for the MeJA and the combined hormone treatments (Figure S1B). The cell membrane stability (CMS) of leaves decreased by 42% under drought stress compared to control plants (Figures 3D and S1A). Exogenously applied MeJA and SA mitigated this effect by increasing the CMS by 52% and 47%, respectively, compared to drought-stressed plants (Figure S1B). The protective effect of MeJA + SA was similar to that observed with application of either hormone alone.

Drought stress triggered a profound increase of 436% in the proline content in French bean leaves (Figures 3E and S1A). Application of MeJA, SA, and MeJA + SA to drought-stressed plants lowered the proline content by 23%, 16%, and 36%, respectively, compared to the drought-stressed plants (Figure S1B).

3.5. MeJA and SA Suppressed the Generation of Oxidative Stress Indicators

Oxidative stress due to the drought stress in French bean plants was determined by measurements of the $O_2^{\bullet-}$ generation rate, the H_2O_2 and MDA levels, and the LOX activity. A sharp increase in $O_2^{\bullet-}$ generation (270%) was observed in drought-stressed plants compared to unstressed controls (Figures 4A and S2A). Application of MeJA, SA, and MeJA + SA significantly lowered (31%, 36%, and 51%, respectively) $O_2^{\bullet-}$ generation (Figures 4A and S2B). The H_2O_2 levels also significantly increased by 55% in drought-stressed plants compared to the unstressed controls (Figures 4B and S2A). Treatment with MeJA, SA, and MeJA + SA lowered the H_2O_2 levels by 14%, 13%, and 19%, respectively, compared to untreated drought-stressed plants (Figure S2B).

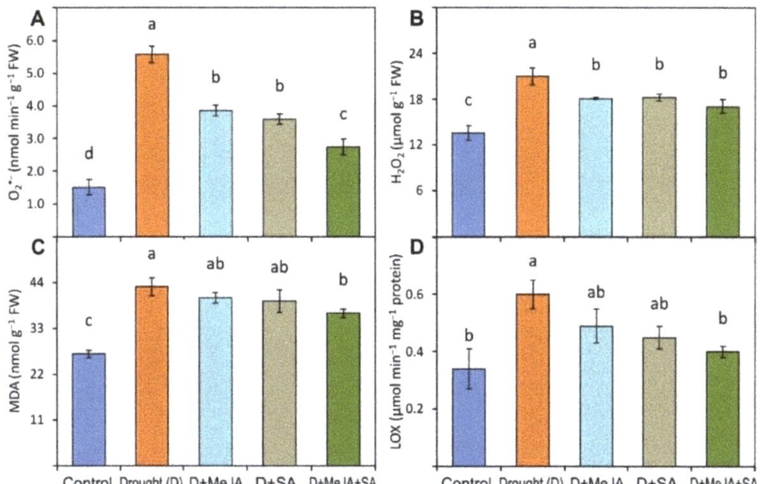

Figure 4. (**A**) superoxide ($O_2^{\bullet-}$) generation rate; (**B**) hydrogen peroxide content (H_2O_2); (**C**) malondi-aldehyde content (MDA); (**D**) lipoxygenase (LOX) activity of French bean plants grown under control, drought (D), and drought-stressed plants treated with methyl jasmonate (D + MeJA), salicylic acid (D + SA), and their combination (D + MeJA + SA). Vertical bars represent +/− SE values. Different letter(s) denote a significant difference at $p \leq 0.05$. FW—Fresh weight. Additional details are shown in Figure 1.

Compared to unstressed controls, the levels of MDA and the LOX activity increased by 60% and 76%, respectively, due to the drought stress (Figures 4C,D and S2A). However, MeJA, SA, and MeJA + SA treatments restrained the production of MDA and the LOX activity in drought-stressed plants. Combined MeJA + SA reduced MDA content and LOX activity by 15% and 33%, respectively.

3.6. Antioxidant Enzyme Activities and MeJA and SA Pre-Treatment

Drought stress promoted a substantial increase in SOD activity by 54% compared to the unstressed control (Figures 5A and S2A). Exogenous application of MeJA or SA further enhanced SOD activity compared to untreated drought-stressed plants, while the combined MeJA + SA treatment caused a statistically significant 83% rise in SOD activity compared to the untreated drought-stressed plants (Figure S2B). Unlike SOD, the activities of CAT and POD substantially decreased (42% and 54%, respectively) in the drought-stressed plants compared to the controls (Figures 5B,C and S2A). Application of MeJA or SA alone to drought-stressed plants did not show any remarkable effect on these enzyme activities compared to the activities in the untreated drought-stressed plants; however, the combined MeJA + SA treatment significantly increased the activities of CAT and POD (79% and 144%, respectively) compared to activities in the untreated drought-stressed plants (Figure S2B).

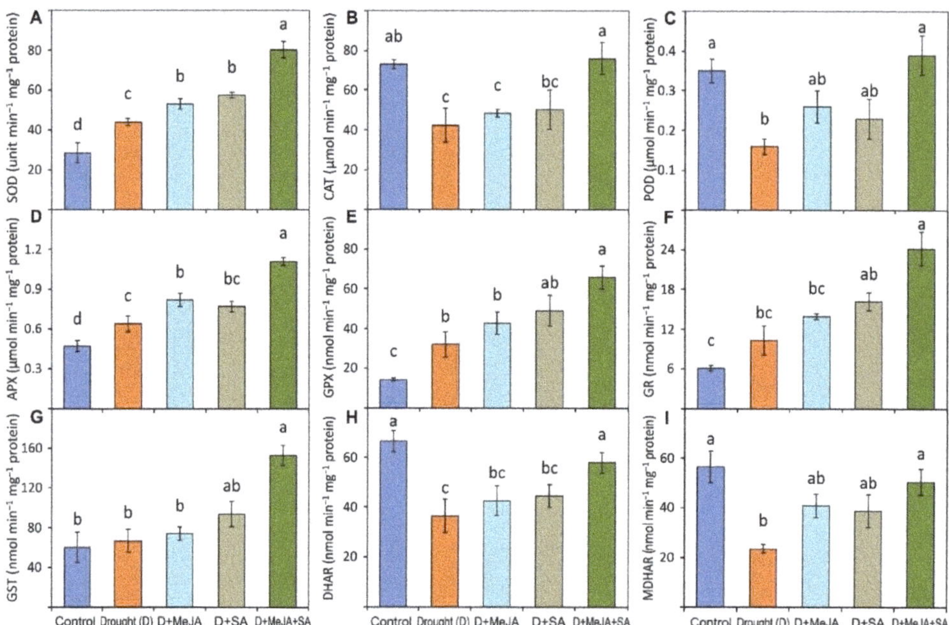

Figure 5. Specific activity of (**A**) superoxide dismutase (SOD); (**B**) catalase (CAT); (**C**) peroxidase (POD); (**D**) ascorbate peroxidase (APX); (**E**) glutathione peroxidase (GPX); (**F**) glutathione reductase (GR); (**G**) glutathione-S-transferase (GST); (**H**) dehydroascorbate reductase (DHAR); (**I**) monodehydroascorbate reductase (MDHAR) of French bean plants grown under control, drought (D), and drought-stressed plants treated with methyl jasmonate (D + MeJA), salicylic acid (D + SA), and their combination (D + MeJA + SA). Vertical bars represent +/− SE values. Different letter(s) denote a significant difference at $p \leq 0.05$. Additional details are shown in Figure 1.

A slight increase was noted in APX and GST activities in the drought-stressed plants (36% and 11%, respectively) compared to the unstressed control (Figures 5D,G and S2A). Application of SA alone to drought-stressed plants did not change the APX and GST activities substantially, while MeJA significantly increased the activity of APX only (Figure 5D,G). However, the combined application of MeJA + SA significantly increased the APX and GST activities (73% and 128%) compared to untreated drought-stressed plants (Figures 5D,G and S2B). A marked increase in GPX activity (126%) was recorded in the drought-stressed plants compared to the unstressed controls, but no significant increases were noted for GR activity (Figures 5E,F and S2A). Treatment with MeJA or SA alone did not cause any marked increases, whereas the combined MeJA + SA treatment significantly increased both GPX and GR activities (105% and 134%, respectively) compared to the untreated drought-stressed plants (Figure S2B).

Compared to the unstressed controls, the activities of DHAR and MDHAR decreased (45% and 58%, respectively) in the drought-stressed plants (Figures 5H,I and S2A). Treatment of drought-stressed plants with MeJA or SA alone did not significantly increase these enzyme activities, whereas the combined MeJA + SA treatment resulted in a statistically significant increase in the activities of both DHAR and MDHAR by 59% and 115%, respectively, compared to the untreated drought-stressed plants (Figures 5H,I and S2B).

3.7. Comparative Assessment of Responses across Treatments

A comparative heatmap analysis revealed three distinct groups among the parameters measured in this study (Figure 6). Group 1 consisted of most of the antioxidant enzymes (SOD, APX, GPX, GR, and GST), root length, and CMS. Oxidative stress indicators and proline content were placed in group 3, and the rest of the parameters, including CAT, POD, DHAR, and MDHAR, were in group 2. A distinct categorization of the growth, physiological processes, and antioxidant scavenging machinery was observed, as drought-stressed plants without pre-treatment and with pre-treatment with either MeJA or SA alone were grouped in a cluster (Figure 6). However, the plants given the combined MeJA + SA treatment showed significantly higher amelioration of drought stress-induced damage and were grouped in the same cluster as the unstressed control plants (Figure 6). A contrasting response between the unstressed controls and the plants receiving the combined MeJA + SA treatment was observed for the ROS scavenging machinery, whereas the ROS levels and other growth and physiological responses were aligned in the same direction in both plant groups (Figure 6).

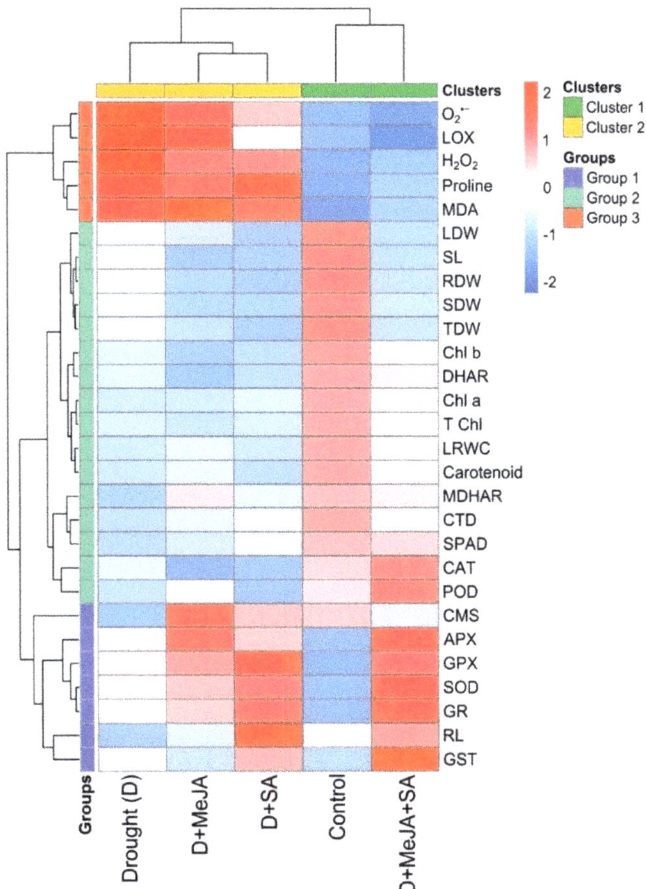

Figure 6. Heatmap and cluster analysis of the growth and physiological attributes, oxidative stress indicators, and antioxidant enzymes in French bean plants grown under control, drought (D), and drought-stressed plants treated with methyl jasmonate (D + MeJA), salicylic acid (D + SA), and their combination (D + MeJA + SA). Additional details are shown in Figure 1.

3.8. Principal Component Analysis (PCA)

The PCA of the responses of the French bean plants to different treatments depicted a total of four principal components (PCs), but only two PCs exhibited eigenvalues > 1 and were significant. Both the PCs explained approximately 97% of the variability in drought response (Figure 7 and Table S2). A PCA biplot showed that PC1 exhibited approximately 75% of the total variability and contributed positively via morpho-physiological traits, CAT, POD, DHAR, and MDHAR; it contributed negatively via proline, SOD, H_2O_2, MDA, and

LOX (Figure 7, Table S2). The second PC accounted for approximately 22% of the total variation and contributed principally by SOD, APX, GPX, GR, GST, and RL and partly by SPAD, CAT, POD, and MDHAR.

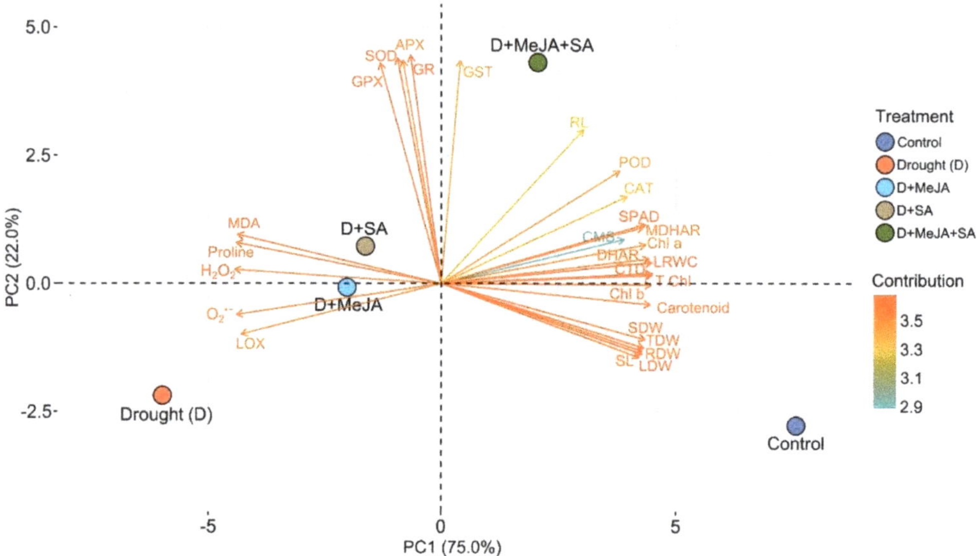

Figure 7. A PCA biplot stipulates the relationship between treatments and observed parameters. Parameters are distributed in different ordinates depending on the divergence among them. In this biplot, the length and color intensity of a vector indicate the quality of representation and the contribution of the traits to the principal components, respectively. The angles between the vectors formed from the middle point of the biplots indicate whether the traits interacted positively or negatively. Additional details are shown in Figure 1.

4. Discussion

The findings from this study revealed how exogenous application of MeJA, SA, and their combination can help to reduce the adverse effects of drought in French bean by maintaining cellular water levels, membrane stability, photosynthetic pigment levels, and antioxidant defenses to ultimately improve morpho-physiological growth. The growth parameters of French bean (shoot and root lengths and dry weights) were reduced under drought stress but were less affected following application of MeJA and SA, separately or in combination. An improvement in drought stress tolerance by the combined application of MeJA and SA has recently been demonstrated in maize by Tayyab et al. [27]; however, the present report is the first to demonstrate similar effects in French beans by significant activation of the plant antioxidant defense system. Previous studies have demonstrated an effective minimization of the damaging effects of drought by exogenous application of MeJA or SA alone in mustard [18] and wheat [28] and by the combination of MeJA and SA in maize [27] through enhancement of various biochemical, morphological, and physiological responses [54] including cell elongation, cell expansion, and cell differentiation [55].

The chlorophyll index, including Chl a, b, and total Chl, and the carotenoid content measured in the same leaves were reduced by drought stress (Table 2 and Figure S1A). Drought stress is known to lower the levels of photosynthetic pigments, including chlorophylls and carotenoids, in various crop plants due to the oxidation and impaired biosynthesis of the pigments [56,57]. The observed decrease in chlorophyll concentration under drought stress could represent a suppression of chlorophyll biosynthesis and/or decreased synthesis and assembly of the PSI and PSII light-harvesting complexes to suppress excess absorption and ROS production [6,58]. The negative impacts on pigments were partly ameliorated in French bean plants treated with MeJA or SA, but the effects were consistently statistically significant for the combined MeJA + SA treatment, and the plant growth and physiology were similar to those of the unstressed control plants (Table 2 and Figure S1B). Increases in leaf chlorophyll and carotenoid contents in drought-stressed plants treated with MeJA, SA, and their combination have been reported previously in soybean [59], soybean [60], and maize [27], respectively.

Canopy temperature depression, LRWC, and CMS are considered effective indicators of drought stress tolerance [6,61–63]. Our findings provide support for earlier results showing that French beans under drought stress have significantly lower CTD, LRWC, and CMS; however, these reductions were suppressed by the application of MeJA or SA and statistically significantly suppressed by the combined MeJA + SA treatment Figures 3 and S1B). CTD is regarded as a reliable indication of plant water status, and a positive CTD means the canopy is cooler than the surrounding air [64].

An improved LRWC was reported in drought-stressed plants following MeJA treatment of soybean [59], SA treatment of soybean [60] and *Ctenanthe setosa* [65], and a combined MeJA and SA treatment of maize [27]. The enhanced CTD, LRWC, and CMS in the drought-stressed French bean plants treated with MeJA and SA in the present study indicates that these hormones, particularly when supplied in combination, have the potential to reduce the harmful effects of drought by maintaining cellular water status and maintaining a cooler leaf temperature.

Proline has been considered an osmotic stress mediator, stabilizer of macromolecules, compatible solutes to preserve enzymes, and capable of storing carbon and nitrogen for usage during drought and other stresses [66]. Proline accumulation in plants is also a marker of stress induction [67]. Although proline content profoundly increased under drought stress in French beans, the MeJA, SA, and MeJA + SA treatments markedly reduced the proline content in drought-stressed plants (Figures 3E and S1B). This was due to the reduction in osmotic stress following the application of hormones [18,24,26].

The effect of exogenous MeJA and SA on the drought-induced oxidative stress was investigated by determining $O_2^{\bullet-}$ generation and the H_2O_2 and MDA levels as well as the specific activity of LOX. Drought stress significantly increased the levels of $O_2^{\bullet-}$, H_2O_2, and MDA and LOX activity compared to the unstressed control plants, indicating that the plants were under severe oxidative stress (Figures 4 and S2A). The drought-induced increases in oxidative stress indicators in our study were consistent with results from previous studies on soybean [60], mustard [24], and maize [27]. Exogenous application of MeJA or SA suppressed the production of the oxidative stressors, while MeJA + SA showed statistically significant effects in drought-stressed French bean plants Figures 4 and S2B). These findings are in accordance with previous studies showing that combined treatments with MeJA and SA reduced the levels of oxidative stress indicators in maize seedlings under drought [27] and in Eureka lemon under chilling stress [30].

In plant cells, SOD renders key protection against $O_2^{\bullet-}$ by converting it to H_2O_2 and, subsequently, neutralizing it to H_2O by CAT and peroxidases (POD, APX, and GPX) [15,16,68]. The plant GSTs are a large and diverse group of enzymes that catalyze the coupling of electrophilic xenobiotic substrates with GSH and are linked with the induction of tolerance to different abiotic stresses [69]. The four enzymes of the ascorbate–glutathione cycle (i.e., APX, MDHAR, DHAR, and GR) played a pivotal role in the systematic detoxification of cellular H_2O_2 produced due to the oxidative stress in wheat [41].

In the present study, increased activity of SOD, APX, GPX, GST, and GR and decreased activity of CAT, POD, DHAR, and MDHAR were observed in drought-stressed plants compared with unstressed the control plants (Figure 5), which is in agreement with earlier findings of similar drought-stress effects in rapeseed [70], mustard [18], and mung bean [71]. Exogenous MeJA, SA, and their combination augmented the antioxidant enzyme activities in French beans under drought stress and reduced drought-induced oxidative damage (Figures 5 and S2B). However, MeJA (20 µM) and SA (2 mM) applied independently did not consistently improve the antioxidant enzyme system under drought; this may be because of a reliance on other stress signals that can activate antioxidant enzymes' activities [55]. MeJA or SA induced an upregulation of the antioxidant enzyme system under drought stress in *Ctenanthe setosa* [65], *Cucumis melo* [72], mustard [18,24], jatropha [22], and *Verbascum sinuatum* [26]. However, a combined MeJA and SA treatment (10 μM + 1 mM) was more effective in this upregulation of the antioxidant enzymes (Figures 5 and S2B). These results are in agreement with Tayyab et al. [27], who observed a synergistically positive role of MeJA + SA in increasing antioxidant enzyme activity under drought-induced oxidative stress in maize seedlings. Our results suggest that the signaling routes of MeJA and SA would operate similarly to trigger defensive responses in French bean plants under drought stress. However, investigating additional mechanisms that lead to an increased effectiveness by combined hormone applications on minimizing drought stress is an intriguing area for future research.

Taken together, our results demonstrated that MeJA or SA reduced oxidative stress by upregulating antioxidant enzyme activities, improving the physiological functions of French bean plants under drought stress; however, the alleviation was further boosted by combined hormone application. Heatmap and PCA biplot analyses clearly showed that the combined application of MeJA and SA mitigated drought-induced oxidative stress principally by the upregulation of SOD, APX, GPX, GR, and GST and partly by CAT, POD, and MDHAR activities (Figures 6 and 7). The results presented in this study add new insights into oxidative stress defense mechanisms triggered by MeJA and SA in French bean plants.

5. Conclusions

French beans pre-treated with exogenous MeJA and SA showed reductions in drought stress responses including improved photosynthetic performance, membrane stability, water status, leaf temperature control, and antioxidant enzyme activities. Interestingly, the combined application of MeJA (10 μM) and SA (1 mM) was more efficacious at alleviating the adverse effects of drought-induced oxidative stress by upregulating antioxidant enzymatic activities. The combined treatment also improved the physiological activities and increased plant biomass compared to drought-stressed plants. Seed and foliar treatment of French bean with a combination of MeJA and SA significantly improved drought stress tolerance by augmenting antioxidant systems. The findings from this study provide the rationale for future research related to the defense signaling pathways employed by French bean plants pre-treated with MeJA and SA to aid in the development of new cultivars with improved adaptation to future drier scenarios.

Supplementary Materials: The following are available online at https://www.mdpi.com/article/10.3390/plants10102066/s1, Table S1: Daily weather data at the experimental site during the study period, Table S2: Extracted Eigenvalues and latent vectors of studied traits associated with the first two principal components, Figure S1: Variations in the studied growth and physiological parameters of French bean plants. A. Percent change due to drought over control and B. percent change over drought due to the application of methyl jasmonate (D + MeJA), salicylic acid (D + SA) and their combination (D + MeJA + SA) on drought-stressed plants, and Figure S2: Variations in the oxidative stress indicators and antioxidant enzymes activity of French bean plants. A. Percent change due to drought over control and B. percent change over drought due to the application of methyl jasmonate (D + MeJA), salicylic acid (D + SA) and their combination (D + MeJA + SA) on drought stressed plants.

Author Contributions: M.M.-U.-D., D.T., J.U.A. and M.R. contributed to the conception of the work, data acquisition, and interpretation of the data; M.M.-U.-D. and D.T. performed the experiments and contributed to the data analysis; M.R. contributed the reagents and materials; M.M.-U.-D. and M.H. contributed to the graphical presentation of the data; M.M.-U.-D., D.T., J.U.A., T.I. and M.H. drafted the manuscript; S.V.K.J., T.I. and M.H. contributed to the critical revision and finalization of the manuscript. All authors have read and agreed to the published version of the manuscript.

Funding: This research was partially funded by the contribution number 21-323-J from the Kansas Agricultural Experiment Station, Kansas State University, Manhattan, KS 66506, USA.

Institutional Review Board Statement: Not applicable.

Informed Consent Statement: Not applicable.

Data Availability Statement: All relevant data are available in this manuscript.

Acknowledgments: We extend our thanks to the Molecular Breeding Laboratory, Bangladesh Agricultural Research Institute, Gazipur, Bangladesh, for providing lab facilities during the research work.

Conflicts of Interest: The authors declare no conflict of interest.

References

1. Graham, P.H.; Ranalli, P. Common bean (*Phaseolus vulgaris* L.). *Field Crop. Res.* **1997**, *53*, 131–146. [CrossRef]
2. Rosales-Serna, R.; Kohashi-Shibata, J.; Acosta-Gallegos, J.A.; Trejo-Lopez, C.; Ortizcereceres, J.; Kelly, J. Biomass distribution, maturity acceleration and yield in drought-stressed common bean cultivars. *Field Crops Res.* **2004**, *85*, 203–211. [CrossRef]
3. Nazrul, M.I.; Shaheb, M.R. Performance of French bean (*Phaseolus vulgaris* L.) genotypes in Sylhet region of Bangladesh. *Bangladesh Agron. J.* **2016**, *19*, 37–44. [CrossRef]
4. Assefa, T.; Wu, J.; Beebe, S.E.; Rao, I.M.; Marcomin, D.; Claude, R.J. Improving adaptation to drought stress in small red common bean: Phenotypic differences and predicted genotypic effects on grain yield, yield components and harvest index. *Euphytica* **2015**, *203*, 477–489. [CrossRef]
5. Jaleel, C.A.; Manivannan, P.; Wahid, A.; Farooq, M.; Somasundaram, R.; Panneerselvam, R. Drought stress in plants: A review on morphological characteristics and pigments composition. *Int. J. Agric. Biol.* **2009**, *11*, 100–105.
6. Ghobadi, M.; Taherabadi, S.; Ghobadi, M.-E.; Mohammadi, G.-R. Jalali-Honarmand, S. Antioxidant capacity, photosynthetic characteristics and water relations of sunflower (*Helianthus annuus* L.) cultivars in response to drought stress. *Ind. Crop. Prod.* **2013**, *50*, 29–38. [CrossRef]
7. Lesznyák, M.; Hunyadi-Borbély, É.; Csajbók, J. The role of nutrient-water-supply and the cultivation in the yield of pea (*Pisum sativum* L.). *Cereal Res. Commun.* **2008**, *36*, 1079–1082.
8. Trenberth, K.E.; Dai, A.; Van Der Schrier, G.; Jones, P.D.; Barichivich, J.; Briffa, K.R.; Sheffield, J. Global warming and changes in drought. *Nat. Clim. Chang.* **2014**, *4*, 17–22. [CrossRef]
9. Schwalm, C.R.; Anderegg, W.R.L.; Michalak, A.M.; Fisher, J.B.; Biondi, F.; Koch, G.; Litvak, M.; Ogle, K.; Shaw, J.D.; Wolf, A.; et al. Global patterns of drought recovery. *Nature* **2017**, *548*, 202–205. [CrossRef]
10. MoEF. *Bangladesh Climate Change Strategy and Action Plan 2009*; Ministry of Environment and Forests: Dhaka, Bangladesh, 2009.
11. Jain, M.; Kataria, S.; Hirve, M.; Prajapati, R. Water deficit stress effects and responses in maize. In *Plant Abiotic Stress Tolerance*; Hasanuzzaman, M., Hakeem, K., Nahar, K., Alharby, H., Eds.; Springer: Cham, Switzerland, 2019; pp. 129–151.
12. Gao, M.; Zhou, J.J.; Liu, H.; Zhang, W.; Hu, Y.; Liang, J.; Zhou, J. Foliar spraying with silicon and selenium reduces cadmium uptake and mitigates cadmium toxicity in rice. *Sci. Total Environ.* **2018**, *631–632*, 1100–1108. [CrossRef]
13. Singh, M.; Kushwaha, B.K.; Singh, S.; Kumar, V.; Singh, V.P.; Prasad, S.M. Sulphur alters chromium (VI) toxicity in *Solanum melongena* seedlings: Role of sulphur assimilation and sulphur-containing antioxidants. *Plant. Physiol. Biochem.* **2017**, *112*, 183–192. [CrossRef]
14. Fazeli, F.; Ghorbanli, M.; Niknam, V. Effect of drought on biomass, protein content, lipid peroxidation and antioxidant enzymes in two sesame cultivars. *Plant. Biol.* **2007**, *51*, 98–103. [CrossRef]
15. Foyer, C.H.; Noctor, G. Ascorbate and glutathione: The heart of the redox hub. *Plant. Physiol.* **2011**, *155*, 2–18. [CrossRef]
16. Gill, S.S.; Tuteja, N. Reactive oxygen species and antioxidant machinery in abiotic stress tolerance in crop plants. *Plant. Physiol. Biochem.* **2010**, *48*, 909–930. [CrossRef]
17. Noctor, G.; Lelarge-Trouverie, C.; Mhamdi, A. The metabolomics of oxidative stress. *Phytochemistry* **2014**, *112*, 33–53. [CrossRef]
18. Alam, M.M.; Hasanuzzaman, M.; Nahar, K.; Fujita, M. Exogenous salicylic acid ameliorates short term drought stress in mustard (*Brassica juncea* L.) seedlings by up-regulating the antioxidant defense and glyoxalase system. *Aust. J. Crop. Sci.* **2013**, *7*, 1053–1063.
19. Hasanuzzaman, M.; Nahar, K.; Bhuiyan, T.F.; Anee, T.I.; Inafuku, M.; Oku, H.; Fujita, M. Salicylic acid: An all-rounder in regulating abiotic stress responses in plants. In *Phytohormones-Signaling Mechanisms and Crosstalk in Plant Development and Stress Responses*; El-Esawi, M.A., Ed.; IntechOpen: London, UK, 2017; pp. 31–75.
20. Mir, M.A.; John, R.; Alyemeni, M.N.; Alam, P.; Ahmad, P. Jasmonic acid ameliorates alkaline stress by improving growth performance, ascorbate glutathione cycle and glyoxylase system in maize seedlings. *Sci. Rep.* **2018**, *8*, 2831. [CrossRef]
21. Lazebnik, J.; Frago, E.; Dicke, M.; van Loon, J.J.A. Phytohormone mediation of interactions between herbivores and plant pathogens. *J. Chem. Ecol.* **2014**, *40*, 730–741. [CrossRef]
22. Soares, A.M.S.; Oliveira, J.T.A.; Gondim, D.M.F.; Domingues, D.P.; Machado, O.L.T.; Jacinto, T. Assessment of stress–related enzymes in response to either exogenous salicylic acid or methyl jasmonate in *Jatropha curcas* L. leaves, an attractive plant to produce biofuel. *South. Afr. J. Bot.* **2016**, *105*, 163–168. [CrossRef]
23. Li, L.; Lu, X.; Ma, H.; Lyu, D. Jasmonic acid regulates the ascorbate-glutathione cycle in *Malus baccata* Borkh roots under low root-zone temperature. *Acta Physiol. Plant.* **2017**, *39*, 174. [CrossRef]

24. Alam, M.M.; Nahar, K.; Hasanuzzaman, M.; Fujita, M. Exogenous jasmonic acid modulates the physiology, antioxidant defense and glyoxalase systems in imparting drought stress tolerance in different *Brassica* species. *Plant. Biotechnol. Rep.* **2014**, *8*, 279–293. [CrossRef]
25. Shyu, C.; Brutnell, T.P. Growth-defense balance in grass biomass production: The role of jasmonates. *J. Exp. Bot.* **2015**, *66*, 4165–4176. [CrossRef]
26. Karamian, R.; Ghasemlou, F.; Amiri, H. Physiological evaluation of drought stress tolerance and recovery in *Verbascum sinuatum* plants treated with methyl jasmonate, salicylic acid and titanium dioxide nanoparticles. *Plant. Biosyst.* **2019**, *154*, 277–287. [CrossRef]
27. Tayyab, N.; Naz, R.; Yasmin, H.; Nosheen, A.; Keyani, R.; Sajjad, M.; Hassan, M.N.; Roberts, T.H. Combined seed and foliar pre–treatments with exogenous methyl jasmonate and salicylic acid mitigate drought induced stress in maize. *PLoS ONE* **2020**, *15*, e0232269. [CrossRef]
28. Anjum, S.A.; Tanveer, M.; Hussain, S.; Tung, S.A.; Samad, R.A.; Wang, L.; Khan, I.; ur Rehman, N.; Shah, A.N.; Shahzad, B. Exogenously applied methyl jasmonate improves the drought tolerance in wheat imposed at early and late developmental stages. *Acta Physiol. Plant.* **2016**, *38*, 25. [CrossRef]
29. Yoshida, C.H.P.; Pacheco, A.C.; Lapaz, A.M.; Gorni, P.H.; Vítolo, H.F.; Bertoli, S.C. Methyl jasmonate modulation reduces photosynthesis and induces synthesis of phenolic compounds in sweet potatoes subjected to drought. *Bragantia* **2020**, *79*, 319–334. [CrossRef]
30. Siboza, X.I.; Bertling, I. The effects of methyl jasmonate and salicylic acid on suppressing the production of reactive oxygen species and increasing chilling tolerance in 'Eureka' lemon [*Citrus limon* (L.) Burm F]. *J. Hortic. Sci. Biotechnol.* **2013**, *88*, 269–276. [CrossRef]
31. Ahmmed, S.; Jahiruddin, M.; Razia, S.; Begum, R.A.; Biswas, J.C.; Rahman, A.S.M.M. *Fertilizer Recommendation Guide-2018*; Bangladesh Agricultural Research Council (BARC): Dhaka, Bangladesh, 2018.
32. Ogbaga, C.C.; Stepien, P.; Johnson, G.N. Sorghum (*Sorghum bicolor*) varieties adopt strongly contrasting strategies in response to drought. *Physiol. Plant.* **2014**, *152*, 389–401. [CrossRef]
33. Arnon, D. Copper enzymes isolated chloroplasts, polyphenol oxidase in *Beta Vulgaris*. *Plant. Physiol.* **1949**, *24*, 1–15. [CrossRef]
34. Fischer, R.A.; Rees, D.; Sayre, K.D.; Lu, Z.M.; Condon, A.G.; Saavedra, L.A. Wheat yield progress associated with higher stomatal conductance and photosynthetic rate, and cooler canopies. *Crop. Sci.* **1998**, *38*, 1467–1475. [CrossRef]
35. Shivakrishna, P.; Reddy, K.A.; Rao, D.M. Effect of PEG-6000 imposed drought stress on RNA content, relative water content (RWC), and chlorophyll content in peanut leaves and roots. *Saudi J. Biol. Sci.* **2018**, *25*, 285–289.
36. Sairam, R.K.; Deshmukh, P.S.; Shukla, D.S. Tolerance of drought and temperature stress in relation to increased antioxidant enzyme activity in wheat. *J. Agron. Crop. Sci.* **1997**, *178*, 171–178. [CrossRef]
37. Bates, L.S.; Waldren, R.P.; Teari, D. Rapid determination of free proline for water stress studies. *Plant. Soil* **1973**, *39*, 205–207. [CrossRef]
38. Elstner, E.F.; Heupel, A. Inhibition of nitrite formation from hydroxylammoniumchloride: A simple assay for superoxide dismutase. *Anal. Biochem.* **1976**, *70*, 616–620. [CrossRef]
39. Yang, S.-H.; Wang, L.-J.; Li, S.-H. Ultraviolet-B irradiation-induced freezing tolerance in relation to antioxidant system in winter wheat (*Triticum aestivum* L.) leaves. *Env. Exp. Bot.* **2007**, *60*, 300–307. [CrossRef]
40. Heath, R.L.; Packer, L. Photoperoxidation in isolated chloroplast: I. Kinetics and stoichiometry of fatty acid peroxidation. *Arch. Biochem. Biophys.* **1968**, *125*, 189–198. [CrossRef]
41. Mohi-Ud-Din, M.; Siddiqui, N.; Rohman, M.; Jagadish, S.V.K.; Ahmed, J.U.; Hassan, M.M.; Hossain, A.; Islam, T. Physiological and biochemical dissection reveals a trade-off between antioxidant capacity and heat tolerance in bread wheat (*Triticum aestivum* L.). *Antioxidants* **2021**, *10*, 351. [CrossRef]
42. Bradford, M.M. A rapid and susceptible method for the quantitation of microgram quantities of protein utilizing the principle of protein-dye binding. *Anal. Biochem.* **1976**, *72*, 248–254. [CrossRef]
43. Doderer, A.; Kokkelink, I.; van der Veen, S.; Valk, B.; Schram, A.; Douma, A. Purification and characterization of two lipoxygenase isoenzymes from germinating barley. *Biochim. Biophys. Acta* **1992**, *112*, 97–104. [CrossRef]
44. Spitz, D.R.; Oberley, L.W. An assay for superoxide dismutase activity in mammalian tissue homogenates. *Anal. Biochem.* **1989**, *179*, 8–18. [CrossRef]
45. Noctor, G.; Mhamdi, A.; Foyer, C.H. Oxidative stress and antioxidative systems: Recipes for successful data collection and interpretation. *Plant Cell Environ.* **2016**, *39*, 1140–1160. [CrossRef] [PubMed]
46. Castillo, F.I.; Penel, I.; Greppin, H. Peroxidase release induced by ozone in sedum album leaves. *Plant. Physiol.* **1984**, *74*, 846–851. [CrossRef] [PubMed]
47. Elia, A.C.; Galarini, R.; Taticchi, M.I.; Dörr, A.J.; Mantilacci, L. Antioxidant responses and bioaccumulation in *Ictalurus melas* under mercury exposure. *Ecotoxicol. Env. Saf.* **2003**, *55*, 162–167. [CrossRef]
48. Hossain, M.Z.; Hossain, M.D.; Fujita, M. Induction of pumpkin glutathione *S*-transferase by different stresses and its possible mechanisms. *Biol. Plant.* **2006**, *50*, 210–218. [CrossRef]
49. Kolde, R. Pheatmap: Pretty Heatmaps, Version 1.0.12. rdrr.io 2019. Available online: https://rdrr.io/cran/pheatmap/ (accessed on 29 May 2021).
50. Lê, S.; Josse, J.; Husson, F. FactoMineR: An R package for multivariate analysis. *J. Stat. Softw.* **2008**, *25*, 1–18. [CrossRef]

51. Wickham, H. *Ggplot2: Elegant Graphics for Data Analysis*; Springer: New York, NY, USA, 2016.
52. Pardossi, A.; Incrocci., L.; Incrocci, G.; Malorgio, F.; Battista, P.; Bacci, L.; Rapi, B.; Marzialetti, P.; Hemming, J.; Balendonck, J. Root zone sensors for irrigation management in intensive agriculture. *Sensors* **2009**, *9*, 2809–2835. [CrossRef]
53. Weng, E.; Luo, Y. Soil hydrological properties regulate grassland ecosystem responses to multifactor global change: A modeling analysis. *J. Geophys. Res.* **2008**, *113*, G03003. [CrossRef]
54. Anjum, N.A.; Umar, S.; Ahmad, A. *Oxidative Stress in Plant: Causes, Consequences and Tolerance*; IK International Publishing House Pvt Ltd.: New Delhi, India, 2012.
55. Huang, H.; Liu, B.; Liu, L.; Song, S. Jasmonate action in plant growth and development. *J. Exp. Bot.* **2017**, *68*, 1349–1359. [CrossRef]
56. Abbaspour, H.; Saeidi–Sar, S.; Afshari, H. Improving drought tolerance of *Pistacia vera* L. seedlings by arbuscular mycorrhiza under greenhouse conditions. *J. Med. Plant. Res.* **2011**, *5*, 7065–7072. [CrossRef]
57. Pandey, H.C.; Baig, M.J.; Bhatt, R.K. Effect of moisture stress on chlorophyll accumulation and nitrate reductase activity at vegetative and flowering stage in *Avena* species. *Agric. Sci. Res. J.* **2012**, *2*, 111–118.
58. Dalal, V.K.; Tripathy, B.C. Water-stress induced downsizing of light-harvesting antenna complex protects developing rice seedlings from photo–oxidative damage. *Sci. Rep.* **2018**, *8*, 5955. [CrossRef]
59. Mohamed, H.I.; Latif, H.H. Improvement of drought tolerance of soybean plants by using methyl jasmonate. *Physiol. Mol. Biol. Plants* **2017**, *23*, 545–556. [CrossRef] [PubMed]
60. Razmi, N.; Ebadi, A.; Daneshian, J.; Jahanbakhsh, S. Salicylic acid induced changes on antioxidant capacity, pigments and grain yield of soybean genotypes in water deficit condition. *J. Plant. Interact.* **2017**, *12*, 457–464. [CrossRef]
61. Royo, C.; Villegas, D.; Del Moral, L.G.; Elhani, S.; Aparicio, N.; Rharrabti, Y.; Araus, J.L. Comparative performance of carbon isotope discrimination and canopy temperature depression as predictors of genotype differences in durum wheat yield in Spain. *Aust. J. Agric. Res.* **2002**, *53*, 561–569. [CrossRef]
62. Ahmed, H.G.M.-D.; Zeng, Y.; Yang, X.; Anwaar, H.A.; Mansha, M.Z.; Hanif, C.M.S.; Ikram, K.; Ullah, A.; Alghanem, S.M.S. Conferring drought-tolerant wheat genotypes through morpho-physiological and chlorophyll indices at seedling stage. *Saudi J. Biol. Sci.* **2020**, *27*, 2116–2123. [CrossRef]
63. Mohi-Ud-Din, M.; Hossain, M.A.; Rohman, M.M.; Uddin, M.N.; Haque, M.S.; Ahmed, J.U.; Hossain, A.; Hassan, M.M.; Mostofa, M.G. Multivariate analysis of morpho-physiological traits reveals differential drought tolerance potential of bread wheat genotypes at the seedling stage. *Plants* **2021**, *10*, 879. [CrossRef]
64. Balota, M.; Payne, W.A.; Evett, S.R.; Peters, T.R. Morphological and physiological traits associated with canopy temperature depression in three closely related wheat lines. *Crop. Sci.* **2008**, *48*, 1897–1910. [CrossRef]
65. Kadioglu, A.; Saruhan, N.; Sağlam, A.; Terzi, R.; Acet, T. Exogenous salicylic acid alleviates effects of long–term drought stress and delays leaf rolling by inducing antioxidant system. *Plant. Growth Regul.* **2011**, *64*, 27–37. [CrossRef]
66. Ashraf, M.; Foolad, M.R. Roles of glycinebetaine and proline in improving plant abiotic resistance. *Env. Exp. Bot.* **2007**, *59*, 206–216. [CrossRef]
67. Rampino, P.; Pataleo, S.; Gerardi, C.; Mita, G.; Perrotta, C. Drought stress response in wheat: Physiological and molecular analysis of resistant and sensitive genotypes. *Plant. Cell Env.* **2006**, *29*, 2143–2152. [CrossRef]
68. Wang, G.P.; Zhang, X.Y.; Li, F.; Luo, Y.; Wang, W. Overaccumulation of glycine betaine enhances tolerance to drought and heat stress in wheat leaves in the protection of photosynthesis. *Photosynthetica* **2010**, *48*, 117–126. [CrossRef]
69. Dixon, D.P.; Skipsey, M.; Edwards, R. Roles for glutathione transferases in plant secondary metabolism. *Phytochemistry* **2010**, *71*, 338–350. [CrossRef] [PubMed]
70. Hasanuzzaman, M.; Fujita, M. Selenium pretreatment upregulates the antioxidant defense and methylglyoxal detoxification system and confers enhanced tolerance to drought stress in rapeseed seedlings. *Biol. Trace Elem. Res.* **2011**, *143*, 1758–1776. [CrossRef] [PubMed]
71. Nahar, K.; Hasanuzzaman, M.; Alam, M.M.; Fujita, M. Glutathione-induced drought stress tolerance in mung bean: Coordinated roles of the antioxidant defence and methylglyoxal detoxification systems. *Aob Plants* **2015**, *7*, plv069. [CrossRef]
72. Nafie, E.; Hathout, T.; Mokadem, A.S.A. Jasmonic acid elicits oxidative defense and detoxification systems in *Cucumis melo* L. cells. *Braz. J. Plant. Physiol.* **2011**, *23*, 161–174. [CrossRef]

Article

Foliar Spray of Alpha-Tocopherol Modulates Antioxidant Potential of Okra Fruit under Salt Stress

Maria Naqve [1], Xiukang Wang [2,*], Muhammad Shahbaz [1], Sajid Fiaz [3], Wardah Naqvi [4], Mehwish Naseer [5], Athar Mahmood [6,*] and Habib Ali [7]

[1] Department of Botany, Faculty of Sciences, University of Agriculture Faisalabad, Faisalabad 38000, Pakistan; marianaqvi26@gmail.com (M.N.); shahbazmuaf@yahoo.com (M.S.)

[2] College of Life Sciences, Yan'an University, Yan'an 716000, China

[3] Department of Plant Breeding and Genetics, The University of Haripur, Haripur 22620, Pakistan; sfiaz@uoh.edu.pk

[4] Institute of Agricultural and Resource Economics, Faculty of Social Sciences, University of Agriculture Faisalabad, Faisalabad 38000, Pakistan; wardahnaqvi5@gmail.com

[5] Department of Botany, Faculty of Science and Technology, Govt. College Women University Faisalabad, Faisalabad 38000, Pakistan; mehwishnaseer@gcwuf.edu.pk

[6] Department of Agronomy, Faculty of Agriculture, University of Agriculture Faisalabad, Faisalabad 38000, Pakistan

[7] Department of Agricultural Engineering, Khawaja Fareed University of Engineering and Information Technology, Rahim Yar Khan, Punjab 64200, Pakistan; habib_ali1417@yahoo.com

* Correspondence: wangxiukang@yau.edu.cn (X.W.); athar.mahmood@uaf.edu.pk (A.M.)

Citation: Naqve, M.; Wang, X.; Shahbaz, M.; Fiaz, S.; Naqvi, W.; Naseer, M.; Mahmood, A.; Ali, H. Foliar Spray of Alpha-Tocopherol Modulates Antioxidant Potential of Okra Fruit under Salt Stress. *Plants* **2021**, *10*, 1382. https://doi.org/10.3390/plants10071382

Academic Editors: Mirza Hasanuzzaman and Masayuki Fujita

Received: 11 May 2021
Accepted: 29 June 2021
Published: 6 July 2021

Publisher's Note: MDPI stays neutral with regard to jurisdictional claims in published maps and institutional affiliations.

Abstract: As an antioxidant, alpha-tocopherol (α-Toc) protects plants from salinity-induced oxidative bursts. This study was conducted twice to determine the effect of α-Toc as a foliar spray (at 0 (no spray), 100, 200, and 300 mg L^{-1}) to improve the yield and biochemical constituents of fresh green capsules of okra (*Abelmoschus esculentus* L. Moench) under salt stress (0 and 100 mM). Salt stress significantly reduced K$^+$ and Ca^{2+} ion concentration and yield, whereas it increased H$_2$O$_2$, malondialdehyde (MDA), Na$^+$, glycine betaine (GB), total free proline, total phenolics, and the activities of catalase (CAT), guaiacol peroxidase (GPX), and protease in both okra varieties (Noori and Sabzpari). Foliar application of α-Toc significantly improved the yield in tested okra varieties by increasing the activity of antioxidants (CAT, GPX, SOD, and ascorbic acid), accumulation of GB, and total free proline in fruit tissues under saline and non-saline conditions. Moreover, α-Toc application as a foliar spray alleviated the adverse effects of salt stress by reducing Na$^+$ concentration, MDA, and H$_2$O$_2$ levels and improving the uptake of K$^+$ and Ca^{2+}. Among the tested okra varieties, Noori performed better than Sabzpari across all physio-biochemical attributes. Of all the foliar-applied α-Toc levels, 200 mg L^{-1} and 300 mg L^{-1} were more effective in the amelioration of salinity-induced adverse effects in okra. Thus, we concluded that higher levels of α-Toc (200 mg L^{-1} and 300 mg L^{-1}) combat salinity stress more effectively by boosting the antioxidant potential of okra plants.

Keywords: antioxidants; alpha-tocopherol; foliar spray; salinity; okra varieties

1. Introduction

Okra [*Abelmoschus esculentus* (L.) Moench] is the most popular mallow crop and a common food crop in Asia. Its finger-like fruits, called capsules, are mainly consumed as a vegetable. These capsules are a rich source of vitamins, minerals and dietary fiber, and are low in calories. The mucilage's properties have medicinal enormous value [1]. The high mucilage content sets okra apart from other vegetables and makes it suitable for various medicinal and industrial applications [2].

A 90% loss (6.5 dSm^{-1}) in okra yield has been reported under high salt levels [3]. Soil salinity is one of the most prominent obstacles suppressing plant productivity. It remarkably affects the production of crops by disrupting the overall cellular metabolism of plants [4]. In Asia, it is expected that increasing levels of salinization could result in a loss of 50% of cultivated land by 2050 [5]. In developing countries like Pakistan, these losses are of considerable attention because its economy relies on agriculture. In Pakistan, 6 Mha of cultivated land is affected by salinity, which is a great threat to future food production [6].

Salinity affects plant growth and yield by reducing the photosynthesis rate, biomass, and water use efficiency [7]. The continued deposition of salts shunts osmotic stress, ionic imbalance, and physiological drought in plants [8]. The combination of these stresses directly influences fruit production in plants due to the considerable adverse effects on the composition of amino acids, proteins, and carbohydrates [9]. The production of reactive oxygen species (ROS) due to oxidation stress under high salt levels is another prominent threat as thesedamage the proteins, nucleic acids, and other biomolecules, thus limiting plant metabolism and yield [10]. This oxidative burst in the form of ROS induces the peroxidation of lipids, resulting in the production of lipid radicles and malondialdehyde (MDA), thereby damaging cellular membranes [11].

Plants have an antioxidant defense system to combat ROS. This system consists of enzymatic and non-enzymatic antioxidants found in all cellular compartments. These antioxidants detoxify cells from oxidative free radicles produced under varying saline regimes [12]. Oxidative damage generated under saline regimes can be alleviated by the exogenous application of these antioxidants [13]. Foliar spraying with α-Toc is one such approach to the improvement of plant growth under salinity stress [14]. Tocopherols are members of the vitamin E family and consist of alpha, beta, gamma, and delta forms [14]. α-Toc is more active than all other categories of vitamin E, as it protects photosystem II and lipid membranes in chloroplasts from salinity-induced damages [13,14].

Tocopherols are non-enzymatic antioxidants which protect plants by quenching ROS and guard cellular membranes against lipid peroxidation [15]. Among these tocopherols, α-Toc is the most active antioxidant as it shields photosystems from photo-inhibition and protects membrane lipids in chloroplasts under salinity stress [9]. As chloroplasts are sensitive to salinity stress [16], to combat salinity-induced ROS, α-Toc works in coordination with other antioxidants, including catalase (CAT), superoxide dismutase (SOD), and guaiacol peroxidase (GPX) [17]. SOD is the first line of defense under stress conditions as it converts singlet oxygen species to H_2O_2, and this H_2O_2 is converted to H_2O by CAT and GPX [18].

Few studies have assessed the impact of foliar spraying with α-Toc in boosting the antioxidant potential of okra fruit under salt stress. Therefore, this study aimed to examine the modulations in antioxidant defense mechanism in response to α-Toc foliar spray on the yield and the related attributes of okra under salt stress conditions.

2. Results

The data revealed that catalase (CAT) activity was significantly increased under salinity stress. Neither of the tested varieties of okra differ significantly in terms of CAT activity. Foliar spraying (300 mg L^{-1}) with α-Toc enhanced CAT activity in okra capsules under saline and non-saline conditions (Table 1; Figure 1A).

Table 1. Analysis of variance (mean squares) for enzymatic and non-enzymatic antioxidants traits of okra treated with α-Toc as foliar spray under saline and non-saline conditions.

Source	df	CAT	POD	SOD	GPX	Protease	Phenolics	Ascorbic Acid
V	1	264.06 ns	0.266 ns	119.52 **	690,058.34 ***	28,532.84 ***	1.22 ***	1388.89 ns
S	1	15,314.06 ***	0.548 ns	54.83 ns	306,132.59 ***	8883.06 ***	293,265,625 ***	41,769.14 ***
α-toc	3	2143.22 **	0.515 ns	16.29 ns	279,930.76 ***	607.43 ns	1.32 ***	4013.78 *
V × S	1	1501.56 ns	0.523 ns	47.23 ns	290,599.16 ***	70.84 ns	3.94 ***	28,110.11 ***
V × α-toc	3	489.06 ns	0.065 ns	21.94 ns	18,401.16 ns	48,183.32 ***	6,912,135.4 ns	3462.61 *
S × α-toc	3	205.72 ns	0.997 **	6.42 ns	4860.57 ns	18,282.91 ***	32,911,354 **	16,191.37 ***
V × S × α-toc	3	801.56 ns	0.109 ns	24.20 ns	57,395.68 *	7053.95 ***	9,296,354.2 ns	13,829.37 ***
Error	48	403.64	0.184	15.23	18,769.04	404.83	6,673,880.2	1109.76

*, ** and *** = significant at 0.05, 0.01 and 0.001 levels respectively, ns = non-significant, V: Varieties, S: Salinity, α-Toc: Alpha-tocopherol, CAT: Catalase, POD: Peroxidase, SOD: Superoxide dismutase, GPX: Guaiacol peroxidase.

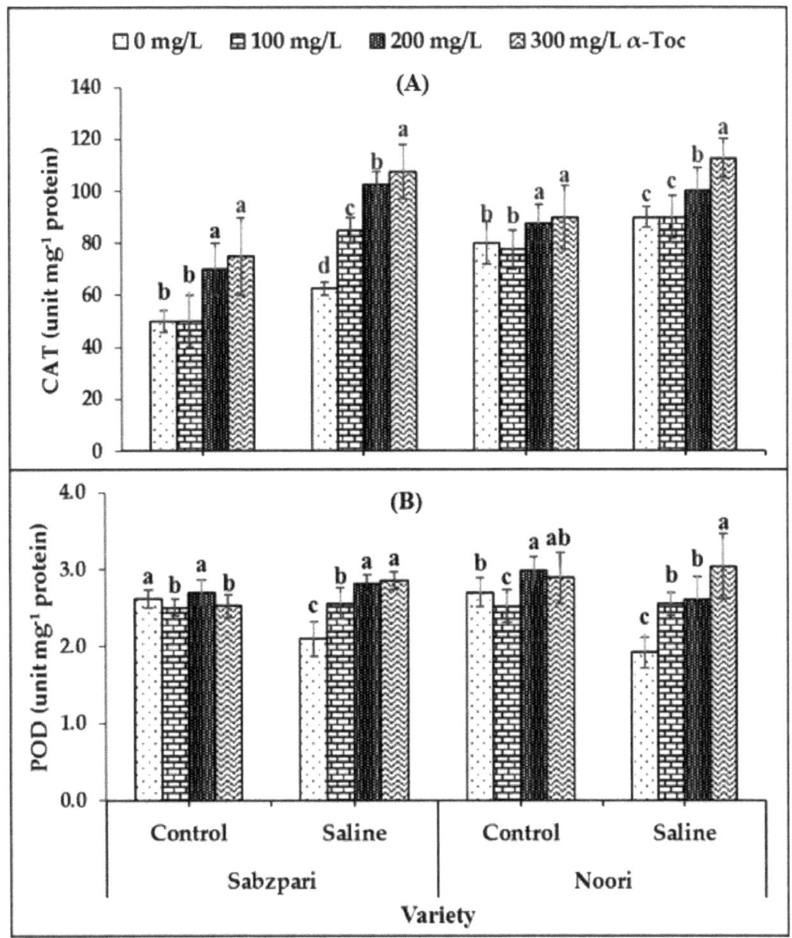

Figure 1. Effect of different levels of α-Toc under saline and non-saline conditions on (**A**) activities of catalase (CAT) and (**B**) peroxidase (POD) of okra varieties sprayed with different levels of α-Toc under saline and non-saline conditions. Values represent means ± S.D. Significant differences among row spacing were measured by the least significant difference (LSD) at *p* > 0.05 and indicated by different letters. S.D. stands for standard deviation.

Both okra varieties demonstrated non-significant performance in term of peroxidase (POD) activity. Neither the application of salt stress nor α-Toc affected the POD activity of the okra capsules. The data showed that root medium salinity had no effect on the activity of protease. However, a significant interaction was recorded between salinity and the α-Toc spray, where 300 mg L^{-1} of the spray proved effective to increase the activity of POD under saline conditions (Table 1; Figure 1B).

The experimental data showed that root medium 100 mM saline stress significantly enhanced the activity of guaiacol peroxidase (GPX) in the fruit tissues of both okra varieties. However, in the salinized Noori plants, GPX activity was slightly higher than Sabzpari. In addition, α-Toc spray markedly improved the activity of GPX in both varieties. Inclusively, the 200 mg L^{-1} application of α-Toc proved to be better at enhancing the activity of GPX (Table 1; Figure 2A).

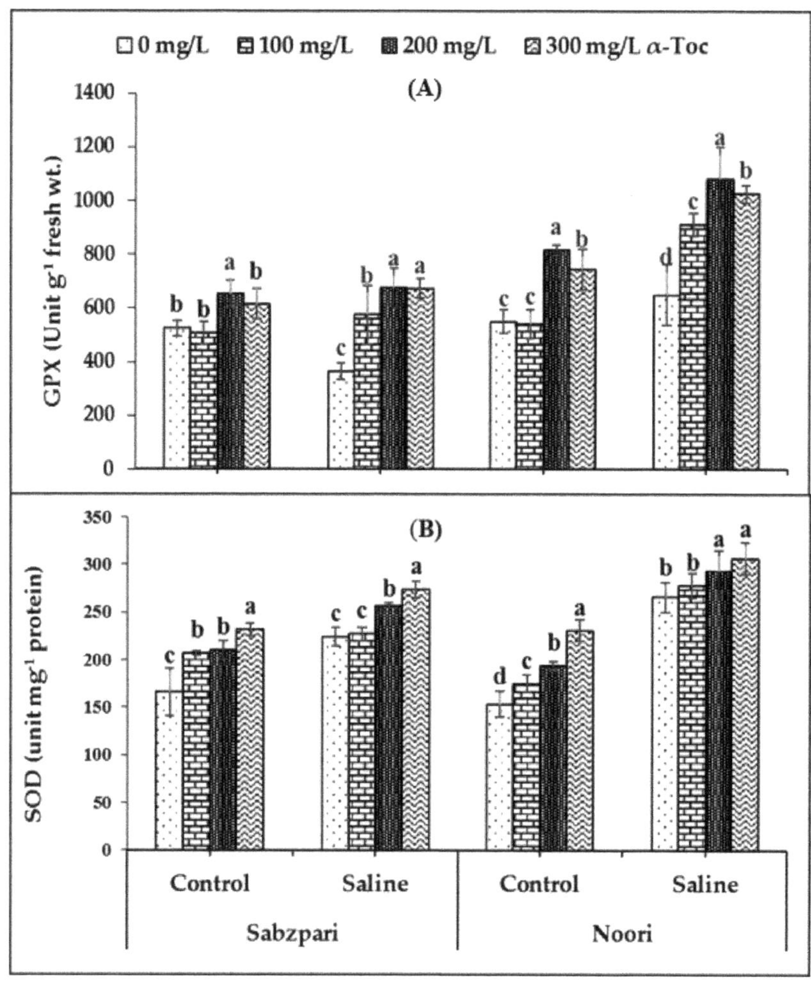

Figure 2. Effect of different levels of α-Toc under saline and non-saline conditions on (**A**) activities of guaiacol peroxidase (GPX) and (**B**) superoxide dismutase (SOD) of okra varieties. Values represent means ± S.D. Significant differences among row spacing were measured by the least significant difference (LSD) at p > 0.05 and indicated by different letters. S.D. stands for standard deviation.

A slight increase in the activity of superoxide dismutase (SOD) in the fruits of both varieties of okra plants was noticed due to salinization. Interestingly, levels of superoxide dismutase (SOD) were higher in Noori than Sabzpari capsules under both stressed and non-stressed conditions. In the current study, foliar-applied α-Toc failed to significantly affect the activity of SOD (Table 1; Figure 2B).

In this study, a significant decrease was recorded in the activity of protease under salinity stress. It was observed from the data that Noori performed better than Sabzpari with respect to this attribute. A significant interaction was also recorded among the tested varieties. The application of 200 mg L^{-1} spray of α-Toc under saline conditions significantly enhanced the protease content of Noori okra in comparison with other treatments (Table 1; Figure 3A).

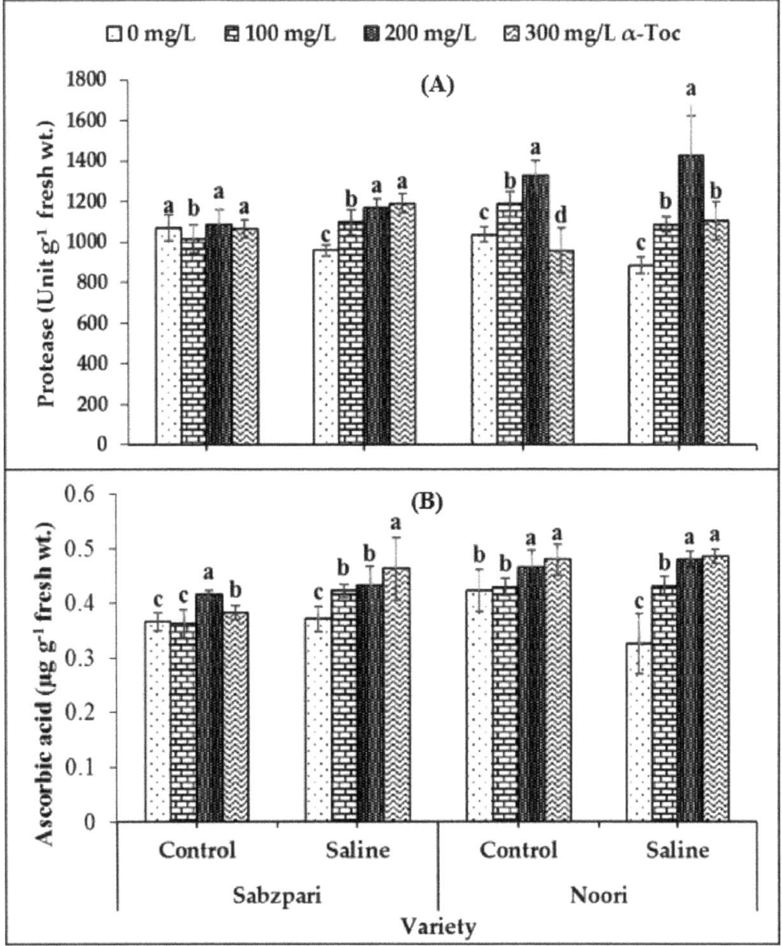

Figure 3. Effect of different levels of α-Toc under saline and non-saline conditions on (**A**) activities of protease and (**B**) ascorbic acid of okra varieties. Values represent means ± S.D. Significant differences among row spacing were measured by the least significant difference (LSD) at $p > 0.05$ and indicated by different letters. S.D. stands for standard deviation.

A significant accumulation of ascorbic acid was recorded under salinity stress in the Noori variety, while it remained the same in Sabzpari under saline and control conditions.

The application of α-Toc significantly enhanced the ascorbic acid content in the Noori variety under saline conditions. Among all treatments, α-Toc concentrations of 200 mg L^{-1} and 300 mg L^{-1} were more effective with respect to this attribute in the case of the Noori variety (Table 1; Figure 3B).

The data revealed that the total phenolics content significantly increased under salt stress. The tested varieties differed significantly in terms of total phenolics content. Noori produced more Sabzpari. Higher levels of applied α-Toc (200 mg L^{-1} and 300 mg L^{-1}) spray exhibited clearly higher total phenolics content under saline regimes (Table 1; Figure 4A).

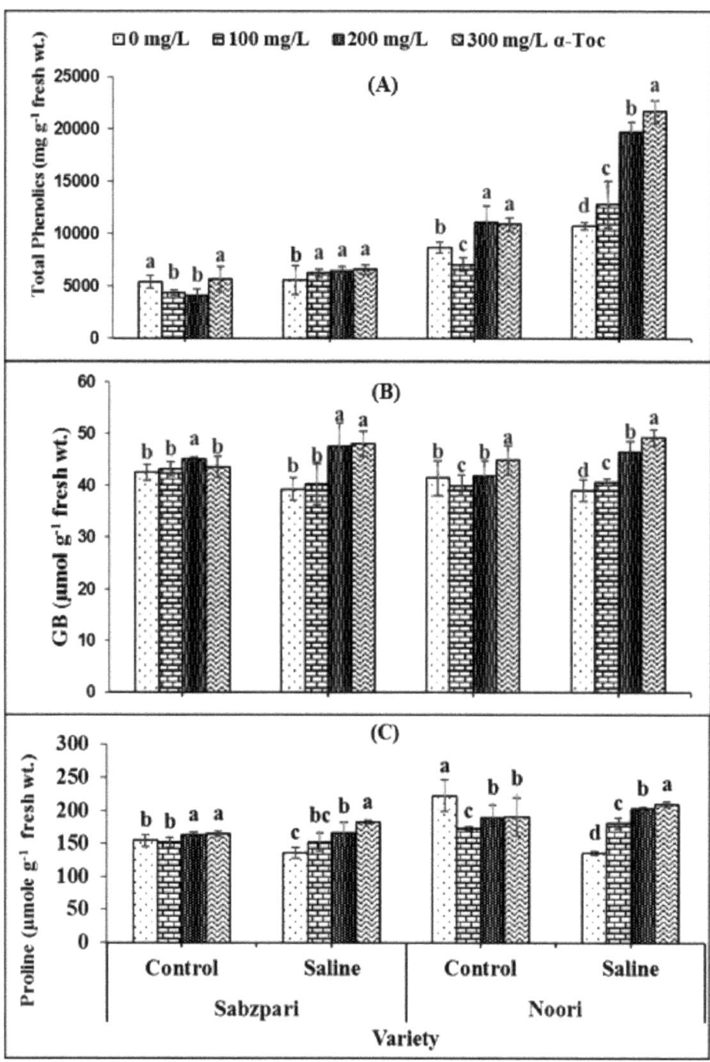

Figure 4. Effect of different levels of α-Toc under saline and non-saline conditions on (**A**) total phenolics (**B**) glycinebetaine (GB), and (**C**) proline content of okra varieties. Values represent means ± S.D. Significant differences among row spacing were measured by the least significant difference (LSD) at $p > 0.05$ and indicated by different letters. S.D. stands for standard deviation.

Both okra varieties have shown non-significant differences in terms of fruit glycine betaine (GB) and free proline contents. Root medium salinity stress non-significantly enhanced GB content (Table 2; Figure 4B,C) while markedly enhancing free proline in both okra varieties. Foliage supplementation with α-Toc did not markedly affect the proline content of the fruit, whereas a significant interaction was observed between α-Toc spray and salt stress with respect to the accumulation of proline content. Higher levels of α-Toc spray produced significantly higher proline content under 100 mM salt stress conditions as compared to other combinations. In addition, foliar spraying with α-Toc at 200 and 300 mg L^{-1} significantly increased GB content in okra fruits of both varieties under controlled and stressed environments (Table 2; Figure 4B,C).

Table 2. Analysis of variance (mean squares) for osmolytes, ROS, ionic traits and yield of okra treated with α-Toc as foliar spray under saline and non-saline conditions.

Source	df	Proline	GB	H_2O_2	MDA	Na$^+$	K$^+$	Ca^{2+}	cap/pl
V	1	992.25 ns	0.33 ns	7284.92 *	228.39 ns	0.19 ns	11.81 ***	8.6 **	43.89 ***
S	1	10,905.3 **	316.19 ns	3519.39 ns	25,656.80 ***	53.47 ***	36.75 ***	30.9 ***	43.88 ***
α-toc	3	1983.47 ns	1097.5 **	1344.55 ns	5587.57 **	1.11 ns	1.19 ns	7.9 ***	9.515 **
V $\times$ S	1	2102.87 ns	5.3 ns	63.87 ns	8050.72 **	18.59 ***	1.72 ns	3.7 *	0.390 ns
V $\times$ α-toc	3	2973.15 *	313.3 *	840.67 ns	12,923.24 ***	6.34 ***	0.84 ns	0.6 ns	0.182 ns
S $\times$ α-toc	3	3381.21 *	548.0 **	1129.41 ns	2793.45 ns	0.79 ns	3.51 **	2.2 *	0.682 ns
V $\times$ S $\times$ α-toc	3	9211.3 ***	9.3 ns	404.06 ns	3888.35 *	0.96 ns	0.60 ns	0.2 ns	0.182 ns
Error	48	1019.95	86.6	1014.53	1044.26	0.98	0.70	0.7	2.234

*, ** and *** = significant at 0.05, 0.01 and 0.001 levels respectively, ns = non-significant, V: Varie Table 2. O$_2$, Hydrogen peroxide, MDA: Malondialdehyde, cap/pl: Capsules per plant.

Interestingly, neither the foliar application of α-Toc nor the application of salt (NaCl) showed a clear impact on hydrogen peroxide (H_2O_2) content. However, both okra varieties differed significantly in terms of hydrogen peroxide concentrations; Sabzpari accumulated more H_2O_2 than Noori under salt stress and non-stressed conditions (Table 2; Figure 5A).

Malondialdehyde (MDA) content was markedly enhanced under salt stress conditions in the Sabzpari variety. Foliar spraying with α-Toc significantly reduced MDA contents under salinity treatment conditions. The experimental data of this study showed that higher levels of α-Toc (200 mg L^{-1} and 300 mg L^{-1}) were more effective at reducing the MDA content of okra (Table 2; Figure 5B).

Root medium applied salt stress was found to significantly enhance the fruit Na$^+$ content of both okra varieties. The response of the two okra varieties was non-significant in terms of fruit Na$^+$ content. Foliar spraying with α-Toc failed to have a significant impact on this ionic attribute, but the interactive effect of both varieties and α-Toc was significant; it was found that 200 and 300 mg L^{-1} of foliar spray decreased Na$^+$ content in both varieties under saline and non-saline regimes. Noori performed better than Sabzpari in lowering the Na$^+$ ion content under salinity stress (Table 2; Figure 6A).

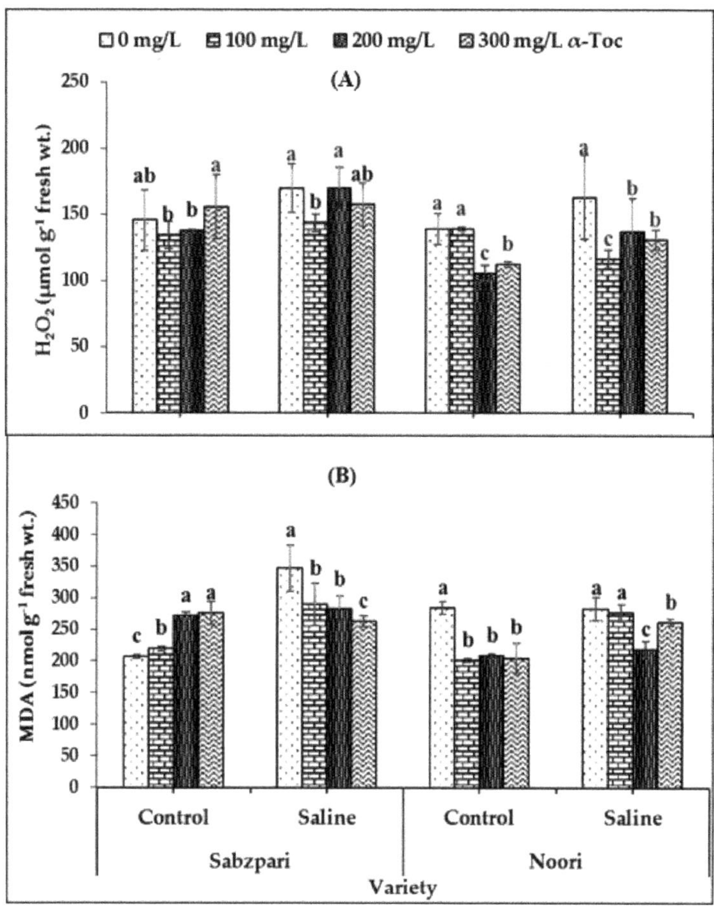

Figure 5. Effect of different levels of α-Toc under saline and non-saline conditions on (**A**) hydrogen peroxide (H$_2$O$_2$) and (**B**) malondialdehyde (MDA) content of okra varieties. Values represent means ± S.D. Significant differences among row spacing were measured by the least significant difference (LSD) at $p > 0.05$ and indicated by different letters. S.D. stands for standard deviation.

In this study, a clear reduction in fruit K$^+$ content was determined in both okra varieties under saline conditions. Although both varieties have shown significant differences in terms of fruit K$^+$ content, the Noori variety proved to be better at accumulating more K$^+$ than Sabzpari under both salinity treatments. However, a significant interactive effect between α-Toc and salinity stress was noted in this study. Higher levels of α-Toc spray enhanced fruit K$^+$ content under salinity stress (100 mM NaCl) conditions in comparison with other combinations (Table 2; Figure 6B).

Repeated experimental data showed that soil salinization remarkably decreased fruit Ca^{2+} content in both okra varieties. Noori had higher fruit Ca^{2+}content than Sabzpari under the salinity-free regime. The interactive effect of both varieties towards salinity was also significant. Fruit Ca^{2+} content increased remarkably after foliar application of α-Toc. The response of α-Toc under salinity stress was also significant. Inclusively, 300 mg L^{-1} increased the fruit Ca^{2+} content in tested varieties under saline and no-saline environments (Table 2; Figure 6C).

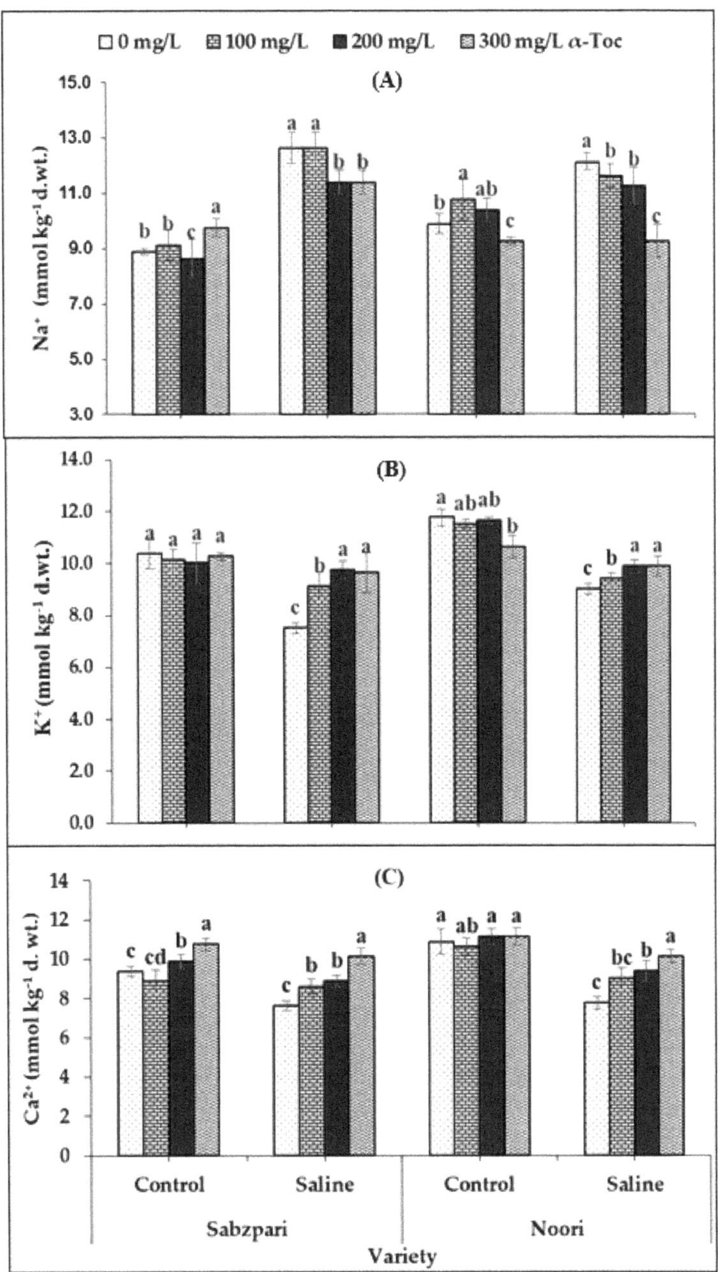

Figure 6. Effect of different levels of α-Toc under saline and non-saline conditions on (**A**) Na$^+$ (**B**) K$^+$ and (**C**) Ca^{2+} ion content of okra varieties. Values represent means ± S.D. Significant differences among row spacing were measured by the least significant difference (LSD) at $p > 0.05$ and indicated by different letters. S.D. stands for standard deviation.

Salt stress significantly reduced the yield of both okra varieties in terms of the number of capsules plant^{-1}. Both varieties differed markedly though; yield reduction was more pronounced in Sabzpari than Noori. However, foliar application of α-Toc significantly enhanced the number of capsules in both okra varieties of stressed and non-stressed plants. In addition to this, enhancement in yield was more pronounced at 200 mg L^{-1} and 300 mg L^{-1} levels of α-Toc application (Table 2; Figure 7).

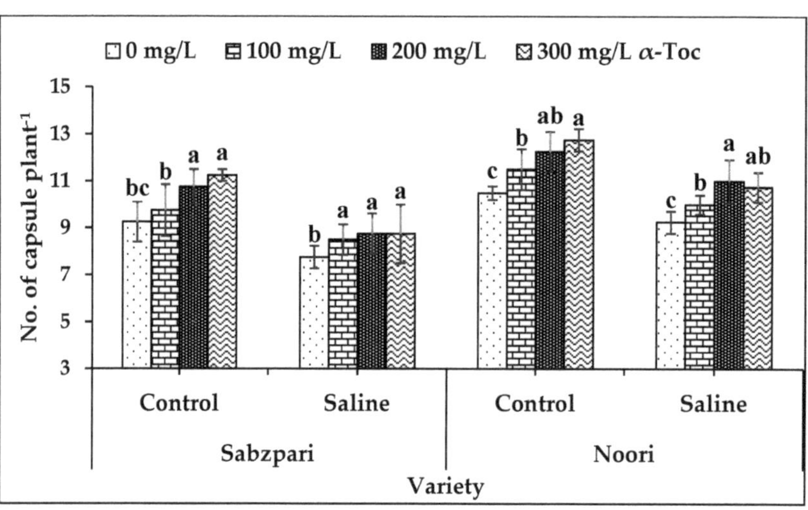

Figure 7. Effect of different levels of α-Toc under saline and non-saline conditions on number of capsule Plant^{-1} of okra varieties. Values represent means ± S.D. Significant differences among row spacing were measured by the least significant difference (LSD) at *p* > 0.05 and indicated by different letters. S.D. stands for standard deviation.

3. Discussion

Salinity is considered an abiotic stress due to its potential to harm various physiological and biochemical processes in plants [19,20]. It influences plant performance, limiting its production and causing water shortages to the plants. Salinity also negatively affects chlorophyll, the photosynthetic apparatus, and chloroplast ultra-structure, and may cause cell death [21,22]. Salinity-induced ROS triggers phytotoxic reactions and adversely affects the cellular processes of plants, causing tissue damage due to the oxidization of macro-molecules such as proteins and lipids. The induction of antioxidants is the prominent activity plants use to combat salinity-induced ROS stress [16]. Antioxidant enzymes are recognized as effective mitigators of the adverse effects of salt stress on cells and tissues [9]. The results of this study suggest that the activities of antioxidants, including CAT, GPX, and protease, were significantly increased due to salt stress in both varieties of okra plants. A slight but non-significant increase was observed in POD activity due to salt stress. The GPX is considered to be a defensive agent against oxidation induced by H_2O_2 [23], while POD is as efficient as GPX at scavenging H_2O_2 under salt stress conditions [24]. Previous research has shown that catalase plays an important role in ROS detoxification, converting 26 million H_2O_2 to H_2O in one minute [25].

In the current study, the activities of these enzymes viz. CAT, POD, GPX, and protease were enhanced by the foliar fertigation of α-Toc in fruit tissues of the tested okra varieties. To combat the overproduction of ROS, α-Toc aids in the coordination of CAT, POD and GPX [26]. SOD activity was also increased in tested varieties of okra plants grown in saline medium, but α-Toc spray did not significantly increase its activity. The enzyme SOD is considered to be the first line of defense against ROS. In a previous study, the activities of SOD, POD, and CAT were increased in two mung bean varieties under stress conditions [27]. Similarly, a study of *Brassica napus* showed a remarkable enhancement in the activity of SOD and POD enzymes, while a decline was noticed in the activity of CAT under stress conditions [28]. It is concluded from the findings of the present study that salt-stress-induced ROS accumulation is scavenged by α-Toc antioxidants as they prevent ROS production by chelating metal ions. Moreover, α-Toc fertigation could compensate for salinity-induced damages by upregulating other enzymatic antioxidants and quenching ROS via its antioxidant properties. Researchers have concluded that the stress-induced ROS accumulation is dependent on antioxidants, including α-Toc [29].

Foliar spraying with α-Toc improved ascorbic acid and total phenolic content in the tested okra plants under salt stress. It is suggested that ascorbic acid protects the plants against salinity effects by acting as a free radical reductant and antioxidant, and salt tolerance may have also been manifested by the upregulation of ascorbic acid, as it induces Toc biosynthesis and total phenols provide membrane stability against salinity-induced ROS. There is a positive correlation between spraying with α-Toc and the acceleration of the content of total phenolics [30].

Glycine betaine (GB) has a significant role in stabilizing the structure and activity of protein complexes and enzymes and also maintains the membrane against devastating impacts caused by salinity [31]. Proline is effective in maintaining the potential of the osmotic process in the cytoplasm and in maintaining proteins and ribosomes against the harmful effects of Na^+ ions. In the current study, salt stress did not significantly affect the proline and GB content in fruit tissues of okra varieties. Contrary to this, previous research has shown a remarkable increase in the accumulation of GB and proline under stress [31]. Foliar spraying with α-Toc significantly enhanced the proline and GB content, indicating its role in shielding the plants from salinity effects by adjusting the osmotic balance.

The ROS can induce the peroxidation of lipids (in MDA), producing aldehydes, of which MDA is a major type. MDA consequently has a role as an indicator of membrane damage induced by ROS. Our findings showed that neither the foliar application of α-Toc nor salt (NaCl) significantly affected H_2O_2 content, whereas MDA contents were increased under salt stress conditions in both okra varieties. In the current study, foliar spraying with α-Toc significantly reduced MDA content under salinity stress in the tested okra capsules. Contrary to the present study, lower levels of H_2O_2 had been observed in *Vicia faba* plants treated with α-tocopherol during salinity stress [14]. This could be due to the variation in genetic traits among different plant species. However, α-Toc helped stabilize lipid membranes by scavenging ROS [32].

In this study, higher concentrations of Na^+ ions had been observed in the fruit tissues of okra plants under saline regimes. Enhanced Na^+ concentration disturbed photosynthesis, plant metabolism, enzymatic activities, and ultimately reduced crop yield [33,34]. In contrast, reduced uptake of K^+ and Ca^{2+} had been observed in the fruit tissues of the tested okra plants under salt stress conditions. The reduced uptake of Ca^{2+} and K^+ is correlated with the enhanced concentration of Na^+. α-Toc spray proved to be successful in minimizing the toxic concentrations of Na^+ and maximizing the concentration of K^+ and Ca^{2+} ion contents in fruit tissues of okra plants, suggesting its pivotal role in osmo-tolerance by lowering Na^+ levels and enhancing the uptake of K^+ and Ca^{2+} ions in plants under salinity conditions.

Plants are the target of abiotic stresses, thus suffering severe yield losses [35]. In the current investigation, reduced numbers of capsules were observed in okra under salinity stress. A reduction in the yield of crop plants under salinity stress is directly linked with inhibited uptake of essential nutrients [36]. However, α-Toc supplementation significantly improved yield under saline and non-saline conditions. Increased yield due to α-Toc spray is linked with the improved uptake of beneficial ions, ionic homeostasis, chloroplast stability, and reduced oxidative damage.

4. Material and Methods

Two-year pot experiments were carried out to examine the possible role of α-Toc in modulating the antioxidant potential of okra capsules under salinity stress. Experiments were conducted in plastic pots (diameter 26 cm, depth 29 cm) each containing 10 kg of well washed river sand. The number of plots was 64 for each experiment. The crop duration was six months from sowing to harvesting. Each experiment was Seeds of two okra varieties (Sabzpari and Noori) were collected from the Ayub Agricultural Research Institute, Faisalabad, Pakistan. Experiment was laid out under completely randomized design (CRD) with four replications. Hoagland's nutrient solution (1 L/pot) was applied weekly after sowing. Five plants were maintained in each pot. Twenty-four days old plants were treated with two salt levels (0 and 100 mM NaCl) in Hoagland's nutrient solution. Concentration of NaCl was maintained in aliquot parts of 50 mM to prevent salt shock. In fact, in order to avoid osmotic shocks to plants, the salinity level was maintained in two phases. In the first phase (at start of salinity) 50 mM NaCl level was maintained, and after two days at this level, the required level of 100 mM was maintained. Foliar spray of each concentration of α-Toc (0, 100, 200 and 300 mg/L) was applied to 36 days old plants. The 50 mL solution of each of α-Toc levels was foliar sprayed to fully saturate the plants. Tween 20 at 0.1% was used to enhance the absorbance of solution as surfactant. All plants were allowed to grow until the complete formation of capsules. Fresh green capsules of okra plants were collected from each plant, weighed, and used for chemical analyses as described below.

4.1. Estimation of Enzymatic Antioxidants

Fresh green capsule material (0.5 g) was homogenized with 10 mL (50 mM) potassium phosphate buffer with pH 7.0, after centrifugation supernatant was used for the determination of enzymatic antioxidants; activity of CAT and POD was determined by following the protocol of Chance and Maehly [37]. The procedure of Giannopolitis and Ries [36] was followed for recording the activities of SOD.

Protocol ascribed by Carlberg and Mannervik [38] was followed for the determination of GPX activities. Protease activities were recorded by following Drapeau et al. [39].

4.2. Estimation of Non-Enzymatic Antioxidants

For the determination of total phenolics content from the fresh green tissue of capsules, the protocol of Ainsworth and Gillespie [40] was followed. Ascorbic acid content from the capsules tissues of okra was determined by following the protocol established Mukherjee and Choudhuri [41]. Similarly, using the protocol established by Grieve and Grattan [42], glycinebetaine content was estimated from fresh fruit tissue.

The protocol ascribed by Bates et al. [43] was followed to estimate the free proline from the fruit material.

4.3. Determination of ROS

The procedure ascribed by Alexieva et al. [44] was adopted for the determination of H_2O_2 level, and by following procedure of Heath and Packer [45] with slight modifications [46] malondialdehyde (MDA) content was estimated.

4.4. Nutrients Analysis for Na^+, K^+ and Ca^{2+}

The protocol proposed by Wolf [47] was followed to determine mineral elements by acid digestion.

4.5. Yield

The plucked fresh capsules were collected, and yield (number of capsules/plants) was recorded.

4.6. Statistical Analysis

By following Snedecor and Cochran [48], analysis of variance (ANOVA) of data for all the parameters with four replicates was calculated under three factor factorial and design of experiment was completely randomized (CRD). The least significance difference (LSD) values at 5% probability were worked out and are presented in each figure.

5. Conclusions

The foliar fertigation of α-Toc was effective in compensating harmful salinity-induced effects in okra by enhancing antioxidant activities (CAT, POD, SOD, GPX, protease, ascorbic acid, total phenolics) and organic osmolytes (GB, free proline), as well as by improving ionic homeostasis and yield, possibly by quenching salinity-induced ROS and protecting chloroplasts with its antioxidant potential. Among the tested okra varieties, Noori showed better tolerance against salinity. α-Toc levels of 200 and 300 mg L^{-1} were more effective. Thus, this study suggests the Noori variety can be grown in salt-affected soils with α-Toc foliar spray (300 mg L^{-1}) to increase okra production. The brutal effects of salinity can be mitigated in okra as well as other crops by foliar spraying with α-Toc.

Author Contributions: M.S. and M.N. (Maria Naqve) conceived the idea; S.F., W.N. and M.N. (Mehwish Naseer) collected the literature review; X.W. and A.M. provided technical expertise to strengthen the basic idea. M.N. (Maria Naqve), M.S., W.N. and A.M. helped in collection of data and its analysis; H.A., M.S., A.M. and S.F. proofread and provided intellectual guidance. All authors have read the first draft, helped in revision and approved the article. All authors have read and agreed to the published version of the manuscript.

Funding: The publication of the present work is supported by the Natural Science Basic Research Program of Shaanxi Province (grant no. 2018JQ5218) and the National Natural Science Foundation of China (51809224), Top Young Talents of Shaanxi Special Support Program.

Institutional Review Board Statement: Not applicable.

Informed Consent Statement: Not applicable.

Data Availability Statement: Not applicable.

Conflicts of Interest: The authors declare that they have no conflict of interest.

References

1. Temam, N.; Mohammed, W.; Aklilu, S.; Belayneh, A.H.; Gebremedhin, K.G.; Egziabher, Y.G. Variability Assessment of Okra (*Abelmoschus esculentus* (L.) Moench) Genotypes Based on Their Qualitative Traits. *Int. J. Agron.* **2021**, *2021*, 6678561. [CrossRef]
2. Haruna, S.; Aliyu, B.S.; Bala, A. Plant gum exudates (Karau) and mucilages, their biological sources, properties, uses and potential applications: A review. *Bayero J. Pure Appl. Sci.* **2016**, *9*, 159–165. [CrossRef]
3. Mushtaq, N.U.; Saleem, S.; Rasool, A.; Shah, W.H.; Hakeem, K.R.; Rehman, R.U. Salt Stress Threshold in Millets: Perspective on Cultivation on Marginal Lands for Biomass. *Phyton* **2020**, *90*, 51. [CrossRef]
4. Mushtaq, Z.; Faizan, S.; Gulzar, B. Salt stress, its impacts on plants and the strategies plants are employing against it: A review. *J. Appl. Biol. Biotechnol.* **2020**, *8*, 81–91.
5. Khan, A.; Khan, A.L.; Muneer, S.; Kim, Y.-H.; Al-Rawahi, A.; Al-Harrasi, A. Silicon and salinity: Crosstalk in crop-mediated stress tolerance mechanisms. *Front. Plant Sci.* **2019**, *10*, 1429. [CrossRef]
6. Naveed, M.; Sajid, H.; Mustafa, A.; Niamat, B.; Ahmad, Z.; Yaseen, M.; Kamran, M.; Rafique, M.; Ahmar, S.; Chen, J.-T. Alleviation of salinity-induced oxidative stress, improvement in growth, physiology and mineral nutrition of canola (*Brassica napus* L.) through calcium-fortified composted animal manure. *Sustainability* **2020**, *12*, 846.
7. Moles, T.M.; de Brito Francisco, R.; Mariotti, L.; Pompeiano, A.; Lupini, A.; Incrocci, L.; Carmassi, G.; Scartazza, A.; Pistelli, L.; Guglielminetti, L. Salinity in autumn-winter season and fruit quality of tomato landraces. *Front. Plant Sci.* **2019**, *10*, 1078. [CrossRef]
8. Maria, N.; Muhammad, S.; Abdul, W.; Waraich, E.A. Seed priming with alpha tocopherol improves morpho-physiological attributes of okra under saline conditions. *Int. J. Agric. Biol.* **2018**, *20*, 2647–2654.
9. Sadiq, M.; Akram, N.A.; Ashraf, M.; Al-Qurainy, F.; Ahmad, P. Alpha-tocopherol-induced regulation of growth and metabolism in plants under non-stress and stress conditions. *J. Plant Growth Regul.* **2019**, *38*, 1325–1340.
10. Soltabayeva, A.; Ongaltay, A.; Omondi, J.O.; Srivastava, S. Morphological, Physiological and Molecular Markers for Salt-Stressed Plants. *Plants* **2021**, *10*, 243. [CrossRef] [PubMed]
11. Demidchik, V. Mechanisms of oxidative stress in plants: From classical chemistry to cell biology. *Environ. Exp. Bot.* **2015**, *109*, 212–228. [CrossRef]
12. Kumar, S.; Li, G.; Yang, J.; Huang, X.; Ji, Q.; Zhou, K.; Khan, S.; Ke, W.; Hou, H. Investigation of an Antioxidative System for Salinity Tolerance in Oenanthe javanica. *Antioxidants* **2020**, *9*, 940. [CrossRef]
13. Surówka, E.; Potocka, I.; Dziurka, M.; Wróbel-Marek, J.; Kurczyńska, E.; Żur, I.; Maksymowicz, A.; Gajewska, E.; Miszalski, Z. Tocopherols mutual balance is a key player for maintaining Arabidopsis thaliana growth under salt stress. *Plant Physiol. Biochem.* **2020**, *156*, 369–383. [CrossRef] [PubMed]

14. Semida, W.; Taha, R.; Abdelhamid, M.; Rady, M. Foliar-applied α-tocopherol enhances salt-tolerance in Vicia faba L. plants grown under saline conditions. *S. Afr. J. Bot.* **2014**, *95*, 24–31. [CrossRef]
15. Fritsche, S.; Wang, X.; Jung, C. Recent advances in our understanding of tocopherol biosynthesis in plants: An overview of key genes, functions, and breeding of vitamin E improved crops. *Antioxidants* **2017**, *6*, 99. [CrossRef]
16. Bose, J.; Rodrigo-Moreno, A.; Shabala, S. ROS homeostasis in halophytes in the context of salinity stress tolerance. *J. Exp. Bot.* **2014**, *65*, 1241–1257. [CrossRef] [PubMed]
17. Laxa, M.; Liebthal, M.; Telman, W.; Chibani, K.; Dietz, K.-J. The role of the plant antioxidant system in drought tolerance. *Antioxidants* **2019**, *8*, 94. [CrossRef]
18. Luis, A.; Corpas, F.J.; López-Huertas, E.; Palma, J.M. Plant superoxide dismutases: Function under abiotic stress conditions. In *Antioxidants and Antioxidant Enzymes in Higher Plants*; Springer: Berlin/Heidelberg, Germany, 2018; pp. 1–26.
19. Dubey, R.S. Photosynthesis in plants under stressful conditions. In *Handbook of Photosynthesis*; CRC Press: Boca Raton, FL, USA, 2018; pp. 629–649.
20. Singh, D. *Stress Physiology*; New Age International (P) limited, Publishers: New Dehli, India, 2003.
21. Stefanov, M.; Biswal, A.; Misra, M.; Misra, A.; Apostolova, E. Responses of Photosynthetic Apparatus to Salt Stress: Structure, Function, and Protection. In *Handbook of Plant and Crop Stress*, 4th ed.; CRC Press: Boca Raton, FL, USA, 2019; pp. 233–250.
22. Muhammad, I.; Shalmani, A.; Ali, M.; Yang, Q.-H.; Ahmad, H.; Li, F.B. Mechanisms regulating the dynamics of photosynthesis under abiotic stresses. *Front. Plant Sci.* **2021**, *11*, 2310. [CrossRef] [PubMed]
23. Wang, W.-B.; Kim, Y.-H.; Lee, H.-S.; Kim, K.-Y.; Deng, X.-P.; Kwak, S.-S. Analysis of antioxidant enzyme activity during germination of alfalfa under salt and drought stresses. *Plant Physiol. Biochem.* **2009**, *47*, 570–577. [CrossRef]
24. Rady, M.; Sadak, M.S.; El-Bassiouny, H.; El-Monem, A. Alleviation the adverse effects of salinity stress in sunflower cultivars using nicotinamide and α-tocopherol. *Aust. J. Basic Appl. Sci.* **2011**, *5*, 342–355.
25. Mehla, N.; Sindhi, V.; Josula, D.; Bisht, P.; Wani, S.H. An introduction to antioxidants and their roles in plant stress tolerance. In *Reactive Oxygen Species and Antioxidant Systems in Plants: Role and Regulation under Abiotic Stress*; Springer: Berlin/Heidelberg, Germany, 2017; pp. 1–23.
26. Hasanuzzaman, M.; Nahar, K.; Fujita, M. Plant response to salt stress and role of exogenous protectants to mitigate salt-induced damages. In *Ecophysiology and Responses of Plants under Salt Stress*; Springer: Berlin/Heidelberg, Germany, 2013; pp. 25–87.
27. Ali, Q.; Javed, M.T.; Noman, A.; Haider, M.Z.; Waseem, M.; Iqbal, N.; Waseem, M.; Shah, M.S.; Shahzad, F.; Perveen, R. Assessment of drought tolerance in mung bean cultivars/lines as depicted by the activities of germination enzymes, seedling's antioxidative potential and nutrient acquisition. *Arch. Agron. Soil Sci.* **2018**, *64*, 84–102. [CrossRef]
28. Abedi, T.; Pakniyat, H. Antioxidant enzymes changes in response to drought stress in ten cultivars of oilseed rape (*Brassica napus* L.). *Czech J. Genet. Plant Breed.* **2010**, *46*, 27–34. [CrossRef]
29. Szarka, A.; Tomasskovics, B.; Bánhegyi, G. The ascorbate-glutathione-α-tocopherol triad in abiotic stress response. *Int. J. Mol. Sci.* **2012**, *13*, 4458. [CrossRef]
30. Li, J.; Jin, H. Regulation of brassinosteroid signaling. *Trends Plant Sci.* **2007**, *12*, 37–41. [CrossRef] [PubMed]
31. Sakamoto, A.; Murata, N. The role of glycine betaine in the protection of plants from stress: Clues from transgenic plants. *Plant Cell Environ.* **2002**, *25*, 163–171. [CrossRef]
32. Tavakkoli, E.; Rengasamy, P.; McDonald, G.K. High concentrations of Na^+ and Cl^- ions in soil solution have simultaneous detrimental effects on growth of faba bean under salinity stress. *J. Exp. Bot.* **2010**, *61*, 4449–4459. [CrossRef]
33. Ashraf, M.; Harris, P.J. Photosynthesis under stressful environments: An overview. *Photosynthetica* **2013**, *51*, 163–190. [CrossRef]
34. Farooq, M.; Gogoi, N.; Barthakur, S.; Baroowa, B.; Bharadwaj, N.; Alghamdi, S.S.; Siddique, K. Drought stress in grain legumes during reproduction and grain filling. *J. Agron. Crop. Sci.* **2017**, *203*, 81–102. [CrossRef]
35. Wang, X.; Wang, G.; Guo, T.; Xing, Y.; Mo, F.; Wang, H.; Fan, J.; Zhang, F. Effects of plastic mulch and nitrogen fertilizer on the soil microbial community, enzymatic activity and yield performance in a dryland maize cropping system. *Eur. J. Soil Sci.* **2021**, *72*, 400–412. [CrossRef]
36. Wang, X.; Fan, J.; Xing, Y.; Xu, G.; Wang, H.; Deng, J.; Wang, Y.; Zhang, F.; Li, P.; Li, Z. The effects of mulch and nitrogen fertilizer on the soil environment of crop plants. *Adv. Agron.* **2019**, *153*, 121–173.
37. Chance, B.; Maehly, A. Assay of catalases and peroxidases. *Methods Enzymol.* **1955**, *2*, 764–775.
38. Giannopolitis, C.N.; Ries, S.K. Superoxide dismutases: I. Occurrence in higher plants. *Plant Physiol.* **1977**, *59*, 309–314. [CrossRef]
39. Carlberg, I.; Mannervik, B. Glutathione reductase. *Methods Enzymol.* **1985**, *113*, 484–490. [PubMed]
40. Drapeau, G. *Protease from Staphylococcus Aureus, Method of Enzymology 45b, L*; Academic Press: New York, NY, USA, 1974.
41. Ainsworth, E.A.; Gillespie, K.M. Estimation of total phenolic content and other oxidation substrates in plant tissues using Folin–Ciocalteu reagent. *Nat. Protoc.* **2007**, *2*, 875–877. [CrossRef] [PubMed]
42. Mukherjee, S.; Choudhuri, M. Implications of water stress-induced changes in the levels of endogenous ascorbic acid and hydrogen peroxide in Vigna seedlings. *Physiol. Plant.* **1983**, *58*, 166–170. [CrossRef]
43. Grieve, C.; Grattan, S. Rapid assay for determination of water soluble quaternary ammonium compounds. *Plant Soil* **1983**, *70*, 303–307. [CrossRef]
44. Bates, L.S.; Waldren, R.P.; Teare, I. Rapid determination of free proline for water-stress studies. *Plant Soil* **1973**, *39*, 205–207. [CrossRef]

45. Alexieva, V.; Sergiev, I.; Mapelli, S.; Karanov, E. The effect of drought and ultraviolet radiation on growth and stress markers in pea and wheat. *Plant Cell Environ.* **2001**, *24*, 1337–1344. [CrossRef]
46. Dhindsa, R.S.; Plumb-Dhindsa, P.; Thorpe, T.A. Leaf senescence: Correlated with increased levels of membrane permeability and lipid peroxidation, and decreased levels of superoxide dismutase and catalase. *J. Exp. Bot.* **1981**, *32*, 93–101. [CrossRef]
47. Wolf, B. A comprehensive system of leaf analyses and its use for diagnosing crop nutrient status. *Commun. Soil Sci. Plant Anal.* **1982**, *13*, 1035–1059. [CrossRef]
48. Snedecor, G.W.; Cochran, W.G. *Statistical Methods*, 7th ed.; Iowa State University Press: Ames, IA, USA, 1980.

Article

Bitter Melon (*Momordica charantia* L.) Rootstock Improves the Heat Tolerance of Cucumber by Regulating Photosynthetic and Antioxidant Defense Pathways

Mei-Qi Tao [1], Mohammad Shah Jahan [1,2], Kun Hou [1], Sheng Shu [1,3], Yu Wang [1], Jin Sun [1,3] and Shi-Rong Guo [1,3,*]

[1] Key Laboratory of Southern Vegetable Crop Genetic Improvement in Ministry of Agriculture, College of Horticulture, Nanjing Agricultural University, Nanjing 210095, China; 2017104116@njau.edu.cn (M.-Q.T.); shahjahansau@gmail.com (M.S.J.); 2017104117@njau.edu.cn (K.H.); shusheng@njau.edu.cn (S.S.); ywang@njau.edu.cn (Y.W.); jinsun@njau.edu.cn (J.S.)
[2] Department of Horticulture, Sher-e-Bangla Agricultural University, Dhaka 1207, Bangladesh
[3] Suqian Academy of Protected Horticulture, Nanjing Agricultural University, Suqian 223800, China
*   Correspondence: srguo@njau.edu.cn; Tel.: +86-25-8439-5267

Received: 3 May 2020; Accepted: 25 May 2020; Published: 29 May 2020

Abstract: High temperature is considered a critical abiotic stressor that is increasing continuously, which is severely affecting plant growth and development. The use of heat-resistant rootstock grafting is a viable technique that is practiced globally to improve plant resistance towards abiotic stresses. In this experiment, we explored the efficacy of bitter melon rootstock and how it regulates photosynthesis and the antioxidant defense system to alleviate heat stress (42 °C/32 °C) in cucumber. Our results revealed that bitter-melon-grafted seedlings significantly relieved heat-induced growth inhibition and photoinhibition, maintained better photosynthesis activity, and accumulated a greater biomass than self-grafted seedlings. We measured the endogenous polyamine and hydrogen peroxide (H_2O_2) contents to determine the inherent mechanism responsible for these effects, and the results showed that heat stress induced a transient increase in polyamines and H_2O_2 in the inner courtyard of grafted seedlings. This increment was greater and more robust in bitter-melon-grafted seedlings. In addition, the use of polyamine synthesis inhibitors MGBG (methylglyoxal bis-guanylhydrazone) and D-Arg (D-arginine), further confirmed that the production of H_2O_2 under heat stress is mediated by the accumulation of endogenous polyamines. Moreover, compared with other treatments, the bitter-melon-grafted seedlings maintained high levels of antioxidant enzyme activity under high temperature conditions. However, these activities were significantly inhibited by polyamine synthesis inhibitors and H_2O_2 scavengers (dimethylthiourea, DMTU), indicating that bitter melon rootstock not only maintained better photosynthetic activity under conditions of high temperature stress but also mediated the production of H_2O_2 through the regulation of the high level of endogenous polyamines, thereby boosting the antioxidant defense system and comprehensively improving the heat tolerance of cucumber seedlings. Taken together, these results indicate that grafting with a resistant cultivar is a promising alternative tool for reducing stress-induced damage.

Keywords: heat stress; grafting; cucumber; bitter-melon rootstock; polyamines; photosynthesis

1. Introduction

During their life cycle, plants face various environmental stimuli, including high temperature stress. In recent years, with the rise in global temperature, the greenhouse effect has continuously increased, and high temperatures have become a major environmental threat that adversely affects crop growth and productivity [1–5]. In China's facility cultivation, especially in southern China, facility cultivation is frequently subjected to high-temperature stress [6], which leads to the suppression of crop growth and seriously inhibits the production and supply of vegetables [7].

Photosynthesis is one of the most heat-sensitive biological processes [8]. The activity of photosynthesis is directly associated with the amount of biomass production in plants [9]. It has been reported in several studies that heat stress can lead to the inhibition of plant photosynthesis and is the main reason for a reduction in crop yield [5,10–12]. The main reason for the reduction of photosynthesis is the inhibition of photosystem II (PSII) [13]. At the same time, PSII is considered to be a key part of high-temperature stress-induced photoinhibition [7].

Recently, a couple of studies showed that polyamines play a direct controlling role in regulating resistance against different types of plant stress, such as salt stress [14–16], heat stress [17], low temperature stress [18], drought stress [19], and flood stress [20,21]. Polyamines participate in complex signaling systems under abiotic stress which, in turn, regulates a series of defense responses in plants, thereby improving plant resistance against different environmental stressors [22]. There is a strong link between polyamines and different signaling molecules, such as H_2O_2, NO, and Ca^{2+}, and these signaling molecules mediate the mitigation effect of polyamines on stress [23–27].

Grafting is a mature technical method that is used to enhance the stress tolerance of plants [28–30]. Grafting roots play a vital role in plants' response to various stressors [31]. The tolerance of grafted rootstocks to adverse conditions directly affects the resistance of grafted seedlings [29]. Some previous studies have confirmed that resistant rootstock grafting can improve the stress tolerance of grafted plants by reducing photosynthesis inhibition [32], regulating osmotic substances [7], enhancing antioxidant defense [33], regulating hormone signaling [34], and mediating with microRNA transcription [35]. Moreover, rootstock grafting can improve the salt tolerance of cucumber seedlings by regulating endogenous polyamine metabolism. However, the specific role of endogenous polyamines and their regulatory networks in grafted plants under stressful conditions, particularly high temperature stress, has still not been fully elucidated.

Cucumber is an important facility horticultural crop with a high level of heat sensitivity. Bitter melon originated in India and is not cold-tolerant but is heat-resistant [6]. Therefore, we extensively studied the effect of heat-resistant bitter melon rootstock on the photosynthesis of grafted cucumber under conditions of high temperature stress and the physiological mechanism by which grafting alleviates the high temperature stress injury of cucumber plants. We also explored the regulatory mechanism of endogenous polyamines and H_2O_2 signaling molecules in grafted plants, and our results provide a theoretical basis for the cultivation of facility crops under a high-temperature regime.

2. Results

2.1. Effects of Bitter Melon Rootstock on Plant Growth under Heat Stress

The growth attributes such as plant height, fresh and dry weight of self-grafted plants showed significantly greater values in control plants than in bitter-melon-grafted plants, except for the stem diameter (Table 1, Figure 1). In contrast, after 7 days of high-temperature treatment, except for the plant height, bitter-melon-grafted seedlings had significantly higher stem diameters and above-ground fresh and dry biomass weights than the self-grafted seedlings. Specifically, compared with the control, the plant height, stem thickness, and fresh and dry biomass weights of the self-grafted seedlings were reduced by 4.03%, 5.52%, 20.30%, and 29.51%, respectively, while the above four indicators were reduced by 0.25%, 4.09%, 6.26%, and 15.98%, respectively, in bitter-melon-grafted seedlings, indicating that bitter-melon-grafted seedlings could maintain greater biomass accumulation after high-temperature stress (Table 1).

Table 1. Interactive effects of bitter melon rootstock on the growth of grafted cucumber seedlings after 7 days of high temperature stress.

Treatments	Plant Height (cm)	Stem Diameter (mm)	Fresh Weight (Plant g^{-1})	Dry Weight (Plant g^{-1})
Cs-28	18.37 ± 0.32 [a]	5.03 ± 0.13 [bc]	17.19 ± 0.01 [a]	1.83 ± 0.03 [a]
Mc-28	16.27 ± 0.12 [b]	5.62 ± 0.14 [a]	16.62 ± 0.15 [b]	1.69 ± 0.01 [b]
Cs-42	17.63 ± 0.23 [a]	4.79 ± 0.12 [c]	13.70 ± 0.21 [d]	1.29 ± 0.04 [d]
Mc-42	16.23 ± 0.54 [b]	5.39 ± 0.09 [ab]	15.58 ± 0.21 [c]	1.42 ± 0.02 [c]

Note. Self-grafted plants treated with 28 °C/18 °C (day/night), Cs-28; bitter-melon-grafted plants treated with 28 °C/18 °C (day/night), Mc-28; self-grafted plants treated with 42 °C/32 °C (day/night), Cs-42; bitter-melon-grafted plants treated with 42 °C/32 °C (day/night), Mc-42. Data are the mean ± standard error (SE) of three independent experiments. Different letters indicate significant differences at $p < 0.05$ according to Duncan's multiple range test.

Figure 1. Interactive effects of bitter melon rootstock and grafted cucumber seedlings on the phenotype of cucumber seedlings after 7 days of heat stress. Self-grafted plants treated with 28 °C/18 °C (day/night), Cs-28; bitter-melon-grafted plants treated with 28 °C/18 °C (day/night), Mc-28; self-grafted plants treated with 42 °C/32 °C (day/night), Cs-42; bitter-melon-grafted plants treated with 42 °C/32 °C (day/night), Mc-42.

2.2. Effects of Bitter Melon Rootstock on Photosystem II (PSII) Photochemistry under Heat Stress Conditions

To explore the protective effect of rootstock grafting on photosynthetic systems, the following photosynthetic parameters were measured. As shown in Figure 2, there were no significant differences in Fv/Fm, F_0, Y(II), and qP between self-grafted plants and the bitter-melon-grafted plants under normal growth conditions. Conversely, heat stress significantly reduced the Fv/Fm, Y(II), and qP in both self-grafted and bitter-melon-grafted plants (Figure 2A,B,D,E), but the above three indicators were, respectively, 43.2%, 96.2%, and 36.8% higher in bitter-melon-grafted plants than in self-grafted plants under heat stress conditions. However, the F_0 value in both grafted plants was increased under heat stress conditions (Figure 2C), and the value of F_0 in the bitter-melon-grafted plant was considerably reduced compared with that of the self-grafted plants.

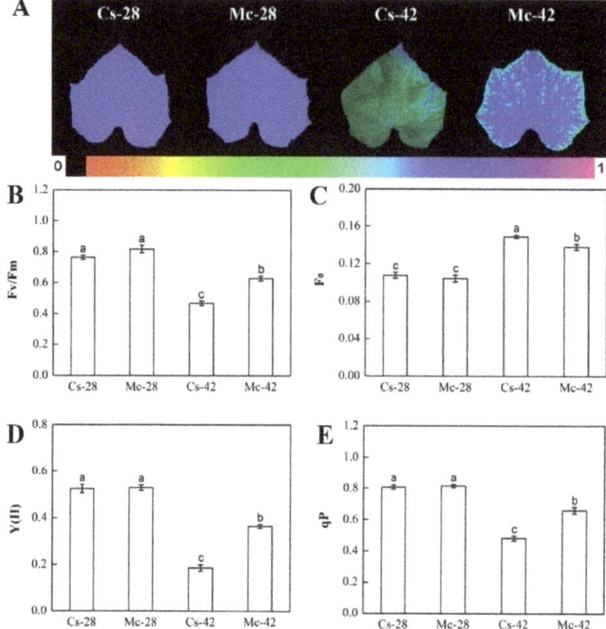

Figure 2. Interactive effects of bitter melon rootstock and grafted cucumber seedlings on chlorophyll fluorescence following exposure to heat stress for 7 days. The maximal photochemical efficiency of photosystem II (PSII, *Fv/Fm*) (**A**,**B**) under the initial fluorescence (F_0) (**C**), the effective photochemical quantum yield of PSII (Y(II)) (**D**) under photochemical quenching (qP) (**E**). Bars represent the mean ± SE of at least three independent experiments. Self-grafted plants exposed to 28 °C/18 °C (day/night), Cs-28; bitter-melon-grafted plants exposed to 28 °C/18 °C (day/night), Mc-28; self-grafted plants exposed to 42 °C/32 °C (day/night), Cs-42; bitter-melon-grafted plants exposed to 42 °C/32 °C (day/night), Mc-42. Different letters indicate significant differences at $p < 0.05$ according to Duncan's multiple range test.

2.3. Effect of Bitter Melon Rootstock on the Endogenous Polyamine Content under Heat Stress Conditions

As displayed in Figure 3, under normal conditions, the endogenous contents of the three free polyamines in the leaves of bitter-melon-grafted seedlings and self-grafted seedlings were not significantly different throughout the whole treatment period. After exposure to heat stress, the accumulation of polyamines in the leaves of the two grafted seedlings was significantly greater than the control level, and the polyamine content in the leaves of bitter-melon-grafted seedlings was significantly higher than that of self-grafted seedlings. The specific performance was as follows: the putrescine (Put) content of the two grafted seedlings began to rise at 4 h after heat stress and became significantly higher than the control level. It began to decline after reaching the peak at 8 h; at this time, the contents of spermidine (Spd) and spermine (Spm) also began to rise slightly. Spd and Spm began to decline after reaching their peak values at 8 and 12 h, respectively. It is important to note that the content of Spm at 8 h was close to that at 12 h.

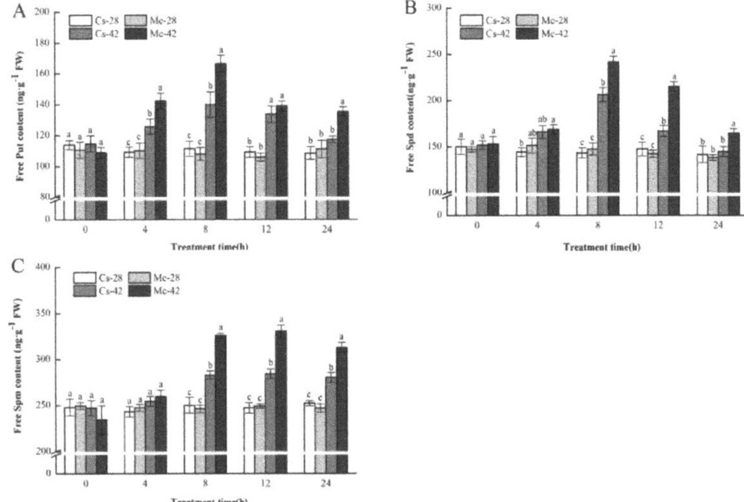

Figure 3. Interactive effects of bitter melon rootstock and grafted cucumber seedlings on the time course change of polyamine contents (**A**): Putrescine (Put), (**B**): Spermidine (Spd), (**C**): Spermine (Spm) in grafted cucumber leaves under heat stress conditions. The bars represent the mean ± SE values of at least three independent experiments. Self-grafted plants exposed to 28 °C/18 °C (day/night), Cs-28; bitter-melon-grafted plants exposed to 28 °C/18 °C (day/night), Mc-28; self-grafted plants exposed to 42 °C/32 °C (day/night), Cs-42; bitter-melon-grafted plants exposed to 42 °C/32 °C (day/night), Mc-42. Different letters indicate significant differences at $p < 0.05$ according to Duncan's multiple range test.

2.4. Effect of Bitter Melon Rootstock on the Accumulation of H_2O_2 under Heat Stress Conditions

As shown in Figure 4, throughout the whole treatment period, there was no significant change in the H_2O_2 content in the leaves of the two grafted seedlings under control conditions; however, high temperature stress significantly increased the H_2O_2 content in the leaves of bitter-melon-grafted seedlings and self-grafted seedlings, to a value significantly higher than that in the control plants. Under heat stress conditions, the changing trend of the H_2O_2 content in the leaves of the two grafted seedlings was roughly the same, that is, the H_2O_2 content increased first and then decreased. However, the increase in the H_2O_2 content of bitter-melon-grafted seedlings was substantially higher than that of self-grafted seedlings at 8, 12, and 24 h. Under heat stress conditions, the H_2O_2 content of the leaves of bitter-melon-grafted seedlings increased rapidly after 4 h, and it reached a peak at 8 h, and then began to decline. The peak content was about 1.28 times more than the control level; the self-grafted seedlings also showed a slow upward trend after 4 h. Furthermore, it began to decline after reaching the peak at 12 h. The peak content was only 1.16 times higher than the control, and it returned to the control level after 24 h.

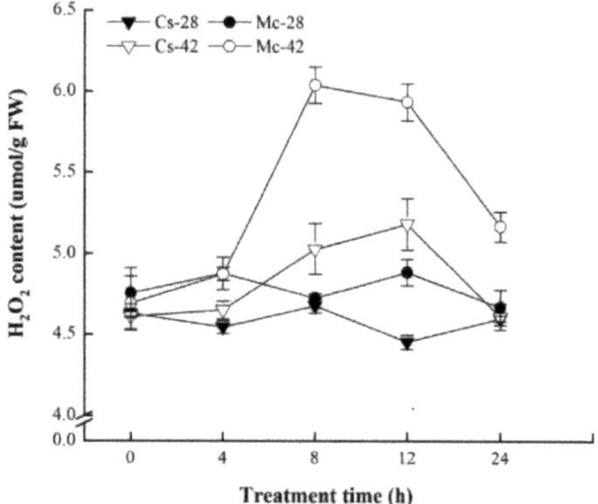

Figure 4. Interactive effects of bitter melon rootstock and grafted cucumber seedlings on the time course change of the H_2O_2 content in grafted cucumber leaves under heat stress conditions. Self-grafted plants exposed to 28 °C/18 °C (day/night), Cs-28; bitter-melon-grafted plants exposed to 28 °C/18 °C (day/night), Mc-28; self-grafted plants exposed to 42 °C/32 °C (day/night), Cs-42; bitter-melon-grafted plants exposed to 42 °C/32 °C (day/night), Mc-42. The values are the means ± SE of three independent experiments. The bars represent the standard error.

2.5. The Production of H_2O_2 is Regulated by Polyamines under Heat Stress Conditions

D-arginine (D-Arg) and methylglyoxal bis-guanylhydrazone (MGBG) are two kinds of polyamine synthesis inhibitor. To explore whether polyamines mediate H_2O_2 production in the leaves of grafted seedling under conditions of high temperature stress, we exogenously applied two kinds of inhibitor. As shown in Figure 5, treatment with both 1mM MGBG and 2 mM D-Arg significantly reduced the high-temperature-induced greater accumulation of H_2O_2 and almost returned to the control level, indicating that polyamines mediate the production of H_2O_2 under conditions of high temperature stress.

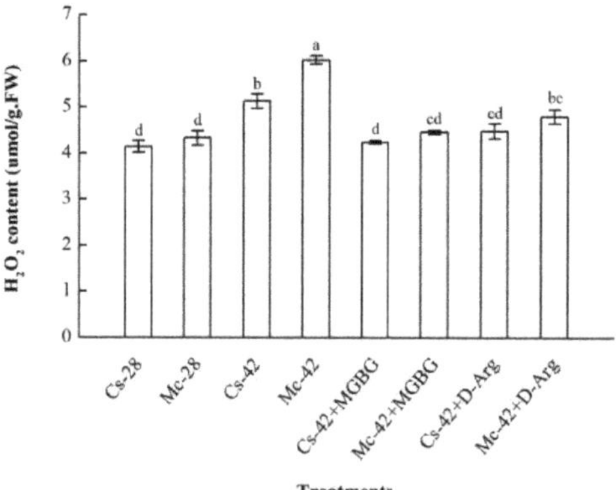

Figure 5. Effects of D-arginine (D-Arg) and methylglyoxal bis-guanylhydrazone (MGBG) on H_2O_2 production in the leaves of self-grafted and bitter-melon-grafted cucumber seedlings under heat stress conditions. Self-grafted plants exposed to 28 °C/18 °C (day/night), Cs-28; bitter-melon-grafted plants exposed to 28 °C/18 °C (day/night), Mc-28; self-grafted plants exposed to 42 °C/32 °C (day/night), Cs-42; bitter-melon-grafted plants exposed to 42 °C/32 °C (day/night), Mc-42. Seedlings were pretreated with 0.2 mM MGBG and 2mM D-Arg for 12 h, respectively, after which they were treated under high temperature conditions for 8 h. Bars represent the mean ± SE of at least three independent experiments. The different letters indicate significant differences at $p < 0.05$ according to Duncan's multiple range test.

2.6. Polyamine and H_2O_2 Accumulation Enhanced the Antioxidant Defense System of Bitter-Melon-Grafted Seedling Leaves under Heat Stress Conditions

As shown in Figure 6, heat stress significantly induced the activity of three antioxidant enzymes: super oxide dismutase (SOD), peroxidase (POD), and ascorbate peroxidase (APX) in grafted seedlings. In particular, the bitter-melon-grafted seedlings exhibited higher enzyme activities. The activities of SOD and POD markedly increased at 8 h after heat treatment (Figure 6A,B), while 4 h later, the activity of APX increased (Figure 6D). Interestingly, the activity of catalase (CAT) showed a clear downward trend from 24 h after heat treatment (Figure 6C). More importantly, it can be clearly seen that the peak point of the antioxidant enzyme activities was significantly lower than the polyamines (PAs) and H_2O_2 contents (Figure 4).

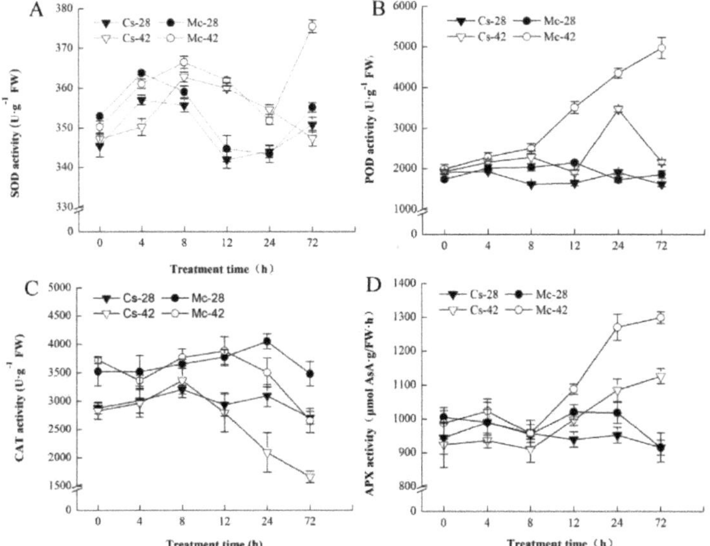

Figure 6. Interactive effects of bitter melon rootstock and grafted cucumber on the time course change of enzyme activity of (**A**) Superoxide dismutase (SOD), (**B**) Peroxidase (POD), (**C**) Catalase (CAT) and (**D**) Ascorbate peroxidase (APX) in cucumber leaves under heat stress conditions. Self-grafted plants exposed to 28 °C/18 °C (day/night), Cs-28; bitter-melon-grafted plants exposed to 28 °C/18 °C (day/night), Mc-28; self-grafted plants exposed to 42 °C/32 °C (day/night), Cs-42; bitter-melon-grafted plants exposed to 42 °C/32 °C (day/night), Mc-42. The values are the mean ± SE of three independent experiments. The bars represent the standard error.

To further elucidate how PAs and H_2O_2 contributed to the change in antioxidant enzyme activities under conditions of high temperature stress, polyamine synthesis inhibitors (MGBG and D-Arg) and hydrogen peroxide scavengers (DMTU) were applied in subsequent experiments. The results are shown in Figure 7. The activity of the three enzymes in the grafted plants sprayed with the three inhibitors was markedly lower than that of the grafted plants under only heat stress conditions. Combined with previous results (Figure 5), it is speculated that polyamines may mediate the production of hydrogen peroxide and further regulate the antioxidant defense.

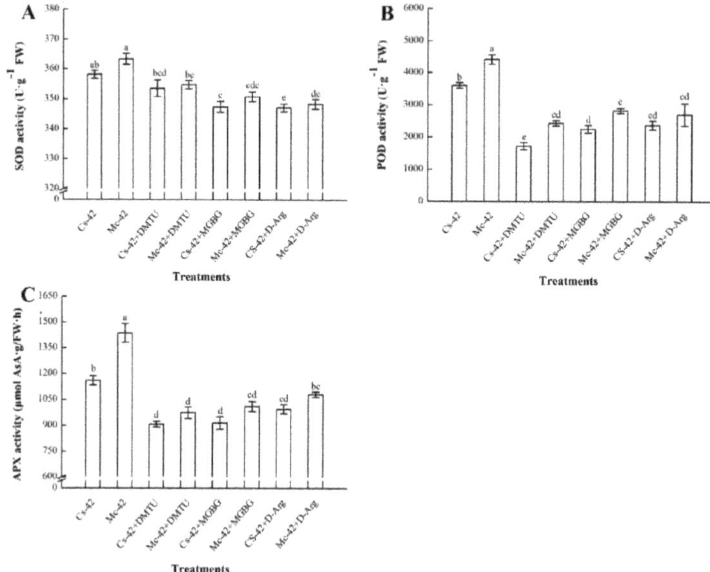

Figure 7. Effects of polyamine synthesis inhibitors D-arginine (D-Arg) and methylglyoxal bis-guanylhydrazone (MGBG) and an H_2O_2 scavenger (DMTU) on enzyme activity of (**A**) Superoxide dismutase (SOD), (**B**) Peroxidase (POD), and (**C**) Ascorbate peroxidase (APX) in grafted cucumber leaves under heat stress conditions. Self-grafted plants treated with 42 °C/32 °C (day/night), Cs-42; bitter-melon-grafted plants treated with 42 °C/32 °C (day/night), Mc-42. All inhibitors were sprayed 12 h before exposure to heat stress. Seedlings were treated for 24 h under heat stress conditions. Bars represent the mean ± SE of at least three independent experiments. The different letters indicate significant differences at $p < 0.05$ according to Duncan's multiple range test.

3. Discussion

In the last few decades, high-temperature stress has become a major environmental stressor that restricts proper growth and yield of crops [36,37], and the promotion of grafting technology provides an opportunity to improve crop resistance to various biotic and abiotic stresses, including high-temperature stress [6]. In our study, heat stress at 42/32 °C (day/night) significantly prevented the growth of grafted seedlings. However, seedlings grafted with bitter melon rootstock showed higher stem thicknesses and above-ground dry/fresh weights than self-grafted seedlings under conditions of high-temperature stress, showing that bitter melon rootstock efficiently ameliorated the growth inhibition caused by heat stress (Table 1, and Figure 1). Similar to the bitter melon rootstock, the grafting of heat-resistant luffa rootstock also alleviated the growth inhibition of cucumber seedlings caused by heat stress [33]. Moreover, in addition to high-temperature stress, rootstock grafting has also shown a mitigating effect on growth inhibition under exposure to other abiotic stresses [38,39]. Leaf photosynthesis performs a vital function in determining crop yield [12], and electron transfer during photosynthesis is very sensitive to heat stress, which is the main limiting factor of photosynthesis under high-temperature stress [40]. PSII is also considered to be more vulnerable to high-temperature damage [7]. Our results indicate that high temperatures induce photoinhibition of grafted seedlings, which is manifested in decreases in *Fv/Fm*, qP, and Y (II) and an increase in F_0. However, bitter melon rootstock significantly alleviated the photoinhibition caused by high-temperature stress. The changes in fluorescence parameters agreed with previous results of studies where cucumbers were grafted onto other rootstocks under heat stress conditions [32,41].

Polyamines have been widely reported to have an important role in the plant response to abiotic stress [42–44]. In our study, the results showed that high temperatures induce a large and rapid increase in the free endogenous PAs contents in the leaves of grafted seedlings, which mostly reached a peak at 8 h (Figure 3). The trend was identical to the time course of the changes in the endogenous PA level in cucumber plants under salt stress [23]. In addition, the increased content of PAs in the leaves of bitter-melon-grafted seedlings was more prominent than that in self-grafted seedlings. All of these results imply that rootstock grafting may produce a stronger response to stress by increasing endogenous polyamine levels.

With the deepening of research, more studies have indicated that polyamines participate in the regulation of complex signal systems to resist stress [22,24,26]. Hydrogen peroxide has also been frequently documented as a signaling molecule that plays a role in different stresses [45–47]. Moreover, in recent years, studies have confirmed that there are interconnections between polyamine and hydrogen peroxide signaling molecules under stress conditions [48,49]. However, it is unclear whether they respond similarly to grafted plants exposed to high-temperature stress. In the present study, our results showed that high-temperature stress induces a rapid response from H_2O_2 (Figure 4), and the changing trend of the content of H_2O_2 is consistent with the results of our previous study [50]. Moreover, grafted seedlings showed an obvious sequence in response to H_2O_2, whereby the bitter-melon-grafted seedlings peaked at 8 h, while the self-grafted seedlings peaked at 12 h, indicating that bitter-melon-grafted seedlings can accumulate hydrogen peroxide more quickly and significantly in response to high-temperature stress. To further explore the correlation between PAs and H_2O_2, starting from the difference between the peaks of PAs and H_2O_2 in the leaves of self-grafted cucumber seedlings, polyamine synthesis inhibitors were used to investigate whether changes in endogenous polyamines under stress have a mediating effect on H_2O_2. The results showed (Figure 5) that pretreatment with D-Arg and MGBG dramatically decreased the H_2O_2 content in the leaves of grafted seedlings under conditions of heat stress treatment. Therefore, it is speculated that the H_2O_2 content in the leaves is mediated by endogenous polyamines. The result is consistent with findings in cucumber seedlings under salt stress [23].

The antioxidant defense system is a key strategy by which plants cope with abiotic stress [4,5,51,52]. Some previous reports have shown that polyamines and H_2O_2 signaling molecules actively participate in antioxidant defense under exposure to a stressful environment [23,26]. In this study, under conditions of high temperature stress, bitter-melon-grafted seedlings maintained a high level of antioxidant enzyme activity (except CAT) (Figure 6), similar to that observed in wheat seedling leaves under a heat stress environment [53], and the peak of antioxidant enzyme activity occurred after the production of PAs and H_2O_2. From this, we speculate that the heat-induced antioxidant defense in bitter-melon-grafted seedlings is triggered by polyamines and involves H_2O_2. In order to verify our conjecture, a follow-up inhibitor test was designed. Our result showed that the polyamine synthesis inhibitors (D-Arg and MGBG) and H_2O_2 scavenger (DMTU) significantly reduced the higher level of antioxidants in the leaves of rootstock-grafted seedlings under conditions of high-temperature stress (Figure 7). This result is very similar to that of exogenous polyamines, and this was confirmed by increasing the content of endogenous polyamines to regulate the hydrogen peroxide signal, thereby improving the antioxidant defense capacity of cucumber seedlings under salt stress [24]. It further shows that under a high-temperature stress environment, both polyamines and H_2O_2 mediate the antioxidant defense of grafted seedlings. Under heat stress conditions, the bitter-melon-grafted seedlings showed greater endogenous polyamine accumulation and a more sensitive H_2O_2 response than the self-rooted grafted seedlings. Thus, they exhibited a stronger antioxidant defense capacity, and the damage to plants caused by heat stress was significantly attenuated.

4. Materials and Methods

4.1. Plant Materials and Treatments

Cucumber (*Cucumis sativus* L., cv. Jinyou No.35) was used as the scion (Cs), and bitter melon (*Momordica Charantia* L., cv. Changlv) was used as the rootstock (Mc). In this study, cleft grafting was used for grafting and self-grafted plants were used as a control. Uniform and germinated seeds of bitter melon were sown in organic substrates (2:2:1 [$v/v/v$] vinegar waste compost/peat/vermiculite; Peilei, Zhenjiang, China). When the cotyledons of cucumber rootstock were flattened, and bitter melon rootstock was outcropped, the cucumber scion was sown into a vermiculite-plated plastic square dish. When the cotyledons of the scions and the second true leaves of the rootstock were fully expanded, cleft grafting was performed. Grafted plants were transferred to a small, plastic, arched shed, where the environment was maintained as follows: a temperature of around 25 °C and a relative humidity of 85–100% for 7 days until the graft union had completely healed. After the third true leaves were fully expanded, grafted plants were shifted to a growth chamber (RDN-560E-4; Dongnan Instrument, Ningbo, China), where a photosynthetic photon flux density (PPFD) of 300 μmol m^{-2} s^{-1}, a relative humidity of 70–75%, and a 12/12 h (day/night) light/dark photoperiod were maintained.

When seedlings attained the fourth leaves stage, the plants underwent different treatments: (1) self-grafted plants were exposed to 28 °C/18 °C (day/night), Cs-28; (2) bitter-melon-grafted plants were exposed to 28 °C/18 °C (day/night), Mc-28; (3) self-grafted plants were exposed to 42 °C/32 °C (day/night), Cs-42; (4) bitter-melon-grafted plants were exposed to 42 °C/32 °C (day/night), Mc-42; (5) self-grafted plants were pretreated with 0.2 mM methylglyoxal bis-guanylhydrazone (MGBG, an inhibitor of SAMDC-S-adenosyl methionine decarboxylase) and then exposed to 42 °C/32 °C (day/night), Cs-42 + MGBG; (6) bitter-melon-grafted plants were pretreated with 1 mM methylglyoxal bis-guanylhydrazone (MGBG, an inhibitor of SAMDC) and then exposed to 42 °C/32 °C (day/night), Mc-42+MGBG; (7) self-grafted plants were pretreated with 2 mM D-Arginine (D-Arg, an inhibitor of ADC) and then exposed to 42 °C/32 °C (day/night), Cs-42 + D-Arg; (8) bitter-melon-grafted plants were pretreated with 2 mM D-Arginine (D-Arg, an inhibitor of ADC), and then exposed to 42 °C/32 °C (day/night), Mc-42 + D-Arg; (9) self-grafted plants were pretreated with 5 mM dimethylthiourea (DMTU, a H$_2$O$_2$ scavenger) and then exposed to 42 °C/32 °C (day/night), Cs-42 + DMTU; (10) bitter-melon-grafted plants were pretreated with 5 mM dimethylthiourea (DMTU, a H$_2$O$_2$ scavenger) and then exposed to 42 °C/32 °C (day/night), Mc-42 + DMTU. The plants were sprayed with inhibitors or scavengers 12 h before the high-temperature treatment. The leaves sampled were collected after different treatment time points (for the durations of the different chemical treatments, please see each figure legend) and stored at −80 °C until subsequent analysis.

4.2. Measurement of Plant Growth

The distance between the graft junctions to the scion growth point was measured by a ruler and taken to be the plant height. The thickness of the scion stem in the direction parallel to the scion cotyledon was determined by an electronic Vernier caliper, and this represented the stem thickness. To determine the fresh weight of the above-ground plant parts, we first washed them with distilled water and then wiped off excess water and measured the fresh weight. After drying at 115 °C for 15 min, the above-ground plant parts were then oven-dried at 75 °C to obtain their constant dry weights.

4.3. Measurement of Chlorophyll Fluorescence

Chlorophyll fluorescence measurement was undertaken with the method described by [54] using the M series chlorophyll fluorescence imaging system (Walz, Effeltrich, Germany), and Imaging-Win software was used to obtain the fluorescence parameter data and image.

4.4. Quantification of Endogenous Polyamines

Free polyamines were determined using the method of [55] with a few modifications. Firstly, for polyamine extraction, 0.5 g of leaves was homogenized with 1.6 mL of cold 5% (*w/v*) perchloric acid (PCA) in an ice bath for 1 h followed by centrifugation at 12,000× g for 20 min at 4 °C. The supernatant was then used to determine the free polyamine content. To 0.7 mL of supernatant, 1.4 mL of NaOH (2M) and 15 μL of benzoyl chloride were added, and the solution was vortexed for 30 s, after which it was incubated for 30 min at 37 °C. After that, to terminate the reaction, 2 mL of saturated NaCl solution was added to the resulting solution. To extract benzoyl polyamines, 2 mL of diethyl ether was added to the mixed solution, followed by centrifugation at 3000× g for 5 min at 4 °C. Finally, the extracted benzoyl PAs were evaporated to dryness and then re-dissolved in 1 mL of 64% (*v/v*) methanol. After passing through a 0.45 μmol filter, they were stored at −20 °C. Ultra-performance liquid chromatography (UPLC) was used to determine the content of polyamines.

4.5. Determination of H_2O_2 Content

The H_2O_2 concentration was determined using the method described by [56] with slight modifications. Firstly, the harvested leaf samples were homogenized in 1.6 mL of 0.1% trichloroacetic acid (TCA) followed by centrifugation at 12,000× g for 20 min at 4 °C. Then, 0.25 mL of 0.1 M potassium phosphate buffer (pH 7.8) and 1 mL of 1 M KI (potassium iodine) were incorporated into 0.2 mL of supernatant and held in a dark place for 1 h. After the reaction was completed, 0.1% of TCA was used as a blank control to zero, and the absorbance was read at 390 nm. Finally, the H_2O_2 content was estimated from a standard curve of known concentrations of H_2O_2.

4.6. Assays of Antioxidant Enzyme Activity

Two hundred milligrams of fresh leaf samples were digested in 1.6 mL of 0.05 M pre-cold phosphate buffer (pH 7.8), followed by centrifugation at 12,000× g for 20 min at 4 °C to obtain the supernatant. The extracted supernatant was used to assay the following antioxidant enzyme activities.

For the estimation of superoxide dismutase (SOD) activity, we used the method developed by [57]. Forty microliters (40 μL) of supernatant were added to a reaction mixture consisting of 14.5 mM methionine, 0.05 M phosphate buffer (pH 7.8), 30 μM EDTA-Na$_2$ (Disodium ethylene diamine tetra acetate dihydrate) solution, 2.25 mM nitro blue tetrazolium NBT solution, and 60 μM riboflavin solution. The SOD activity absorbance was measured at 560 nm.

For the measurement of peroxidase (POD) activity, the procedure described by [58] was used with slight modifications. In short, 40 μL of enzyme extract was incorporated into a reaction mixture consisting of 0.2 M phosphate buffer (pH 6.0), 30% H_2O_2 solution, and 50 mM guaiacol, and the absorbance was recorded at 470 nm.

Catalase (CAT) activity was determined by [59]. Briefly, 0.1 mL of enzyme extract was mixed with a reaction solution that contained 0.15 M phosphate buffer (pH 7.0) and 30% H_2O_2 solution. Measurement of the change in absorbance was conducted within 40 s at 240 nm.

For the estimation of ascorbate peroxidase (APX) activity, the method developed in [60] was used. In short, 0.1 mL of enzyme extract was mixed in a complex mixture made up of 50 mM phosphate buffer (pH 6.0), 0.1 mM EDTA–Na$_2$, 5 mM AsA, and 20 mM H_2O_2 solution. The measurement of the change of absorbance was conducted within 40 s at 290 nm.

4.7. Statistical Analysis

For each measurement, at least three (3) independent biological replicates were tested. All data were statistically analyzed with the SPSS 20.0 software program (SPSS, Chicago, IL, USA) using Duncan's multiple range test at the $p < 0.05$ level of significance.

5. Conclusions

In summary, our findings suggest that bitter melon rootstock improves the heat resistance of grafted seedlings by alleviating the photoinhibition induced by heat stress and improving the antioxidant defense capacity of leaves by regulating the changes in endogenous polyamines and H_2O_2 in leaves under conditions of high temperature stress. However, further study is needed to determine how the contents of polyamines and hydrogen peroxide act on the antioxidant defense system.

Author Contributions: S.-R.G. conceived and planned the experiments. M.-Q.T. performed the study and wrote the original manuscript. M.S.J. provided supplied materials, collected data, examined data and edited the original manuscript. K.H. contacted the software and analyzed data. S.S., Y.W. and J.S. supervised and revised the manuscript. All authors have read and agreed to the published version of the manuscript.

Funding: This research work was supported by the China Agriculture Research System (CARS-23-B12) and fund collected by Shi-Rong Guo.

Acknowledgments: This work was supported by the China Agriculture Research System (CARS-23-B12).

Conflicts of Interest: The authors declared that there are no potential conflicts of interest that need to be disclosed.

References

1. Hasanuzzaman, M.; Nahar, K.; Alam, M.; Roychowdhury, R.; Fujita, M. Physiological, Biochemical, and Molecular Mechanisms of Heat Stress Tolerance in Plants. *Int. J. Mol. Sci.* **2013**, *14*, 9643–9684. [CrossRef]
2. Dahal, K.; Li, X.-Q.; Tai, H.; Creelman, A.; Bizimungu, B. Improving Potato Stress Tolerance and Tuber Yield Under a Climate Change Scenario—A Current Overview. *Front. Plant Sci.* **2019**, *10*. [CrossRef]
3. Song, Q.; Yang, F.; Cui, B.; Li, J.; Zhang, Y.; Li, H.; Qiu, N.; Wang, F.; Gao, J. Physiological and molecular responses of two Chinese cabbage genotypes to heat stress. *Biol. Plant.* **2019**, *63*, 548–555. [CrossRef]
4. Jahan, M.S.; Shu, S.; Wang, Y.; Chen, Z.; He, M.; Tao, M.; Sun, J.; Guo, S. Melatonin alleviates heat-induced damage of tomato seedlings by balancing redox homeostasis and modulating polyamine and nitric oxide biosynthesis. *BMC Plant Biol.* **2019**, *19*. [CrossRef]
5. Jahan, M.S.; Wang, Y.; Shu, S.; Zhong, M.; Chen, Z.; Wu, J.; Sun, J.; Guo, S. Exogenous salicylic acid increases the heat tolerance in Tomato (*Solanum lycopersicum* L) by enhancing photosynthesis efficiency and improving antioxidant defense system through scavenging of reactive oxygen species. *Sci. Hortic.* **2019**, *247*, 421–429. [CrossRef]
6. Xu, Y.; Yuan, Y.; Du, N.; Wang, Y.; Shu, S.; Sun, J.; Guo, S. Proteomic analysis of heat stress resistance of cucumber leaves when grafted onto Momordica rootstock. *Hortic. Res.* **2018**, *5*. [CrossRef]
7. Wei, Y.; Wang, Y.; Wu, X.; Shu, S.; Sun, J.; Guo, S. Redox and thylakoid membrane proteomic analysis reveals the Momordica (*Momordica charantia* L.) rootstock-induced photoprotection of cucumber leaves under short-term heat stress. *Plant Physiol. Biochem.* **2019**, *136*, 98–108. [CrossRef] [PubMed]
8. Ding, X.; Jiang, Y.; Hao, T.; Jin, H.; Zhang, H.; He, L.; Zhou, Q.; Huang, D.; Hui, D.; Yu, J. Effects of Heat Shock on Photosynthetic Properties, Antioxidant Enzyme Activity, and Downy Mildew of Cucumber (*Cucumis sativus* L.). *PLoS ONE* **2016**, *11*. [CrossRef] [PubMed]
9. Yamori, W.; Shikanai, T. Physiological functions of cyclic electron transport around photosystem I in sustaining photosynthesis and plant growth. *Annu. Rev. Plant Biol.* **2016**, *67*, 81–106. [CrossRef] [PubMed]
10. Neves, L.H.; Nunes Santos, R.I.; dos Santos Teixeira, G.I.; de Araujo, D.G.; Duarte Silvestre, W.V.; Pinheiro, H.A. Leaf gas exchange, photochemical responses and oxidative damages in assai (*Euterpe oleracea* Mart.) seedlings subjected to high temperature stress. *Sci. Hortic.* **2019**, *257*. [CrossRef]
11. Posch, B.C.; Kariyawasam, B.C.; Bramley, H.; Coast, O.; Richards, R.A.; Reynolds, M.P.; Trethowan, R.; Atkin, O.K. Exploring high temperature responses of photosynthesis and respiration to improve heat tolerance in wheat. *J. Exp. Bot.* **2019**, *70*, 5051–5069. [CrossRef]
12. Rotundo, J.L.; Tang, T.; Messina, C.D. Response of maize photosynthesis to high temperature: Implications for for modeling the impact of global warming. *Plant Physiol. Biochem.* **2019**, *141*, 202–205. [CrossRef]
13. Makonya, G.M.; Ogola, J.B.O.; Muasya, A.M.; Crespo, O.; Maseko, S.; Valentine, A.J.; Ottosen, C.-O.; Rosenqvist, E.; Chimphango, S.B.M. Chlorophyll fluorescence and carbohydrate concentration as field selection traits for heat tolerant chickpea genotypes. *Plant Physiol. Biochem.* **2019**, *141*, 172–182. [CrossRef] [PubMed]

14. Parvin, S.; Lee, O.R.; Sathiyaraj, G.; Khorolragchaa, A.; Kim, Y.-J.; Yang, D.-C. Spermidine alleviates the growth of saline-stressed ginseng seedlings through antioxidative defense system. *Gene* **2014**, *537*, 70–78. [CrossRef]

15. Shen, J.-l.; Wang, Y.; Shu, S.; Jahan, M.S.; Zhong, M.; Wu, J.-Q.; Sun, J.; Guo, S.-R. Exogenous putrescine regulates leaf starch overaccumulation in cucumber under salt stress. *Sci. Hortic.* **2019**, *253*, 99–110. [CrossRef]

16. Nahar, K.; Hasanuzzaman, M.; Rahman, A.; Alam, M.M.; Jubayer-Al, M.; Suzuki, T.; Fujita, M. Polyamines Confer Salt Tolerance in Mung Bean (*Vigna radiata* L.) by Reducing Sodium Uptake, Improving Nutrient Homeostasis, Antioxidant Defense, and Methylglyoxal Detoxification Systems. *Front. Plant Sci.* **2016**, *7*. [CrossRef] [PubMed]

17. Mostofa, M.G.; Yoshida, N.; Fujita, M. Spermidine pretreatment enhances heat tolerance in rice seedlings through modulating antioxidative and glyoxalase systems. *Plant Growth Regul.* **2014**, *73*, 31–44. [CrossRef]

18. Song, Y.; Diao, Q.; Qi, H. Putrescine enhances chilling tolerance of tomato (*Lycopersicon esculentum* Mill.) through modulating antioxidant systems. *Acta Physiol. Plant.* **2014**, *36*, 3013–3027. [CrossRef]

19. Sanchez-Rodriguez, E.; Romero, L.; Ruiz, J.M. Accumulation of free polyamines enhances the antioxidant response in fruits of grafted tomato plants under water stress. *J. Plant Physiol.* **2016**, *190*, 72–78. [CrossRef]

20. Jia, Y.X.; Sun, J.; Guo, S.R.; Li, J.; Hu, X.H.; Wang, S.P. Effect of root-applied spermidine on growth and respiratory metabolism in roots of cucumber (*Cucumis sativus*) seedlings under hypoxia. *Russ. J. Plant Physiol.* **2010**, *57*, 648–655. [CrossRef]

21. Reggiani, R.; Aurisano, N.; Mattana, M.; Bertani, A. Effect of K^+ ions on polyamine levels in wheat seedlings under anoxia. *J. Plant Physiol.* **1993**, *142*, 94–98. [CrossRef]

22. Pal, M.; Szalai, G.; Janda, T. Speculation: Polyamines are important in abiotic stress signaling. *Plant Sci.* **2015**, *237*, 16–23. [CrossRef] [PubMed]

23. Wu, J.; Shu, S.; Li, C.; Sun, J.; Guo, S. Spermidine-mediated hydrogen peroxide signaling enhances the antioxidant capacity of salt-stressed cucumber roots. *Plant Physiol. Biochem.* **2018**, *128*, 152–162. [CrossRef] [PubMed]

24. Agurla, S.; Gayatri, G.; Raghavendra, A.S. Polyamines increase nitric oxide and reactive oxygen species in guard cells of Arabidopsis thaliana during stomatal closure. *Protoplasma* **2018**, *255*, 153–162. [CrossRef]

25. Li, Z.; Zhang, Y.; Peng, D.; Wang, X.; Peng, Y.; He, X.; Zhang, X.; Ma, X.; Huang, L.; Yan, Y. Polyamine regulates tolerance to water stress in leaves of white clover associated with antioxidant defense and dehydrin genes via involvement in calcium messenger system and hydrogen peroxide signaling. *Front. Physiol.* **2015**, *6*. [CrossRef]

26. Gong, B.; Wan, X.; Wei, M.; Li, Y.; Wei, M.; Shi, Q. Overexpression of S-adenosylmethionine synthetase 1 enhances tomato callus tolerance to alkali stress through polyamine and hydrogen peroxide cross-linked networks. *Plant Cell Tissue Organ Cult.* **2016**, *124*, 377–391. [CrossRef]

27. Zepeda-Jazo, I.; Maria Velarde-Buendia, A.; Enriquez-Figueroa, R.; Bose, J.; Shabala, S.; Muniz-Murguia, J.; Pottosin, I.I. Polyamines Interact with Hydroxyl Radicals in Activating Ca^{2+} and K^+ Transport across the Root Epidermal Plasma Membranes. *Plant Physiol.* **2011**, *157*, 2167–2180. [CrossRef]

28. Petropoulos, S.A.; Khah, E.M.; Passam, H.C. Evaluation of rootstocks for watermelon grafting with reference to plant development, yield and fruit quality. *Int. J. Plant Prod.* **2012**, *6*, 481–491.

29. Rouphael, Y.; Cardarelli, M.; Rea, E.; Colla, G. Improving melon and cucumber photosynthetic activity, mineral composition, and growth performance under salinity stress by grafting onto Cucurbita hybrid rootstocks. *Photosynthetica* **2012**, *50*, 180–188. [CrossRef]

30. Tripodi, G.; Condurso, C.; Cincotta, F.; Merlino, M.; Verzera, A. Aroma compounds in mini-watermelon fruits from different grafting combinations. *J. Sci. Food Agric.* **2020**, *100*, 1328–1335. [CrossRef]

31. Mudge, K.; Janick, J.; Scofield, S.; Goldschmidt, E. A History of Grafting. *Hortic. Rev.* **2009**, *35*. [CrossRef]

32. Li, H.; Ahammed, G.J.; Zhou, G.; Xia, X.; Zhou, J.; Shi, K.; Yu, J.; Zhou, Y. Unraveling Main Limiting Sites of Photosynthesis under Below- and Above-Ground Heat Stress in Cucumber and the Alleviatory Role of Luffa Rootstock. *Front. Plant Sci.* **2016**, *7*. [CrossRef] [PubMed]

33. Li, H.; Wang, F.; Chen, X.-J.; Shi, K.; Xia, X.-J.; Considine, M.J.; Yu, J.-Q.; Zhou, Y.-H. The sub/supra-optimal temperature-induced inhibition of photosynthesis and oxidative damage in cucumber leaves are alleviated by grafting onto figleaf gourd/luffa rootstocks. *Physiol. Plant.* **2014**, *152*, 571–584. [CrossRef] [PubMed]

34. Li, H.; Liu, S.-S.; Yi, C.-Y.; Wang, F.; Zhou, J.; Xia, X.-J.; Shi, K.; Zhou, Y.-H.; Yu, J.-Q. Hydrogen peroxide mediates abscisic acid-induced HSP70 accumulation and heat tolerance in grafted cucumber plants. *Plant Cell Environ.* **2014**, *37*, 2768–2780. [CrossRef]

35. Khraiwesh, B.; Zhu, J.-K.; Zhu, J. Role of miRNAs and siRNAs in biotic and abiotic stress responses of plants. *Biochim. Biophys. Acta Gene Regul. Mech.* **2012**, *1819*, 137–148. [CrossRef]

36. Xiang, N.; Hu, J.; Wen, T.; Brennan, M.A.; Brennan, C.S.; Guo, X. Effects of temperature stress on the accumulation of ascorbic acid and folates in sweet corn (*Zea mays* L.) seedlings. *J. Sci. Food Agric.* **2019**. [CrossRef]

37. Hsuan, T.-P.; Jhuang, P.-R.; Wu, W.-C.; Lur, H.-S. Thermotolerance evaluation of Taiwan Japonica type rice cultivars at the seedling stage. *Bot. Stud.* **2019**, *60*. [CrossRef]

38. Liu, Z.; Bie, Z.; Huang, Y.; Zhen, A.; Niu, M.; Lei, B. Rootstocks improve cucumber photosynthesis through nitrogen metabolism regulation under salt stress. *Acta Physiol. Plant.* **2013**, *35*, 2259–2267. [CrossRef]

39. Li, L.; Xing, W.-W.; Shao, Q.-S.; Shu, S.; Sun, J.; Guo, S.-R. The effects of grafting on glycolysis and the tricarboxylic acid cycle in Ca(NO$_3$)($_2$)-stressed cucumber seedlings with pumpkin as rootstock. *Acta Physiol. Plant.* **2015**, *37*. [CrossRef]

40. Huve, K.; Bichele, I.; Tobias, M.; Niinemets, U. Heat sensitivity of photosynthetic electron transport varies during the day due to changes in sugars and osmotic potential. *Plant Cell Environ.* **2006**, *29*, 212–228. [CrossRef]

41. Li, H.; Wang, Y.; Wang, Z.; Guo, X.; Wang, F.; Xia, X.-J.; Zhou, J.; Shi, K.; Yu, J.-Q.; Zhou, Y.-H. Microarray and genetic analysis reveals that csa-miR159b plays a critical role in abscisic acid-mediated heat tolerance in grafted cucumber plants. *Plant Cell Environ.* **2016**, *39*, 1790–1804. [CrossRef] [PubMed]

42. Alcazar, R.; Altabella, T.; Marco, F.; Bortolotti, C.; Reymond, M.; Koncz, C.; Carrasco, P.; Tiburcio, A.F. Polyamines: Molecules with regulatory functions in plant abiotic stress tolerance. *Planta* **2010**, *231*, 1237–1249. [CrossRef]

43. Nahar, K.; Hasanuzzaman, M.; Alam, M.M.; Fujita, M. Exogenous Spermidine Alleviates Low Temperature Injury in Mung Bean (*Vigna radiata* L.) Seedlings by Modulating Ascorbate-Glutathione and Glyoxalase Pathway. *Int. J. Mol. Sci.* **2015**, *16*, 30117–30132. [CrossRef] [PubMed]

44. Liu, J.-H.; Wang, W.; Wu, H.; Gong, X.; Moriguchi, T. Polyamines function in stress tolerance: From synthesis to regulation. *Front. Plant Sci.* **2015**, *6*. [CrossRef]

45. Saxena, I.; Srikanth, S.; Chen, Z. Cross Talk between H$_2$O$_2$ and Interacting Signal Molecules under Plant Stress Response. *Front. Plant Sci.* **2016**, *7*. [CrossRef]

46. Wang, Y.; Zhang, J.; Li, J.-L.; Ma, X.-R. Exogenous hydrogen peroxide enhanced the thermotolerance of Festuca arundinacea and Lolium perenne by increasing the antioxidative capacity. *Acta Physiol. Plant.* **2014**, *36*, 2915–2924. [CrossRef]

47. Liu, T.; Hu, X.; Zhang, J.; Zhang, J.; Du, Q.; Li, J. H$_2$O$_2$ mediates ALA-induced glutathione and ascorbate accumulation in the perception and resistance to oxidative stress in *Solanum lycopersicum* at low temperatures. *BMC Plant Biol.* **2018**, *18*. [CrossRef]

48. Gupta, K.; Sengupta, A.; Chakraborty, M.; Gupta, B. Hydrogen Peroxide and Polyamines Actas Double Edged Swords in Plant Abiotic Stress Responses. *Front. Plant Sci.* **2016**, *7*. [CrossRef]

49. Diao, Q.; Song, Y.; Shi, D.; Qi, H. Interaction of Polyamines, Abscisic Acid, Nitric Oxide, and Hydrogen Peroxide under Chilling Stress in Tomato (*Lycopersicon esculentum* Mill.) Seedlings. *Front. Plant Sci.* **2017**, *8*. [CrossRef]

50. Shu, S.; Gao, P.; Li, L.; Yuan, Y.; Sun, J.; Guo, S. Abscisic Acid-Induced H$_2$O$_2$ Accumulation Enhances Antioxidant Capacity in Pumpkin-Grafted Cucumber Leaves under Ca(NO$_3$)($_2$) Stress. *Front. Plant Sci.* **2016**, *7*. [CrossRef]

51. Jahan, M.S.; Guo, S.; Baloch, A.R.; Sun, J.; Shu, S.; Wang, Y.; Ahammed, G.J.; Kabir, K.; Roy, R. Melatonin alleviates nickel phytotoxicity by improving photosynthesis, secondary metabolism and oxidative stress tolerance in tomato seedlings. *Ecotoxicol. Environ. Saf.* **2020**, *197*, 110593. [CrossRef] [PubMed]

52. Sarwar, M.; Saleem, M.F.; Ullah, N.; Rizwan, M.; Ali, S.; Shahid, M.R.; Alamri, S.A.; Alyemeni, M.N.; Ahmad, P. Exogenously applied growth regulators protect the cotton crop from heat-induced injury by modulating plant defense mechanism. *Sci. Rep.* **2018**, *8*. [CrossRef]

53. Chen, X.; Li, L.; Niu, H.; Yi, Y. Effects of high temperature stress on antioxidant enzyme activities in leaves of different wheat varieties. *Henan Agric. Sci.* **2008**, 38–40.

54. Lu, C.M.; Qiu, N.W.; Wang, B.S.; Zhang, J.H. Salinity treatment shows no effects on photosystem II photochemistry, but increases the resistance of photosystem II to heat stress in halophyte *Suaeda salsa*. *J. Exp. Bot.* **2003**, *54*, 851–860. [CrossRef] [PubMed]
55. Shu, S.; Guo, S.-R.; Sun, J.; Yuan, L.-Y. Effects of salt stress on the structure and function of the photosynthetic apparatus in *Cucumis sativus* and its protection by exogenous putrescine. *Physiol. Plant.* **2012**, *146*, 285–296. [CrossRef]
56. Alexieva, V.; Sergiev, I.; Mapelli, S.; Karanov, E. The effect of drought and ultraviolet radiation on growth and stress markers in pea and wheat. *Plant Cell Environ.* **2001**, *24*, 1337–1344. [CrossRef]
57. Giannopolitis, C.N.; Ries, S.K. Superoxide dismutases: I. Occurrence in higher plants. *Plant Physiol.* **1977**, *59*, 309–314. [CrossRef]
58. Tan, W.; Liu, J.; Dai, T.; Jing, Q.; Cao, W.; Jiang, D. Alterations in photosynthesis and antioxidant enzyme activity in winter wheat subjected to post-anthesis water-logging. *Photosynthetica* **2008**, *46*, 21–27. [CrossRef]
59. Dhindsa, R.; Plumb-Dhindsa, P.; Reid, D. Leaf senescence and lipid peroxidation: Effects of some phytohormones, and scavengers of free radicals and singlet oxygen. *Physiol. Plant.* **2006**, *56*, 453–457. [CrossRef]
60. Nakano, Y.; Asada, K. Hydrogen Peroxide is Scavenged by Ascorbate-specific Peroxidase in Spinach Chloroplasts. *Plant Cell Physiol.* **1980**, *22*. [CrossRef]

 *plants*

Article

Peptone-Induced Physio-Biochemical Modulations Reduce Cadmium Toxicity and Accumulation in Spinach (*Spinacia oleracea* L.)

Naila Emanuil [1], Muhammad Sohail Akram [1,*], Shafaqat Ali [2,3,*], Mohamed A. El-Esawi [4], Muhammad Iqbal [1] and Mohammed Nasser Alyemeni [5]

[1] Department of Botany, Government College University, Faisalabad 38000, Pakistan; naila60jb@gmail.com (N.E.); iqbalm@gcuf.edu.pk (M.I.)

[2] Department of Environmental Sciences and Engineering, Government College University, Faisalabad 38000, Pakistan

[3] Department of Biological Sciences and Technology, China Medical University, Taichung 40402, Taiwan

[4] Botany Department, Faculty of Science, Tanta University, Tanta 31527, Egypt; mohamed.elesawi@science.tanta.edu.eg

[5] Botany and Microbiology Department, College of Science, King Saud University, Riyadh 11451, Saudi Arabia; mnyemeni@ksu.edu.sa

* Correspondence: msohailakram@gcuf.edu.pk (M.S.A.); shafaqataligill@gcuf.edu.pk (S.A.)

Received: 12 November 2020; Accepted: 16 December 2020; Published: 19 December 2020

Abstract: The accumulation of cadmium (Cd) in edible plant parts and fertile lands is a worldwide problem. It negatively influences the growth and productivity of leafy vegetables (e.g., spinach, *Spinacia oleracea* L.), which have a high tendency to radially accumulate Cd. The present study investigated the influences of peptone application on the growth, biomass, chlorophyll content, gas exchange parameters, antioxidant enzymes activity, and Cd content of spinach plants grown under Cd stress. Cd toxicity negatively affected spinach growth, biomass, chlorophyll content, and gas exchange attributes. However, it increased malondialdehyde (MDA), hydrogen peroxide (H_2O_2), electrolyte leakage (EL), proline accumulation, ascorbic acid content, Cd content, and activity of antioxidant enzymes such as superoxide dismutase (SOD), peroxidase (POD), and catalase (CAT) in spinach plants. The exogenous foliar application of peptone increased the growth, biomass, chlorophyll content, proline accumulation, and gas exchange attributes of spinach plants. Furthermore, the application of peptone decreased Cd uptake and levels of MDA, H_2O_2, and EL in spinach by increasing the activity of antioxidant enzymes. This enhancement in plant growth and photosynthesis might be due to the lower level of Cd accumulation, which in turn decreased the negative impacts of oxidative stress in plant tissues. Taken together, the findings of the study revealed that peptone is a promising plant growth regulator that represents an efficient approach for the phytoremediation of Cd-polluted soils and enhancement of spinach growth, yield, and tolerance under a Cd-dominant environment.

Keywords: antioxidants; cadmium; oxidative stress; peptone; spinach

---

1. Introduction

Cadmium (Cd) is a very toxic element and has a high mobility rate from the soil to plants. Cd has been extensively studied in agriculture systems due to its toxic effects on plants and humans [1]. The concentration of Cd is being increased in the soil due to its uses in sludge applications, sewage irrigation, chemical fertilizers, and atmospheric depositions [2,3]. Recently, the contamination of agricultural soils with Cd has become one of the most severe environmental issues in Asian countries [4]. In Pakistan, inorganic fertilizers have been extensively used to increase the yield of crops, resulting in the accumulation of higher concentrations of Cd in the soil. Cd can readily enter into plant roots and translocate to shoots. With increasing the intracellular Cd concentrations, various plant metabolic processes, including cell division, mineral nutrition, nitrogen metabolism, photosynthesis, and respiration, are severely affected [5]. Exposure of plants to Cd stress results in leaf rolling, chlorophyll content decrease, and photosynthesis inhibition via suppressing the chlorophyll biosynthesis [6]. Cd decreases the uptake of micro- and macro-nutrients, which in turn affects the plant transport activities [7]. It also affects nitrate absorption and its translocation from plant roots to shoots via inhibiting nitrate reductase. Moreover, it alters the composition of lipids and causes changes in the membrane functionality. Higher Cd concentrations, within the plant body, generate reactive oxygen species (ROS), which cause oxidative stress [8], leading to the degradation of pigments, nucleic acid, and proteins. The radial absorption capability of Cd by plant roots and its translocation toward the aerial parts make it of great concern regarding its accumulation in the food chain [9]. Vegetables cultivated in contaminated soils accumulate high concentrations of Cd, leading to increases in food contamination risks. Spinach (*Spinacia oleracea* L.) is a leafy vegetable with green and broad large-surface-area leaves, high growth rate, and metal-accumulating ability [10]. Spinach accumulates Cd via plant roots and translocate it to the leaves, resulting in increased food contamination risks. Therefore, there is an urgent need to minimize the accumulation of Cd in spinach plants.

Amino acids are building blocks for protein synthesis and play a critical role in plant metabolism. Amino acids are very important to stimulate cell growth as they can act as a buffer to maintain the optimum pH for plant cells. They eliminate ammonia from plant cells and protect them against ammonia toxicity. They are also involved in the biosynthesis of different organic compounds such as hormones, vitamins, pigments, alkaloids, terpenoids, purine, pyrimidine, and enzymes [11]. Moreover, they are involved in the stimulation of physiological and metabolic functions of plants [12]. Several studies revealed the positive effects of amino acids on plant growth [13,14]. However, very little is known about the amino acids that induce impacts on plants grown under heavy metal stress, particularly Cd.

Biostimulants can regulate natural plant processes to increase the nutrients intake, environmental stress tolerance, and crop yield [15]. It has been suggested that biostimulants could interact with signaling processes and help plants to cope with the environmental stress. Biostimulants are derived from various inorganic and biological compounds, such as fulvic and humic substances, animal feeds, protein hydrolysates, and industrial discards [16–18]. Protein hydrolysates increase the activity of enzymes and improve the uptake of nutrients [19]. Hydrolyzed proteins might contain short-chain bioactive peptides, which have immunological and hormonal activities and could act as plant growth regulators [20,21]. Peptone is a mixture of short peptides and amino acids. It is a biostimulant in the form of macro-granules and is produced from the enzymatic hydrolysis of animal or plant proteins. It is mainly involved in the stimulation of plant growth and may provide protection against pathogens. Acid hydrolysate of peptone from vegetable sources is a rich source of the nitrogenous material and contains a high concentration of free amino acids. The optimal concentration of available amino acids can play diverse roles in plant metabolism and growth under stress conditions [22]. Thus, we hypothesized that the physio-biochemical mechanisms of spinach could be significantly regulated through the foliar application of peptone, leading to the alleviation of the harmful impacts of Cd. Therefore, the main objective of the present study was to evaluate the harmful impacts of Cd on spinach growth attributes, biomass, Cd accumulation, and oxidative mechanisms. The present study also examined the possible role of peptone in enhancing the growth and production of spinach plants grown in Cd-contaminated areas via upregulating the defense systems as well as reducing Cd accumulation in edible plant parts.

2. Materials and Methods

2.1. Soil Analysis

Soil used in the study was taken from agricultural fields and was characterized at the Institute of Soil and Environmental Sciences, University of Agriculture Faisalabad. The soil was cleaned and sieved before using for filling the pots. The methods of Bouyoucos [23] and Page et al. [24] were used for particles size characterization and sodium absorption ratio determination. Soil electrical conductivity (EC) and pH was determined using EC/pH meter. The methodology of Soltanpour [25] was used for extracting the trace elements at pH 7.6 using the ammonium bicarbonate diethylene triamine penta-acetic acid (AB-DTPA) solution. The properties of the soil used for the experiment are shown in Table 1.

Table 1. The characteristics of the soil used in the experiment of the present study.

Soil Characteristics	Values
Texture	Sandy clay
Sand %	48
Silt %	16
Clay %	36
pH	7.65
EC dS m^{-1}	0.811
Soluble ions	
Ca^{2+} mmol L^{-1}	7.71
Na mmol L^{-1}	9.46
K mmol L^{-1}	1.34
Metal concentrations	
Cd mg kg^{-1}	0.83
Pb mg kg^{-1}	6.64
Zn mg kg^{-1}	4.93
Mn mg kg^{-1}	6.37
Fe mg kg^{-1}	50.34

2.2. Experimental Design

Seeds of spinach (*Spinacia oleracea* L.) were obtained from Ayub Agricultural Research Institute (AARI), Faisalabad. The seeds were sown in plastic pots filled with 5 kg soil/pot. The pots were arranged in a completely randomized design. In order to maintain five seedlings per pot, thinning was performed after 7 days and an appropriate fertilizer was applied. Water polluted with cadmium nitrate (0, 50, and 100 μM) was used for the irrigation of plants. Peptone (Millipore Sigma, 51841) solution (500 and 1000 mg L^{-1}) was prepared using ddH$_2$O. A 0.5 L peptone solution (or ddH$_2$O) was sprayed on 20 plants (four replicates for each treatment with five plants in each pot) twice after an interval of one week. Hence, a plant received 0, 25, or 50 mg of peptone after two spray treatments using 0, 500, and 1000 mg L^{-1} peptone solution, respectively.

2.3. Plant Harvesting and Sampling

The plants were harvested after 40 days and separated into shoots and roots to measure the growth parameters. Plants were then washed with a distilled water to remove adhered soil particles and were then air-dried. The plant fresh weight was then measured using an analytical balance. The plants were oven dried at 70 °C for 48 h to be used for estimating the dry weight of root and shoot, discretely.

2.4. Determination of Chlorophyll and Gas Exchange Parameters

The method of Arnon [26] was followed to measure the plant chlorophyll content. A 0.5 g leaf sample was dipped in 80% acetone at 4 °C in dark. The reading was then taken using a spectrophotometer at the wavelengths of 645, 663, and 480 nm in order to calculate the chlorophyll content. An IRGA (infra-red gas analyzer) was used to measure the stomatal conductance and photosynthetic rate at a high intensity of light between 10:00 to 12:00 a.m. on a full-sun day.

2.5. Determination of Electrolyte Leakage, Hydrogen Peroxide (H_2O_2), and Malondialdehyde (MDA)

The protocol of Dionisio-Sese and Tobita [27] was used to measure the electrolyte leakage (EL) level by dipping small pieces of leaves in deionized water. The first EC_1 reading was taken after the samples' incubation at 32 °C for 2 h, and the second EC_2 was determined after the same samples were subsequently incubated at 121 °C for 20 min. The level of EL was calculated as follows:

$$EL = (EC_1/EC_2) \times 100$$

The methods of Velikova et al. [28] and Carmak and Horst [29] were used to measure the levels of MDA and H_2O_2 in plants, respectively.

2.6. Determination of Superoxide Dismutase (SOD), Peroxidase (POD), and Catalase (CAT) Activity

The method of Zhang [30] was used to determine SOD and POD activities in spinach plants by grinding leaf samples in liquid nitrogen followed by their standardization with 0.5 M phosphate buffer of pH 7.8. The activity of CAT was measured following the protocol of Aebi [31].

2.7. Determination of the Contents of Ascorbic Acid and Proline

Ascorbic acid in plant samples was measured following the method of Mukherjee and Choudhuri [32]. The methodology of Bates et al. [33] was used to measure proline content in plant samples.

2.8. Determination of Cd Content

Cadmium content in plant samples was measured by digesting 0.5 g of plant samples in HNO_3 and $HCLO_4$ (3:1, *v/v*) in a conical digestion flask using a hot plate (350 °C) for about 8–10 h. The digested samples were run on an atomic absorption spectrophotometer to measure Cd concentration [34].

2.9. Statistical Analysis

Statistical analysis for the sample data was performed using SPSS for windows. The results were subjected to the two-way analysis of variance (ANOVA) using Tukey's test at a probability level of 5%, and the *p* value at 0.05 was considered significant.

3. Results

3.1. Plant Growth

Plant growth parameters were recorded to measure Cd toxicity in spinach. The results indicated that the foliar application of peptone exhibited positive effects on plant growth and morphology. Under Cd stress, the application of peptone enhanced the plant growth attributes such as fresh and dry biomass, root length and leaf area. By the application of 1000 mgL^{-1} of peptone, the leaf fresh weight exhibited significant increases of 23, 34, and 46% under Cd stress of 0, 50, and 100 µM, respectively, as compared to control. Application of 500 mgL^{-1} of peptone significantly increased the leaf fresh weight at 11%, 13%, and 24% under Cd stress of 0, 50, and 100 µM, respectively, as compared to control (Figure 1A). By the application of 500 mgL^{-1} of peptone, the leaf dry mass was significantly increased up to 15%, 24%, and 27% under Cd stress of 0, 50, and 100 µM, respectively, in comparison to control. A maximum significant increase of 54% in the leaf dry mass was recorded in plants treated with 1000 mgL^{-1} of peptone under 100 µM Cd stress (Figure 1B). The higher level of peptone spray (1000 mgL^{-1}) significantly ($p < 0.05$) increased the leaf area up to 21%, 34%, and 44% under Cd stress of 0, 50, and 100 µM, respectively, as compared to control (Figure 1F).

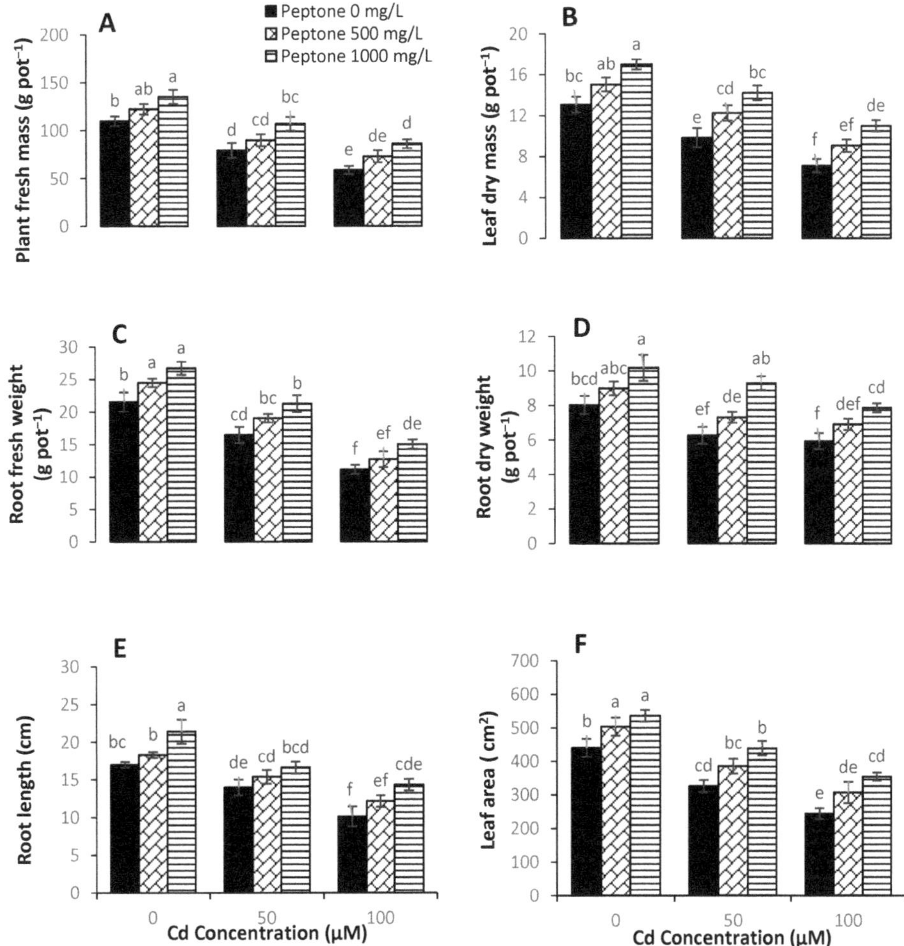

Figure 1. The influence of the foliar application of peptone (0, 500, and 1000 mgL^{-1}) on the leaf fresh weight (**A**), leaf dry weight (**B**), root fresh weight (**C**), root dry weight (**D**), taproot length (**E**), and leaf area (**F**) of spinach plants grown under Cd stress. Values represent the means of four replicates with standard deviations. Letters on the bar indicate the significant differences between given treatments at $p < 0.05$.

3.2. Photosynthetic Pigments

Photosynthesis is considered a major process involved in plant growth and biomass production. This process is very sensitive to heavy metal uptake/accumulation. Decreases in the activity of the photosynthetic system is pronounced under Cd stress. This might be due to the decreases in chlorophyll biosynthesis and photosynthetic enzymes' activity as well as the disturbance in the water and nutrient balance in plants. The current study revealed that the application of peptone for Cd-treated and non-treated plants amended the quantity of photosynthetic pigments (Figure 2). Plants treated with peptone (500 mgL^{-1}) exhibited a significantly increased chlorophyll *a* content of 9%, 43%, and 38% under Cd stress of 0, 50, and 100 μM), respectively, as compared to control. In comparison with control, a higher significant increase in chlorophyll *a* content of 69% was noted in plants treated with 1000 mgL^{-1} of peptone under 100 μM Cd application (Figure 2A). Plants treated with 1000 mgL^{-1} of peptone significantly increased chlorophyll *b* contents under Cd stress (Figure 2B). Moreover, the exogenous application of 500 mgL^{-1} of peptone significantly enhanced chlorophyll *a/b* ratio under Cd stress in comparison to control. A maximum increase of 11% was recorded in chlorophyll *a/b* ratio of plants treated with 1000 mgL^{-1} of peptone and 100 μM of Cd (Figure 2D).

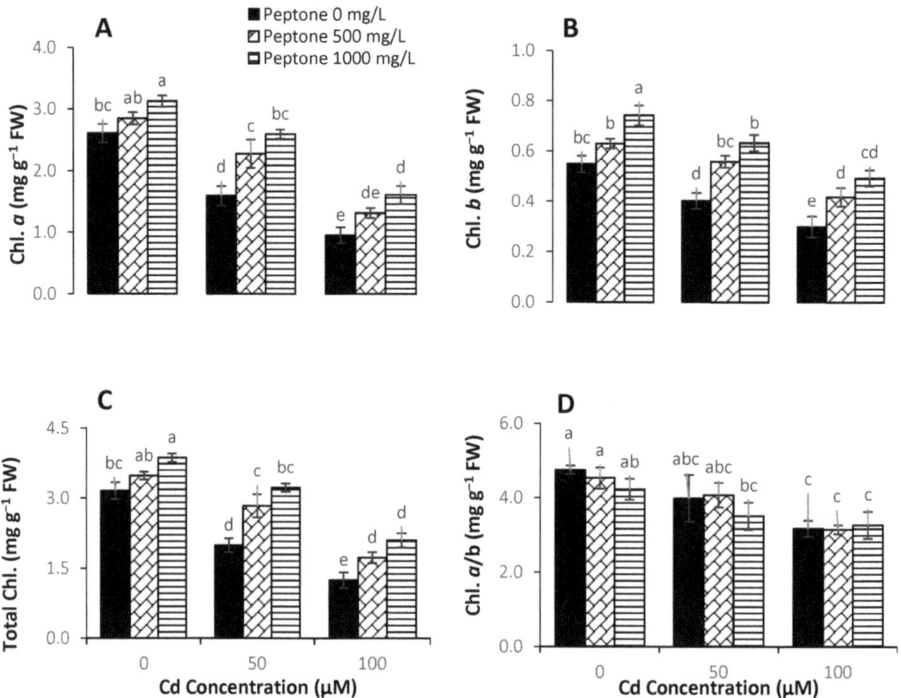

Figure 2. The influence of the foliar application of peptone (0, 500, and 1000 mgL^{-1}) on chlorophyll *a* (**A**), chlorophyll *b* (**B**), total chlorophyll (**C**), and chlorophyll *a/b* ratio (**D**) of spinach plants grown under Cd stress. Values represent the means of four replicates with standard deviations. Letters on the bar indicate the significant differences between given treatments at *p* < 0.05.

3.3. Gas Exchange Parameters

The plant photosynthetic rate was significantly affected by Cd stress, which inhibits chlorophyll synthesis and results in pigment degradation. Cd stress results in the decrease of the photosynthetic rate through increasing the stomatal resistance and reducing the density of stomata, leading to reductions in the gas exchange rate. In the present study, the application of peptone resulted in the modulation of gas exchange parameters such as the photosynthetic rate and stomatal conductance of spinach plants grown under Cd stress. The photosynthetic rate and stomatal conductance were significantly enhanced upon the foliar application of peptone (Figure 3). Significant increases of 35%, 46%, and 56% in the photosynthetic rate and 32%, 40%, and 51% in the stomatal conductance were recorded in plants treated with 1000 mgL^{-1} of peptone under Cd stress of 0, 50, and 100 μM, respectively. Moreover, the application of 500 mgL^{-1} of peptone significantly increased the photosynthetic rate up to 19%, 27%, and 41% and the stomatal conductance up to 15%, 14%, and 28% under Cd stress of 0, 50, and 100 μM, respectively, as compared to control.

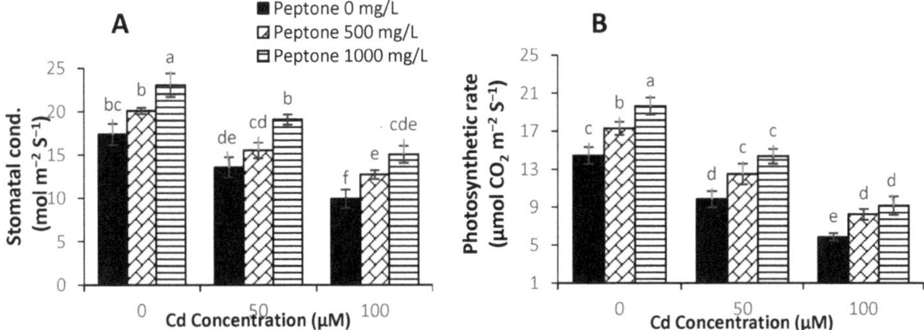

Figure 3. The influence of the foliar application of peptone (0, 500, and 1000 mgL^{-1}) on the stomatal conductance (**A**) and photosynthetic rate (**B**) of spinach plants grown under Cd stress. Values represent the means of four replicates with standard deviations. Letters on the bar indicate the significant differences between given treatments at $p < 0.05$.

3.4. Levels of Hydrogen Peroxide, Malondialdehyde, and Electrolyte Leakage

In order to reveal the impact of peptone application on oxidative damage caused by Cd stress, contents of malondialdehyde (MDA), hydrogen peroxide (H_2O_2), and electrolyte leakage (EL) were recorded in spinach plants (Figure 4). The exogenous application of peptone (500 mgL^{-1}) significantly decreased H_2O_2 content by 23%, 16%, and 11% under Cd levels of 0, 50, and 100 μM, respectively, as compared to control. The highest significant decrease of 24% in H_2O_2 content was noted at the higher level of both peptone (1000 mgL^{-1}) and Cd (100 μM) relative to control. Plants treated with 1000 mgL^{-1} peptone at all levels of Cd (0, 50, and 100 μM) showed significant decreases in MDA content. However, the application of 500 mg/L^{-1} of peptone significantly reduced MDA content by 32%, 11%, and 11% under Cd levels of 0, 50, and 100 μM, respectively, as compared to control. Furthermore, spinach plants treated with peptone showed significant decreases in electrolyte leakage level at all Cd stress levels.

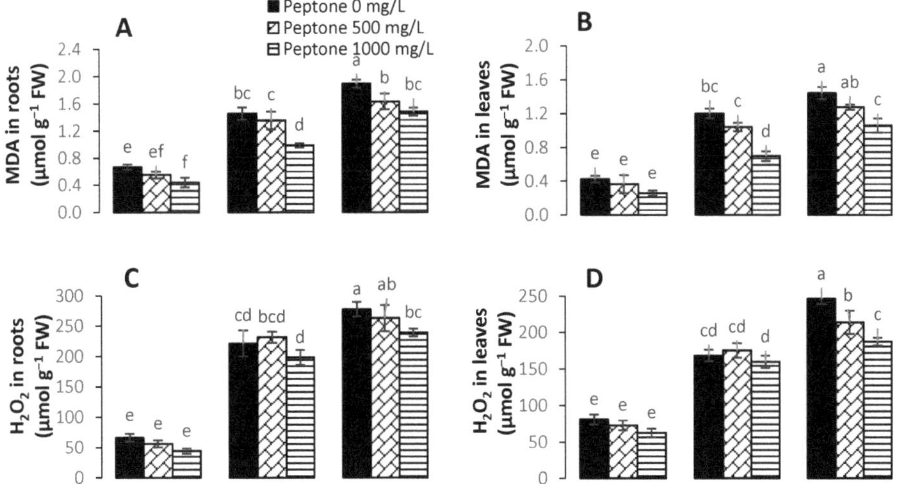

Figure 4. *Cont.*

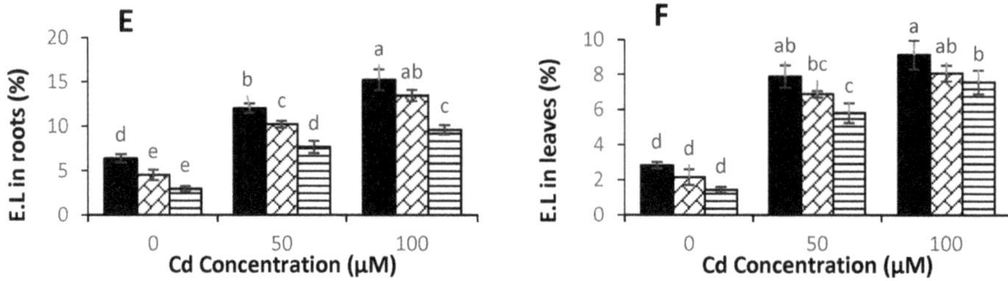

Figure 4. The influence of the foliar application of peptone (0, 500, and 1000 mgL^{-1}) on MDA in roots (**A**), MDA in leaves (**B**), H$_2$O$_2$ in roots (**C**), H$_2$O$_2$ in leaves (**D**), EL in roots (**E**), and EL in leaves (**F**) of spinach plants grown under Cd stress. Values represent the means of four replicates with standard deviations. Letters on the bar indicate the significant differences between given treatments at $p < 0.05$.

3.5. Activity of Antioxidant Enzymes

In order to survive against adverse environmental conditions, plants have developed certain defense systems. Antioxidant enzymes such as superoxide dismutase (SOD), peroxidase (POD), and catalase (CAT) detoxify ROS and decrease the toxic effect of stress. The results of the current study revealed significant increases in the activity of antioxidant enzymes upon the foliar application of peptone (Figure 5). The foliar application of peptone significantly enhanced the activities of SOD and CAT (Figure 3). Moreover, significant increases of 76%, 40%, and 54% for SOD and 59%, 48%, and 27% for CAT were recorded in plants treated with 1000 mgL^{-1} of peptone. The application of 500 mgL^{-1} of peptone significantly enhanced the activity of SOD by 46%, 48%, and 26% and the activity of CAT by 42%, 25%, and 16% under the Cd stress levels of 0, 50, and 100 µM, respectively, as compared to control. The application of 500 mgL^{-1} of peptone significantly enhanced the activity of POD by 29%, 19%, and 24% under the Cd stress levels of 0, 50, and 100 µM, respectively, as compared to control.

3.6. Proline and Ascorbic Acid Content

Non-enzymatic antioxidant molecules, such as proline, could assist plants in surviving under harmful environmental conditions. The results indicated that the application of 500 mgL^{-1} of peptone significantly enhanced proline contents by 24%, 36%, and 21%, while the application of 1000 mgL^{-1} of peptone exhibited significant increases of 40%, 47%, and 35% under Cd stress levels of 0, 50, and 100 µM, respectively, as compared to control (Figure 6). The higher concentration of peptone (1000 mgL^{-1}) significantly enhanced the level of ascorbic acid by 44%, 29%, and 29%, while the application of 500 mgL^{-1} of peptone resulted in significant increases of 22%, 25%, and 17% at Cd stress levels of 0, 50, and 100 µM, respectively, as compared to control.

3.7. Cd Concentration

The application of 500 mgL^{-1} of peptone significantly decreased the Cd concentration in plant tissues under all Cd stress levels, as compared to control. The maximum dose of peptone (1000 mL^{-1}) caused the highest decreases in Cd accumulation, as compared to that in control plants (Figure 7).

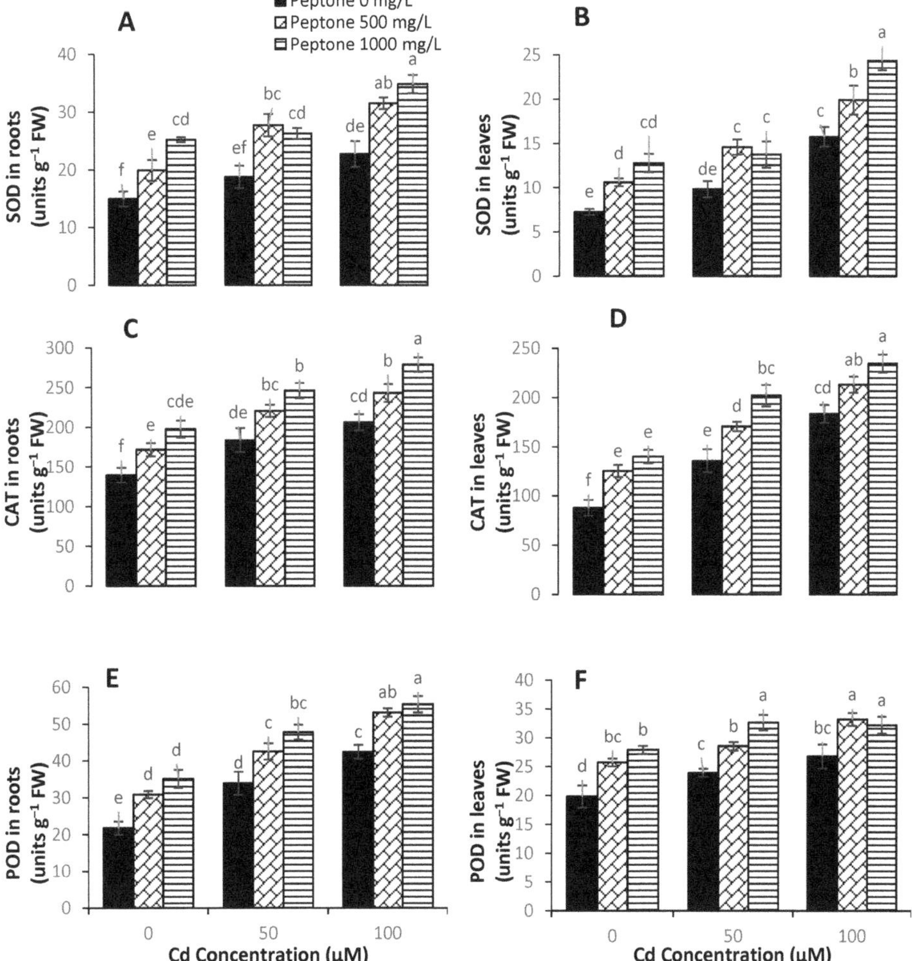

Figure 5. The influence of the foliar application of peptone (0, 500, and 1000 mgL^{-1}) on superoxide dismutase (SOD) in roots (**A**), SOD in leaves (**B**), catalase (CAT) in roots (**C**), CAT in leaves (**D**), peroxidase (POD) in roots (**E**) and POD in leaves (**F**) of spinach plants grown under Cd stress. Values represent the means of four replicates with standard deviations. Letters on the bar indicate the significant differences between given treatments at $p < 0.05$.

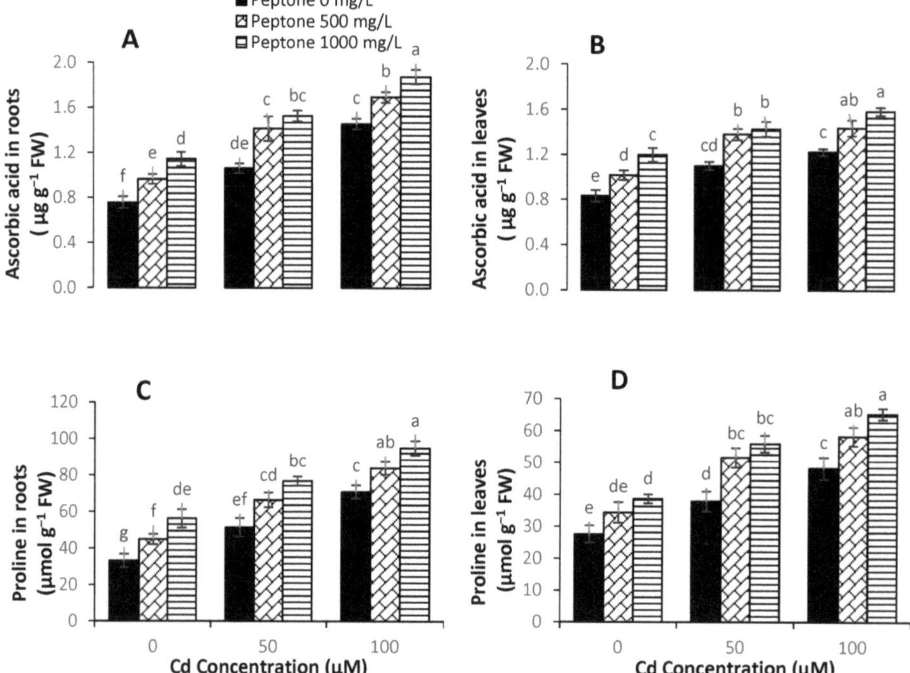

Figure 6. The influence of the foliar application of peptone (0, 500, and 1000 mgL^{-1}) on the levels of ascorbic acid in root (**A**), ascorbic acid in leaves (**B**), proline in roots (**C**), and proline in leaves (**D**) of spinach plants grown under Cd stress. Values represent the means of four replicates with standard deviations. Letters on the bar indicate the significant differences between given treatments at $p < 0.05$.

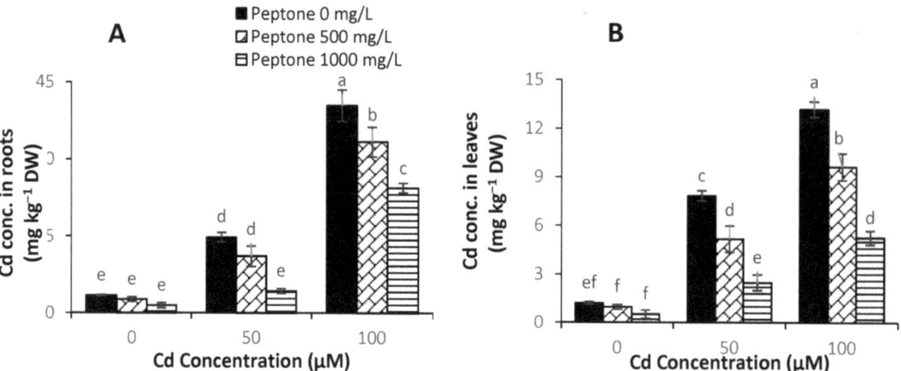

Figure 7. The influence of the foliar application of peptone (0, 500, and 1000 mgL^{-1}) on Cd concentration in roots (**A**) and Cd concentration in leaves (**B**) of spinach plants grown under Cd stress. Values represent the means of four replicates with standard deviations. Letters on the bar indicate the significant differences between given treatments at $p < 0.05$.

4. Discussion

The results of the present study showed that Cd stress significantly reduced the growth and biomass of spinach plants. This decrease in the plant growth might be due to the high concentration of Cd in the plant shoot and roots as well as the higher translocation rate of Cd from plant roots to shoots. It has been well documented that the growth and biomass of different plant species were negatively affected by Cd toxicity, depending upon the type of plant species and the exposure duration to Cd. Previous studies have clearly reported these Cd-stress-induced toxic impacts on the growth traits of various plant crops [35,36]. Like the other abiotic stresses [37,38], Cd could reduce the plant growth via affecting the ultrastructure and normal functioning of plant [39]. On the other hand, the application of amino acids regulated the growth, dry weight, quality, nitrogen contents, and yield of food crops [40]. The current study showed that the increases in plant growth might be due to the direct or indirect positive effects of peptone application for spinach plants grown under Cd stress. Similarly, the application of a mixture of amino acids enhanced the growth and nutrient contents of faba bean plants in seawater-stressed environment [14]. Moreover, the foliar application of peptone enhanced the fresh and dry weight, leaf area, and growth of *Helichrysum bracteatum* L. [12]. Amino acids are rich sources of energy and carbon that help in promoting plant growth [41]. Amino acids can serve as immediate sources of nitrogen, which are usually taken up by plants more rapidly than organic nitrogen. The results of this study are in a coordination with the earlier studies conducted on lemon basil [42], basil plant [43], and *Pelargonium graveolens* L. [44]. As building blocks of proteins, amino acids also play key roles in the regulation of metabolism and nitrogen storage.

Previous studies reported the adverse effects of environmental stresses on the growth, chlorophyl content, and gas exchange attributes of different plant species [45–53]. Similarly, in the current study, Cd stress decreased the chlorophyll content and gas exchange parameters in spinach plants. These decreases might be due to the change in chloroplast structure under Cd stress [54]. Cd stress results in a reduction in chlorophyll contents by causing lipid peroxidation, which results in a disturbance in the structure and function of the thylakoid membrane [55]. Additionally, the decrease in chlorophyll content under Cd stress might be due to the overproduction of ROS, a naturally reactive compound that damages pigments and other biomolecules. On the other hand, in the present study, the foliar application of peptone reduced the damaging effects of Cd stress on chlorophyll content of spinach plants. Previous studies also reported the positive roles of amino acids in increasing the photosynthetic pigments in *Foeniculum vulgare* L. [56], *Salvia farinacea* L. [57], and faba bean [13]. In the same way, three different concentrations of peptone (250, 500, and 1000 mg/L^{-1}) enhanced chlorophyl *a* and chlorophyl *b* contents in *Helichrysum bracteatum* L. [12]. It is likely that the application of peptone, being a source of free amino acids, reduced Cd toxicity in spinach plants as amino acids play key roles in stress signaling and secondary metabolism [58]. In the current study, the foliar spray of peptone enhanced the gas exchange parameters such as stomatal conductance and photosynthetic rate in spinach plants under Cd stress. These results were in agreement with that previously reported [35]. Previous reports have revealed that the environmental stresses considerably induce oxidative stress as well as the levels of electrolyte leakage (EL) and malondialdehyde (MDA) in different plant species, causing damages in plant tissues [59–66]. Similarly, in the current study, spinach plants also exhibited higher levels of EL and MDA in the absence of peptone treatment. The increased levels of EL and MDA in the roots and shoots of spinach plants indicated the Cd-stress-induced oxidative stress. However, in the present study, the foliar spray of peptone decreased EL and MDA contents in plant shoots and roots as compared to control. This decrease in the level of oxidative stress might be due to the protective role of peptone against lipid peroxidation in spinach plants grown in a Cd-stressed environment. The antioxidant enzymes play a key role in ROS scavenging in order to enhance the plant capability to cope with stress conditions [38,67]. It was reported that peptone application provided an optimum concentration of radially available free amino acid molecules, which in turn reduced the free radicals associated with enhanced osmoprotection [68]. Similar results were previously reported in rice plants grown under Cd stress [35]. In addition, the availability of amino acids (e.g., glutamate) from a biostimulant (e.g., peptone) may alleviate the oxidative stress via increasing the synthesis of stress-responsive amino acids such as arginine and proline [22].

The current study also revealed decreases in the activity of antioxidant enzymes in spinach plants grown under Cd stress. Such decreases in antioxidant enzymes activity might be due to the higher concentration of Cd in plants, which in turn increased the levels of EL and MDA. Additionally, the increase or decrease in the activity of antioxidant enzymes depends upon the level of metal stress and plant species [69,70]. In the present study, the application of peptone promoted the antioxidant enzymes activities. The enhanced activities of SOD, POD, and CAT due to peptone application might be due to the lower concentrations of Cd in these plants. Furthermore, the mobility of Cd in the active plant parts might be reduced due to the amino acid ability to form a complex with Cd, resulting in decreases in Cd toxicity [71]. This enhancement in enzymes activity promoted spinach plants' tolerance to Cd stress via reducing the levels of EL and MDA in plant tissues.

The regulation of osmotic adjustment substances such as proline could maintain the cell membrane integrity and osmotic balance in plants [72]. The results of the current study revealed an increase in the contents of ascorbic acid and proline. The increase in the concentration of osmotic adjustment substances in plants suppresses the toxic effects of Cd stress [73]. In the present study, the application of peptone decreased Cd contents in the shoot and roots of spinach. This decrease in Cd contents might be because peptone application increased the amino acids' availability and modulated the intracellular environment, which in turn restricted the Cd uptake and translocation [22].

5. Conclusions

Cd stress exerts negative impacts on the growth and physio-biochemical attributes of spinach plants. Foliar spray of peptone significantly promoted the plant growth and biomass. It also enhanced the activity of antioxidant enzymes, which promoted the plants' defense against Cd stress. The application of peptone also reduced the levels of H_2O_2 and MDA. The amino acids have the ability to bind with Cd and immobilize it. The use of peptone application against metal stress is not well-studied, hence further field studies on other food crops are suggested.

Author Contributions: Conceptualization, N.E., M.S.A., S.A. and M.A.E.-E.; data curation, N.E., M.S.A. and M.I.; formal analysis, N.E., M.S.A., M.I. and M.A.E.-E.; investigation, N.E., M.S.A., M.I., S.A., M.A.E.-E. and M.N.A.; methodology, N.E., M.S.A., M.I., S.A., M.A.E.-E. and M.N.A.; resources, S.A., M.A.E.-E. and M.N.A.; software, N.E., M.S.A., M.I., S.A., M.A.E.-E. and M.N.A.; validation, N.E., M.S.A., M.I., S.A., M.A.E.-E. and M.N.A.; visualization, N.E., M.S.A., M.I., M.A.E.-E. and S.A.; writing—original draft, N.E., M.S.A., S.A. and M.A.E.-E.; writing—review and editing, N.E., M.S.A., S.A., M.A.E.-E., M.I. and M.N.A. All authors have read and agreed to the published version of the manuscript.

Funding: The authors highly acknowledge the Government College University, Faisalabad, Pakistan for its support. The authors would like to extend their sincere appreciation to Researchers Supporting Project Number (RSP-2020/180), King Saud University, Riyadh, Saudi Arabia.

Acknowledgments: The authors highly acknowledge the Higher Education Commission Islamabad, Pakistan for its support. The authors would like to extend their sincere appreciation to Researchers Supporting Project Number (RSP-2020/180), King Saud University, Riyadh, Saudi Arabia.

Conflicts of Interest: The authors declare that they have no conflict of interest.

References

1. Liao, Q.L.; Liu, C.; Wu, H.Y.; Jin, Y.; Hua, M.; Zhu, B.W.; Chen, K.; Huang, L. Association of soil cadmium contamination with ceramic industry: A case study in a Chinese town. *Sci. Total Environ.* **2015**, *514*, 26–32. [CrossRef] [PubMed]

2. Liang, J.; Feng, C.; Zeng, G.; Zhong, M.; Gao, X.; Li, X.; He, X.; Li, X.; Fang, Y.; Mo, D. Atmospheric deposition of mercury and cadmium impacts on topsoil in a typical coal mine city, Lianyuan, China. *Chemosphere* **2017**, *189*, 198–205. [CrossRef] [PubMed]

3. Dharma-Wardana, M.W.C. Fertilizer usage and cadmium in soils, crops and food. *Environ. Geochem. Health* **2018**, *40*, 2739–2759. [CrossRef] [PubMed]

4. Mori, M.; Kotaki, K.; Gunji, F.; Kubo, N.; Kobayashi, S.; Ito, T.; Itabashi, H. Suppression of cadmium uptake in rice using fermented bark as a soil amendment. *Chemosphere* **2016**, *148*, 487–494. [CrossRef]

5. Gallego, S.M.; Pena, L.B.; Barcia, R.A.; Azpilicueta, C.E.; Iannone, M.F.; Rosales, E.P.; Zawoznik, M.S.; Groppa, M.D.; Benavides, M.P. Unravelling cadmium toxicity and tolerance in plants: Insight into regulatory mechanisms. *Environ. Exp. Bot.* **2012**, *83*, 33–46. [CrossRef]

6. Martinez-Penalver, A.; Grana, E.; Reigosa, M.J.; Sanchez-Moreiras, A.M. The early response of Arabidopsis thaliana to cadmium-and copper-induced stress. *Environ. Exp. Bot.* **2012**, *78*, 1–9. [CrossRef]

7. Rochayati, S.; Du Laing, G.; Rinklebe, J.; Meissner, R.; Verloo, M. Use of reactive phosphate rocks as fertilizer on acid upland soils in Indonesia: Accumulation of cadmium and zinc in soils and shoots of maize plants. *J. Plant. Nutr. Soil Sci.* **2011**, *174*, 186–194. [CrossRef]

8. Garnier, J.; Cebron, A.; Tallec, G.; Billen, G.; Sebilo, M.; Martinez, A. Nitrogen behaviour and nitrous oxide emission in the tidal Seine River estuary (France) as influenced by human activities in the upstream watershed. *Biogeochemistry* **2006**, *77*, 305–326. [CrossRef]

9. Mojiri, A. The potential of corn (Zea mays) for phytoremediation of soil contaminated with cadmium and lead. *J. Biol. Environ. Sci.* **2011**, *5*, 17–22.

10. Alia, N.; Sardar, K.; Said, M.; Salma, K.; Sadia, A.; Sadaf, S.; Miklas, S. Toxicity and bioaccumulation of heavy metals in spinach (*Spinacia oleracea*) grown in a controlled environment. *Int. J. Environ. Res. Public Health* **2015**, *12*, 7400–7416. [CrossRef]

11. Melquist, S.; Luff, B.; Bender, J. Arabidopsis PAI gene arrangements, cytosine methylation and expression. *Genetics* **1999**, *153*, 401–413. [PubMed]

12. Soad, M.M.; Lobna, S.T.; Farahat, M.M. Influence of foliar application of pepton on growth, flowering and chemical composition of *Helichrysum bracteatum* L. plants under different irrigation intervals. *Ozean J. Appl. Sci.* **2010**, *3*, 143–155.

13. Sadak, M.S.; Abdelhamid, M.T. Influence of amino acids mixture application on some biochemical aspects, antioxidant enzymes and endogenous polyamines of *Vicia faba* L.plant grown under seawater salinity stress. *Gesunde Pflanz.* **2015**, *67*, 119–129. [CrossRef]

14. Sadak, M.S.H.; Abdelhamid, M.T.; Schmidhalter, U. Effect of foliar application of aminoacids on plant yield and some physiological parameters in bean plants irrigated with seawater. *Acta Biol. Colomb.* **2015**, *20*, 141–152.

15. Calvo, P.; Nelson, L.; Kloepper, J.W. Agricultural uses of plant biostimulants. *Plant Soil* **2014**, *383*, 3–41. [CrossRef]

16. Ali, Q.; Perveen, R.; El-Esawi, M.A.; Ali, S.; Hussain, S.M.; Amber, M.; Iqbal, N.; Rizwan, M.; Alyemeni, M.N.; El-Serehy, H.A.; et al. Low Doses of Cuscuta reflexa Extract Act as Natural Biostimulants to Improve the Germination Vigor, Growth, and Grain Yield of Wheat Grown under Water Stress: Photosynthetic Pigments, Antioxidative Defense Mechanisms, and Nutrient Acquisition. *Biomolecules* **2020**, *10*, 1212. [CrossRef]

17. Canellas, L.P.; Olivares, F.L.; Aguiar, N.O.; Jones, D.L.; Nebbioso, A.; Mazzei, P.; Piccolo, A. Humic and fulvic acids as biostimulants in horticulture. *Sci. Hortic.* **2015**, *196*, 15–27. [CrossRef]

18. Soliman, M.H.; Abdulmajeed, A.M.; Alhaithloul, H.; Alharbi, B.M.; El-Esawi, M.A.; Hasanuzzaman, M.; Elkelish, A. Saponin biopriming positively stimulates antioxidants defense, osmolytes metabolism and ionic status to confer salt stress tolerance in soybean. *Acta Physiol. Plant.* **2020**, *42*, 114. [CrossRef]

19. Cerdan, M.; Sanchez-Sanchez, A.; Oliver, M.; Juarez, M.; Sanchez-Andreu, J.J. Effect of foliar and root applications of amino acids on iron uptake by tomato plants. *Acta Hortic.* **2009**, *830*, 481–488. [CrossRef]

20. Colla, G.; Rouphael, Y.; Canaguier, R.; Svecova, E.; Cardarelli, M. Biostimulant action of a plant-derived protein hydrolysate produced through enzymatic hydrolysis. *Front. Plant. Sci.* **2014**, *5*, 448. [CrossRef]

21. Lachhab, N.; Sanzani, S.M.; Adrian, M.; Chiltz, A.; Balacey, S.; Boselli, M. Soybean and casein hydrolysates induce grapevine immune responses and resistance against *Plasmopara viticola*. *Front. Plant. Sci.* **2014**, *5*, 716. [CrossRef] [PubMed]

22. Teixeira, W.F.; Fagan, E.B.; Soares, L.H.; Umburanas, R.C.; Reichardt, K.; Neto, D.D. Foliar and seed application of amino acids affects the antioxidant metabolism of the soybean crop. *Front. Plant. Sci.* **2017**, *8*, 327. [CrossRef] [PubMed]

23. Bouyoucos, G.J. Hydrometer method improved for making particle size analyses of soils. *Agron. J.* **1962**, *54*, 464–465. [CrossRef]

24. Page, A.L.; Miller, R.H.; Keeny, D.R. Methods of soil analysis. In *Chemical and Microbiological Properties*; American Society of Agronomy: Madison, WI, USA, 1982; p. 9.

25. Soltanpour, P.N. Use of AB-DTPA soil test to evaluate elemental availability and toxicity. *Commun. Soil Sci. Plant. Anal.* **1985**, *16*, 323–338. [CrossRef]

26. Arnon, D.I. Copper enzymes in isolated chloroplasts, polyphenoxidase in *Beta vulgaris* L. *Plant Physiol.* **1949**, *2*, 1–15. [CrossRef]

27. Dionisio-Sese, M.L.; Tobita, S. Antioxidant responses of rice seedlings to salinity stress. *Plant Sci.* **1998**, *135*, 1–9. [CrossRef]

28. Velikova, V.; Yordanov, I.; Edreva, A. Oxidative stress and some antioxidant systems in acid rain-treated bean plants: Protective role of exogenous polyamines. *Plant Sci.* **2000**, *151*, 59–66. [CrossRef]

29. Carmak, I.; Horst, J.H. Effect of aluminium on lipid peroxidation, superoxide dismutase, catalase, and peroxidase activities in root tips of soybean (*Glycine max* L.). *J. Plant Physiol.* **1991**, *83*, 463–468. [CrossRef]

30. Zhang, X.Z. Themeasurement and mechanism of lipid peroxidation and SOD, POD and CAT activities in biological system. In *Research Methodology of Crop Physiology*; Agriculture Press: Beijing, China, 1992; pp. 208–211.

31. Aebi, H. Catalase in vitro. *Methods Enzymol.* **1984**, *105*, 121–126.

32. Mukherjee, S.P.; Choudhuri, M.A. Implications of water stress-induced changes in the levels of endogenous ascorbic acid and hydrogen peroxide in Vigna seedlings. *Physiol. Plant.* **1983**, *58*, 166–170. [CrossRef]

33. Bates, L.S.; Waldren, R.P.; Teare, I.D. Rapid determination of free proline for water-stress studies. *Plant Soil* **1973**, *39*, 205–207. [CrossRef]

34. Rehman, M.Z.U.; Rizwan, M.; Ghafoor, A.; Naeem, A.; Ali, S.; Sabir, M.; Qayyum, M.F. Effect of inorganic amendments for in situ stabilization of cadmium in contaminated soils and its phyto-availability to wheat and rice under rotation. *Environ. Sci. Pollut. Res.* **2015**, *22*, 16897–16906. [CrossRef] [PubMed]

35. Hina, K.; Kanwal, S.S.; Arshad, M.; Gul, I. Effect of cadmium (Cd) stress on spinach (*Spinacea oleracea*) and its retention kinetics in soil in response to organic amendments. *Pak. J. Agric. Sci.* **2019**, *56*, 179–185.

36. Mahnoor, A.; Arshid, P.; Usman, I.; Qaisar, M.; Rafiq, A. Melatonin and plant growth-promoting rhizobacteria alleviate the cadmium and arsenic stresses and increase the growth of *Spinacia oleracea* L. *Plant. Soil Environ.* **2020**, *66*, 234–241.

37. Ahanger, M.A.; Qin, C.; Begum, N.; Maodong, Q.; Dong, X.X.; El-Esawi, M.; El-Sheikh, M.A.; Alatar, A.A.; Zhang, L. Nitrogen availability prevents oxidative effects of salinity on wheat growth and photosynthesis by up-regulating the antioxidants and osmolytes metabolism, and secondary metabolite accumulation. *BMC Plant Biol.* **2019**, *19*, 1–12. [CrossRef]

38. Soliman, M.H.; Alnusaire, T.S.; Abdelbaky, N.F.; Alayafi, A.A.; Hasanuzzaman, M.; Rowezak, M.M.; El-Esawi, M.; Elkelish, A. Trichoderma-Induced Improvement in Growth, Photosynthetic Pigments, Proline, and Glutathione Levels in Cucurbita pepo Seedlings under Salt Stress. *Phyton* **2020**, *89*, 473. [CrossRef]

39. Shanying, H.E.; Xiaoe, Y.A.N.G.; Zhenli, H.E.; Baligar, V.C. Morphological and physiological responses of plants to cadmium toxicity: A review. *Pedosphere* **2017**, *27*, 421–438.

40. Saeed, M.R.; Kheir, A.M.; Al-Sayed, A.A. Supperssive effect of some amino acids against Meloidogyne incognita on soybeans. *J. Agric. Sci. Mansoura Univ.* **2005**, *30*, 1097–1103.

41. Goss, J.A. *Physiology of Plants and Their Cells*; Pergamon Biological Sciences Series; Pergamon: New York, NY, USA, 1973.

42. Youssef, A.A.; Khattab, M.E.; Omer, E.A. Effect of spraying of molybdenum and tyrosine on growth, yield and chemical composition of lemon basil plant. *Egypt Pharm. J.* **2004**, *3*, 87–106.

43. Talaat, I.M.; Youssef, A.A. The role of the amino acids lysine and ornithine in growth and chemical constituents of Basil plants. *Egypt J. Appl. Sci.* **2002**, *17*, 83–95.

44. Mona, H.M.; Talaat, I.M. Physiological response of Rose Geranium (*Pelargonium gravealens* L.) to phenylalanine and nicatonic Acid. *Annal. Agr. Sci. Moshtohor* **2005**, *43*, 807–822.

45. El-Esawi, M.A.; Elkelish, A.; Soliman, M.; Elansary, H.O.; Zaid, A.; Wani, S.H. *Serratia marcescens* BM1 enhances cadmium stress tolerance and phytoremediation potential of soybean through modulation of osmolytes, leaf gas exchange, antioxidant machinery, and stress-responsive genes expression. *Antioxidants* **2020**, *9*, 43. [CrossRef] [PubMed]

46. Abdelaal, K.A.; EL-Maghraby, L.M.; Elansary, H.; Hafez, Y.M.; Ibrahim, E.I.; El-Banna, M.; El-Esawi, M.; Elkelish, A. Treatment of Sweet Pepper with Stress Tolerance-Inducing Compounds Alleviates Salinity Stress Oxidative Damage by Mediating the Physio-Biochemical Activities and Antioxidant Systems. *Agronomy* **2020**, *10*, 26. [CrossRef]

47. Alhaithloul, H.A.; Soliman, M.H.; Ameta, K.L.; El-Esawi, M.A.; Elkelish, A. Changes in Ecophysiology, Osmolytes, and Secondary Metabolites of the Medicinal Plants of *Mentha piperita* and *Catharanthus roseus* Subjected to Drought and Heat Stress. *Biomolecules* **2020**, *10*, 43. [CrossRef]

48. El-Esawi, M.A.; Alayafi, A.A. Overexpression of *StDREB2* Transcription Factor Enhances Drought Stress Tolerance in Cotton (*Gossypium barbadense* L.). *Genes* **2019**, *10*, 142. [CrossRef]

49. Soliman, M.; Alhaithloul, H.A.; Hakeem, K.R.; Alharbi, B.M.; El-Esawi, M.; Elkelish, A. Exogenous nitric oxide mitigates nickel-induced oxidative damage in eggplant by upregulating antioxidants, osmolyte metabolism, and glyoxalase systems. *Plants* **2019**, *8*, 562. [CrossRef]

50. Ali, Q.; Ali, S.; El-Esawi, M.A.; Rizwan, M.; Azeem, M.; Hussain, A.I.; Perveen, R.; El-Sheikh, M.A.; Alyemeni, M.N.; Wijaya, L. Foliar spray of Fe-Asp confers better drought tolerance in sunflower as compared with FeSO$_4$: Yield traits, osmotic adjustment, and antioxidative defense mechanisms. *Biomolecules* **2020**, *10*, 1217. [CrossRef]

51. El-Esawi, M.A.; Alaraidh, I.A.; Alsahli, A.A.; Alzahrani, S.M.; Ali, H.M.; Alayafi, A.A.; Ahmad, M. *Serratia liquefaciens* KM4 improves salt stress tolerance in maize by regulating redox potential, ion homeostasis, leaf gas exchange and stress-related gene expression. *Int. J. Mol. Sci.* **2018**, *19*, 3310. [CrossRef]

52. Vwioko, E.; Adinkwu, O.; El-Esawi, M.A. Comparative physiological, biochemical and genetic responses to prolonged waterlogging stress in okra and maize given exogenous ethylene priming. *Front. Physiol.* **2017**, *8*, 632. [CrossRef]

53. El-Esawi, M.A.; Al-Ghamdi, A.A.; Ali, H.M.; Ahmad, M. Overexpression of *AtWRKY30* transcription factor enhances heat and drought stress tolerance in wheat (*Triticum aestivum* L.). *Genes* **2019**, *10*, 163. [CrossRef]

54. De Araujo, R.P.; de Almeida, A.A.F.; Pereira, L.S.; Mangabeira, P.A.; Souza, J.O.; Pirovani, C.P.; Ahnert, D.; Baligar, V.C. Photosynthetic, antioxidative, molecular and ultrastructural responses of young cacao plants to Cd toxicity in the soil. *Ecotoxicol. Environ. Saf.* **2017**, *144*, 148–157. [CrossRef] [PubMed]

55. Gu, C.S.; Yang, Y.H.; Shao, Y.F.; Wu, K.W.; Liu, Z.L. The effects of exogenous salicylic acid on alleviating cadmium toxicity in *Nymphaea tetragona* Georgi. *S. Afr. J. Bot.* **2018**, *114*, 267–271. [CrossRef]

56. Hassanein, R.A.M. Effect of Some Amino Acids, Trace Elements and Irradiation on Fennel (*Foeniculum vulgare* L.). Ph.D. Thesis, Faculty of Agriculture, Cairo University, Giza, Egypt, 2003.

57. Nahed, G.A.; Balbaa, L.K. Influence of tyrosine and zinc on growth, flowering and chemical constituents of *Salvia farinacea* plants. *J. Appl. Sci. Res.* **2007**, *3*, 1479–1489.

58. Tatjana, M.H.; Nesi, A.N.; Wagner, L.A.; Braun, H.P. Amino acid catabolism in plants. *Mol. Plant.* **2015**, *8*, 1563–1579.

59. Zafar-ul-Hye, M.; Naeem, M.; Danish, S.; Khan, M.J.; Fahad, S.; Datta, R.; Brtnicky, M.; Kintl, A.; Hussain, G.S.; El-Esawi, M.A. Effect of cadmium-tolerant rhizobacteria on growth attributes and chlorophyll contents of bitter gourd under cadmium toxicity. *Plants* **2020**, *9*, 1386. [CrossRef]

60. El-Esawi, M.A.; Alayafi, A.A. Overexpression of rice *Rab7* gene improves drought and heat tolerance and increases grain yield in rice (*Oryza sativa* L.). *Genes* **2019**, *10*, 56. [CrossRef]

61. Naveed, M.; Bukhari, S.S.; Mustafa, A.; Ditta, A.; Alamri, S.; El-Esawi, M.A.; Rafique, M.; Ashraf, S.; Siddiqui, M.H. Mitigation of nickel toxicity and growth promotion in sesame through the application of a bacterial endophyte and zeolite in nickel contaminated soil. *Int. J. Environ. Res. Public Health* **2020**, *17*, 8859. [CrossRef]

62. El-Esawi, M.A.; Alaraidh, I.A.; Alsahli, A.A.; Ali, H.M.; Alayafi, A.A.; Witczak, J.; Ahmad, M. Genetic variation and alleviation of salinity stress in barley (*Hordeum vulgare* L.). *Molecules* **2018**, *23*, 2488. [CrossRef]

63. Imran, M.; Hussain, S.; El-Esawi, M.A.; Rana, M.S.; Saleem, M.H.; Riaz, M.; Ashraf, U.; Potcho, M.P.; Duan, M.; Rajput, I.A.; et al. Molybdenum supply alleviates the cadmium toxicity in fragrant rice by modulating oxidative stress and antioxidant gene expression. *Biomolecules* **2020**, *10*, 1582. [CrossRef]

64. El-Esawi, M.A.; Al-Ghamdi, A.A.; Ali, H.M.; Alayafi, A.A.; Witczak, J.; Ahmad, M. Analysis of genetic variation and enhancement of salt tolerance in French pea (*Pisum Sativum* L.). *Int. J. Mol. Sci.* **2018**, *19*, 2433. [CrossRef]

65. Ali, Q.; Shahid, S.; Ali, S.; El-Esawi, M.A.; Hussain, A.I.; Perveen, R.; Iqbal, N.; Rizwan, M.; Nasser Alyemeni, M.; El-Serehy, H.A.; et al. Fertigation of Ajwain (*Trachyspermum ammi* L.) with Fe-Glutamate confers better plant performance and drought tolerance in comparison with FeSO$_4$. *Sustainability* **2020**, *12*, 7119. [CrossRef]

66. El-Esawi, M.A.; Al-Ghamdi, A.A.; Ali, H.M.; Alayafi, A.A. Azospirillum lipoferum FK1 confers improved salt tolerance in chickpea (*Cicer arietinum* L.) by modulating osmolytes, antioxidant machinery and stress-related genes expression. *Environ. Exp. Bot.* **2019**, *159*, 55–65. [CrossRef]

67. Elkelish, A.; Qari, S.H.; Mazrou, Y.S.A.; Abdelaal, K.A.A.; Hafez, Y.M.; Abu-Elsaoud, A.M.; Batiha, G.-S.; El-Esawi, M.A.; El Nahhas, N. Exogenous Ascorbic Acid Induced Chilling Tolerance in Tomato Plants Through Modulating Metabolism, Osmolytes, Antioxidants, and Transcriptional Regulation of Catalase and Heat Shock Proteins. *Plants* **2020**, *9*, 431. [CrossRef] [PubMed]

68. Raza, S.H.; Athar, H.R.; Ashraf, M.; Hameed, A. Glycinebetaine-induced modulation of antioxidant enzymes activities and ion accumulation in two wheat cultivars differing in salt tolerance. *Environ. Exp. Bot.* **2007**, *60*, 368–376. [CrossRef]

69. Zaheer, I.E.; Ali, S.; Saleem, M.H.; Noor, I.; El-Esawi, M.A.; Hayat, K.; Rizwan, M.; Abbas, Z.; El-Sheikh, M.A.; Alyemeni, M.N.; et al. Iron–Lysine Mediated Alleviation of Chromium Toxicity in Spinach (*Spinacia oleracea* L.) Plants in Relation to Morpho-Physiological Traits and Iron Uptake When Irrigated with Tannery Wastewater. *Sustainability* **2020**, *12*, 6690. [CrossRef]

70. Nazli, F.; Mustafa, A.; Ahmad, M.; Hussain, A.; Jamil, M.; Wang, X.; Shakeel, Q.; Imtiaz, M.; El-Esawi, M.A. A Review on Practical Application and Potentials of Phytohormone-Producing Plant Growth-Promoting Rhizobacteria for Inducing Heavy Metal Tolerance in Crops. *Sustainability* **2020**, *12*, 9056. [CrossRef]

71. Sharma, S.S.; Dietz, K.J. The significance of amino acids and amino acid-derived molecules in plant responses and adaptation to heavy metal stress. *J. Exp. Bot.* **2006**, *57*, 711–726. [CrossRef]

72. Wu, S.; Hu, C.; Tan, Q.; Nie, Z.; Sun, X. Effects of molybdenum on water utilization, antioxidative defense system and osmotic-adjustment ability in winter wheat (*Triticum aestivum* L.) under drought stress. *Plant Physiol. Biochem.* **2014**, *83*, 365–374. [CrossRef]

73. Zhang, C.Y.; Wang, N.N.; Zhang, Y.H.; Feng, Q.Z.; Yang, C.W.; Liu, B. DNA methylation involved in proline accumulation in response to osmotic stress in rice (*Oryza sativa* L.). *Genet. Mol. Res.* **2013**, *12*, 1269–1277. [CrossRef]

 *plants*

Article

Zinc Oxide Nanoparticles Application Alleviates Arsenic (As) Toxicity in Soybean Plants by Restricting the Uptake of as and Modulating Key Biochemical Attributes, Antioxidant Enzymes, Ascorbate-Glutathione Cycle and Glyoxalase System

Parvaiz Ahmad [1,*], Mohammed Nasser Alyemeni [1], Asma A. Al-Huqail [1], Moneerah A. Alqahtani [1], Leonard Wijaya [1], Muhammad Ashraf [2], Cengiz Kaya [3] and Andrzej Bajguz [4]

1. Botany and Microbiology Department, College of Science, King Saud University, 11451 Riyadh, Saudi Arabia; mnyemeni@ksu.edu.sa (M.N.A.); aalhuqail@ksu.edu.sa (A.A.A.-H.); 439203051@student.ksu.edu.sa (M.A.A.); leon077@gmail.com (L.W.)
2. University of Agriculture Faisalabad, Faisalabad 38040, Pakistan; ashrafbot@yahoo.com
3. Soil Science and Plant Nutrition Department, Agriculture Faculty, Harran University, 63210 Sanliurfa, Turkey; c_kaya70@yahoo.com
4. Department of Biology and Ecology of Plants, Faculty of Biology, University of Bialystok, 15-245 Bialystok, Poland; abajguz@uwb.edu.pl
* Correspondence: parvaizbot@yahoo.com or pahmad@ksu.edu.sa

Received: 28 May 2020; Accepted: 20 June 2020; Published: 30 June 2020

Abstract: Accumulation of arsenic (As) in soils is increasing consistently day-by-day, which has resulted in increased toxicity of this element in various crop plants. Arsenic interferes with several plant metabolic processes at molecular, biochemical and physiological levels, which result in reduced plant productivity. Hence, the introduction of novel ameliorating agents to combat this situation is the need of the hour. The present study was designed to examine the effect of zinc oxide nanoparticles (ZnO–NPs) in As-stressed soybean plants. Various plant growth factors and enzymes were studied at varying concentrations of As and ZnO–NPs. Our results showed that with the application of ZnO–NPs, As concentration declined in both root and shoot of soybean plants. The lengths of shoot and root, net photosynthetic rate, transpiration, stomatal conductance, photochemical yield and other factors declined with an increase in external As level. However, the application of ZnO–NPs to the As-stressed soybean plants resulted in a considerable increase in these factors. Moreover, the enzymes involved in the ascorbate–glutathione cycle including superoxide dismutase (SOD), catalase (CAT), ascorbate peroxidase (APX) and glutathione reductase (GR) showed a significant increase in their activity with the application of ZnO–NPs to the As-stressed plants. Hence, our study confirms the significance of ZnO–NPs in alleviating the toxicity of As in soybean plants.

Keywords: arsenic stress; soybean; growth; antioxidant enzymes; ascorbate–glutathione cycle; glyoxalase system

---

1. Introduction

Arsenic (As) is present in lithosphere and is considered as one of the major environmental contaminants [1]. The main As pollution is due to the anthropogenic activities like, over-use of herbicides and pesticides, combustion of coal and preservation of timber (Sharma, 2013). Soil parameters such as redox state and pH have a foremost impact on As toxicity due to its altered accessibility (solubility and mobility) to plants [2]. Uptake of As is believed to be affected by some factors such as pH, nutrient supply, soil type and mugineic acid excreted by some grassy plants [2]. Arsenic within plant cells alters normal cellular metabolic activities mostly by way of binding to the enzymes and changing their course of action, interfering with carbon and sulfur metabolism and disruption of nitrogen assimilation [3]. Arsenic (As III) can react with sulfhydryl groups of proteins and enzymes which can cause loss of function and even cell death [4]. Under higher concentrations of arsenic, the plant fails to balance between the resistance and toxicity, resulting in plant death. [5]. Arsenic stress can significantly reduce the rate of photosynthesis, transpiration as well as stomatal conductance in plants [4] which could be ascribed to metal-induced reduction of water transport [6]. Arsenic toxicity also leads to oxidative stress by the overproduction of reactive oxygen species (ROS) in plants [7–9]. Arsenic stress also enhances the production of methyl glyoxalate (MG) which can damage the normal functioning of the plant cell [10]. According to Yadav, et al. [11], MG and ROS attack the biomolecules and alter their basic structure as well as functions. To mitigate this heavy metal-induced toxicities, plants have developed different stress tolerance mechanisms. The synthesis of osmolytes (proline and glycine betaine (GB) was reported to be enhanced in most plants under heavy metal toxicity, thereby protecting the cells from dehydration stress [8]. The enzymatic [superoxide dismutase (SOD), ascorbate peroxidase (APX), catalase (CAT), glutathione reductase (GR), monodehydroascorbate reductase (MDHAR) and dehydroascorbate reductase (DHAR)] as well as non-enzymatic antioxidants [ascorbic acid (AsA), glutathione (GSH), α-tocopherols and phenolic compounds] can effectively diminish the production of ROS [12,13]. For the detoxification of MG, plants possess the glyoxalase system consisting of glyoxalase I (GlyI) and glyoxalase II (GlyII) which regulates the production of MG. Plants make the full use of their defense system under stress, however, external application of some agents like mineral elements, chemicals, etc. could strengthen the defense system. Hence, the use of nanoparticles provides a new and sustainable approach to mitigate As stress.

In recent times, the use of micro- and macronutrients in nanoparticles (NPs) form has been considered as an effective approach for improving the growth and production of most crops [14]. These nanoparticle supplementations may help to reduce the loss of nutrients and enhance crop production in a sustainable manner [15]. Among metal-based NPs, there is an increasing interest in the application of zinc oxide (ZnO) NPs in agricultural sciences [15,16]. Still, very few studies have reported the impact of ZnO–NPs in the soil-plant system. The plant species, type of the soil and soil pH are the main components which influence the availability of Zn in soils and ZnO–NPs toxicity in plants [17]. Interestingly, low availability of Zn in soils possibly supports the foliar application of ZnO–NPs. Many studies have reported nanoparticle formulations of zinc and their application as a foliar spray to be efficient in decreasing the accumulation of heavy metals in plants [18]. However, at higher concentration, ZnO–NPs may prove to be toxic for plants and act as an obstacle for using it as a nano-fertilizer, but at the same time, at a lower concentration it may prove to be favorable for plants depending upon their growth environment and types of species [19,20]. This study was thus designed to explore the effect of ZnO–NPs application on As-stressed soybean plants by elucidating its impact on various growth and metabolic factors.

2. Results

2.1. Effect of ZnO–NP and As on the Lengths of Shoot and Root and Biomass Yield

Shoot length declined by 26.76% and 49.0% at 10-μM- and 20-μM-As concentration, respectively, with respect to the control (non-arsenic-treated plants). Application of 50-mg/L ZnO–NP on the plants increased the shoot length by 17.59% and 31.12% under 10-μM- and 20-μM-As stress, respectively, over the controls (only As-treated plants). Supplementation of ZnO–NP (100 mg/L) enhanced the shoot length by 37.36% and 63.02% with 10- and 20-μM-As stress, respectively, relative to the controls (only-As-treated plants) (Figure 1A).

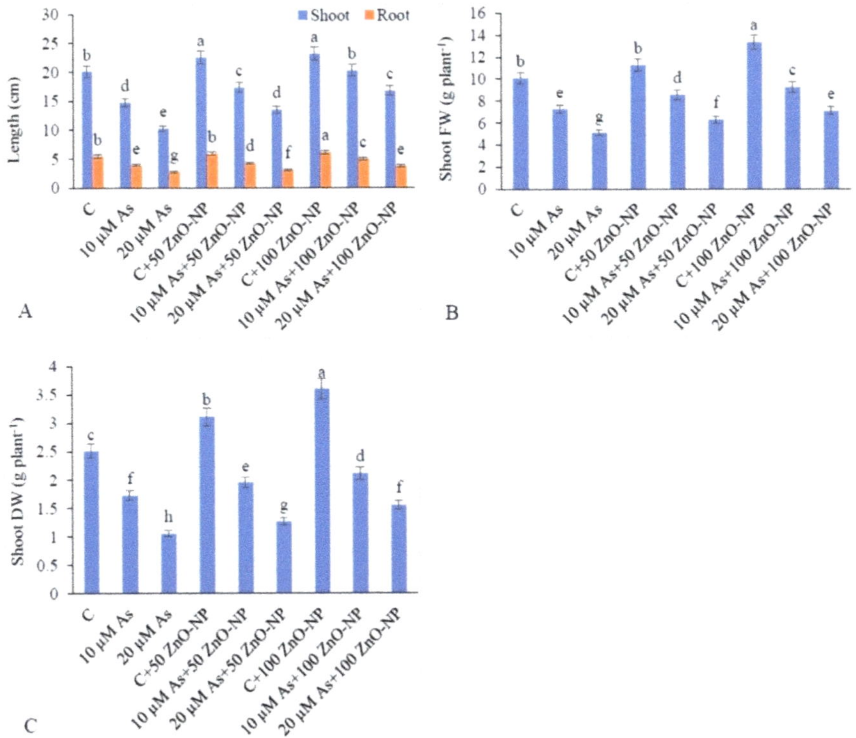

Figure 1. Effect of different concentrations of As and ZnO–NP on (**A**) shoot and root length, (**B**) shoot fresh weight and (**C**) shoot dry weight in soybean plants. Different letters indicate significant difference between the treatments. The data are the means ± SE (n = 5).

Similarly, in case of roots, at zero concentration of ZnO–NP (0 mg/L), the root length declined by 39.27% and 49.26% at 10-µM and 20-µM As, respectively, over the controls (non-As-treated plants). Application of 50-mg/L ZnO–NP on the plants increased the root length by 8.14% and 10.10% under 10 and 20-µM-As stress, respectively, with respect to the controls (non-As-treated plants). Application of ZnO–NP (100 mg/L) enhanced the root length by 25.44% and 33.93% under 10- and 20-µM-As stress, respectively, relative to the plants treated with As only and 50-mg/L ZnO–NP-treated plants (Figure 1A).

Arsenic stress decreased the shoot fresh and dry weights (Figure 1B,C). Shoot dry weight (DW) declined by 31.47% and 58.16%, with 10-µM- and 20-µM-As stress, respectively, with reference to those in the non-As-treated plants. Application of 50-mg/L ZnO–NP to 10-µM- and 20-µM-As-treated plants enhanced the shoot DW by 13.37% and 19.99%, respectively, relative to those in the plants treated with As alone. However, application of 100-mg/L ZnO–NP to 10-µM- and 20-µM-As-treated plants enhanced the shoot DW by 22.67% and 47.61%, respectively, over those in the plants treated with As alone (Figure 1C).

2.2. Effect of As and ZnO–NP on Pigment Content

Supplementation of 10-µM As decreased the amount of total chlorophyll and carotenoids by 12.34% and 21.95%, respectively. Similarly, supplementation of 20-µM As decreased the amount of total chlorophyll by 32.09% and carotenoids by 41.46% relative to the those in the non-As-treated plants. Application of 25-mg/L ZnO–NP to 20-µM-As-stressed plants enhanced the total chlorophyll by

13.63% and carotenoids by 12.50% over those in the plants treated with As only. Application 100-mg/L ZnO–NP to 20-µM-As-stressed plants enhanced the total chlorophyll by 32.72% and carotenoids by 25% with respect to those in the plants treated with As only (Figure 2A).

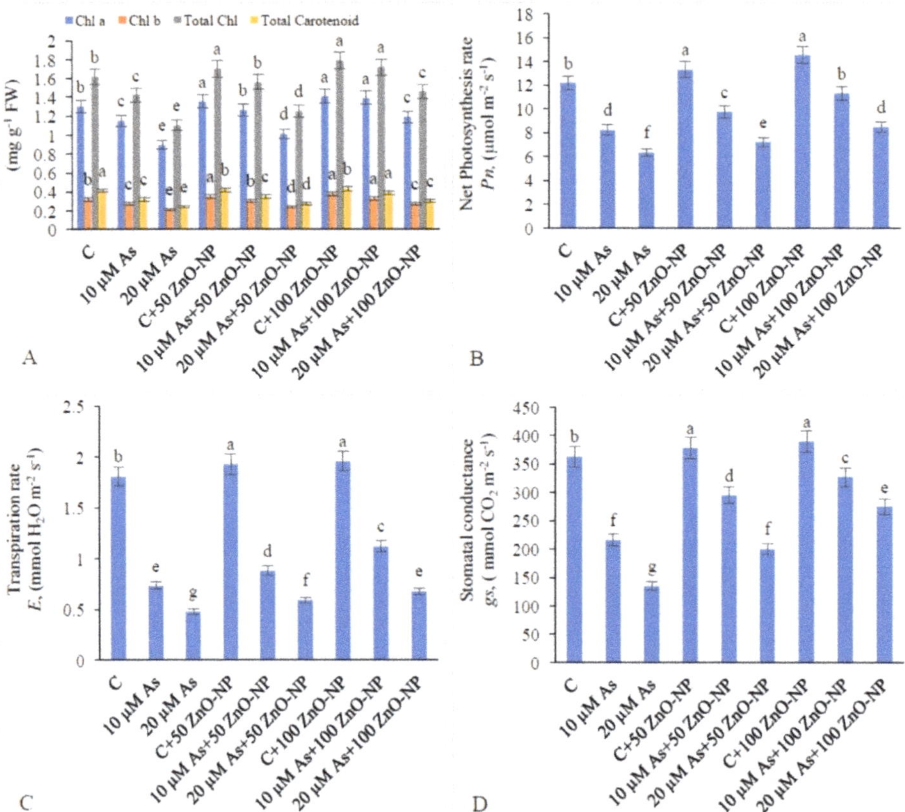

Figure 2. Effect of different concentrations of As and ZnO–NP on (**A**) pigment content, (**B**) net photosynthesis rate, (**C**) transpiration rate and (**D**) stomatal conductance in soybean plants. Different letters indicate significant difference between the treatments. The data are the means ± SE ($n = 5$).

2.3. Effect of As and ZnO–NP on Gas Exchange Attributes

Gas exchange parameters i.e., net photosynthesis rate (Pn), transpiration rate (E) and stomatal conductance (g_s) declined significantly with an increase in As stress. Supplementation of 50-mg/L ZnO–NP to 10-µM-As-stressed plants enhanced Pn by 18.18%, E by 20.54% and g_s by 36.57% and that to 20-µM-treated plants' Pn by 13.70%, E by 22.91% and g_s by 48.14%. Application of 100-mg/L ZnO–NP to 20-µM-As-stressed plants enhanced Pn, E and g_s by 34.01%, 41.66% and 103.70%, respectively, over those in the As-alone-stressed plants (Figure 2B–D).

2.4. Effect of As and ZnO–NP on Chlorophyll Fluorescence

The maximum quantum efficiency (Fv/Fm) and photochemical yield (ΦPSII) of photosystem II declined with an increase in As stress than those in the non-As-treated plants. When 10-µM-As-stressed plants were supplemented with 50-mg/L and 100-mg/L ZnO–NP, both Fv/Fm and ΦPSII increased by 13.33%, 19.99% and 32.14%, 21.05%, respectively, with respect to those in the As-alone-treated plants.

In addition, when 20-µM-As-stressed plants were supplemented with 50-mg/L and 100-mg/L ZnO–NP, both Fv/Fm and ΦPSII further increased by 28.88% and 39.99% and 52.63% and 57.14%, respectively, relative to those in the As-alone-treated plants (Figure 3A).

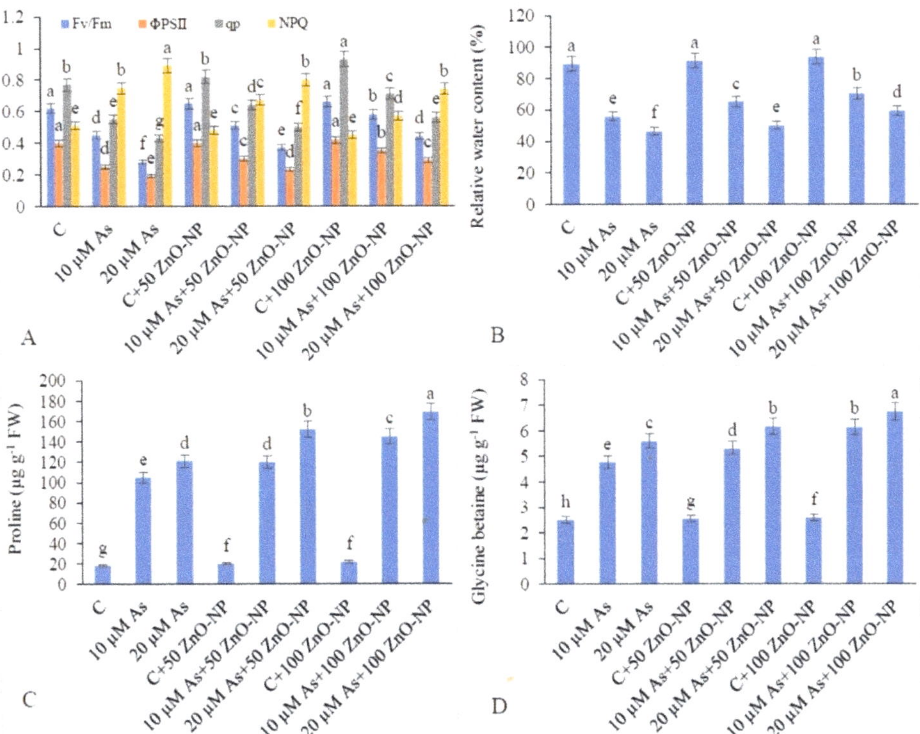

Figure 3. Effect of different concentrations of As and ZnO–NP on (**A**) chlorophyll fluorescence, (**B**) relative water content, (**C**) proline content and (**D**) glycine betaine content in soybean plants. Different letters indicate significant difference between the treatments. The data are the means ± SE (n = 5).

The photochemical quenching (qp) declined in the As-stressed plants and non-photochemical quenching (NPQ) increased in the As-stressed plants with respect to that in the non-As-fed plants (Figure 3A). Application of 50-mg/L and 100-mg/L ZnO–NP increased the qp by 16.27% and 30.23% in the 20-µM-As-stressed plants. In addition, application of 50-mg/L and 100-mg/L ZnO–NP declined the NPQ by 10.11% and 15.73% in the 20-µM-As-stressed plants. (Figure 3A).

2.5. Effect of As and ZnO–NP on Relative Water Content (RWC)

Leaf relative water content (RWC) decreased by 37.71% and 48.07% in the As-stressed plants with reference to that in the non-As-treated plants. However, the supplementation of 50-mg/L and 100-mg/L ZnO–NP to the 10-µM-As-stressed plants increased the water content by 17.14% and 26.26%, respectively, over those in the As-alone-treated plants. In addition, supplementation of 50-mg/L and 100-mg/L ZnO–NP to the 20-µM-As-stressed plants further increased the RWC by 7.87% and 27.79% (Figure 3B).

2.6. Appraisal of Proline and Glycine Betaine (GB) by ZnO–NP in As-Treated Plants

The amount of proline increased by 478.51% and 566.66%, in 10- and 20-µM-As-stressed plants, respectively, over that in the non-As-treated plants. Supplementation of 50-mg/L and 100-mg/L ZnO–NP further increased the proline content by 14.28% and 38.09%, respectively, in the plants exposed to 10 and 20-µM As. In addition, when the 20-µM-As-stressed plants were treated with 50-mg/L and 100-mg/L ZnO–NP, the proline content further increased by 25.61% and 39.66%, respectively, compared to those in the plants treated with As only (Figure 3C).

The amount of glycine betaine (GB) increased by 90.43% and 122.70%, in the 10- and 20-µM-As-stressed plants, respectively, with respect to that in the non-As-fed plants. Supplementation of 50-mg/L and 100-mg/L ZnO–NP further increased the GB content by 11.08% and 28.66% in the plants exposed to 10- and 20-µM As. Moreover, when the 20-µM-As-stressed plants were treated with 50-mg/L and 100-mg/L ZnO–NP, the GB content increased by 10.19% and 20.75% (Figure 3D).

2.7. Measurement of H_2O_2, Lipid Peroxidation (MDA) and Electrolyte Leakage (EL)

The amount of H_2O_2 increased by 126.97% and 206.51%, in the 10- and 20-µM-As-stressed plants, respectively, with respect to that in the non-As-treated plants. However, exogenous application of 50-mg/L ZnO–NP enhanced the H_2O_2 content by 70.69% and of 100-mg/L ZnO–NP by 30.69% only in the 10-µM-As-stressed plants as compared to the respective controls. In plants treated with 20-µM As and additionally supplied with 50-mg/L and 100-mg/L ZnO–NP, the H_2O_2 content increased by 153.02% and 128.37%, respectively, compared with those of As-treated plants which received no ZnO–NP (Figure 4A).

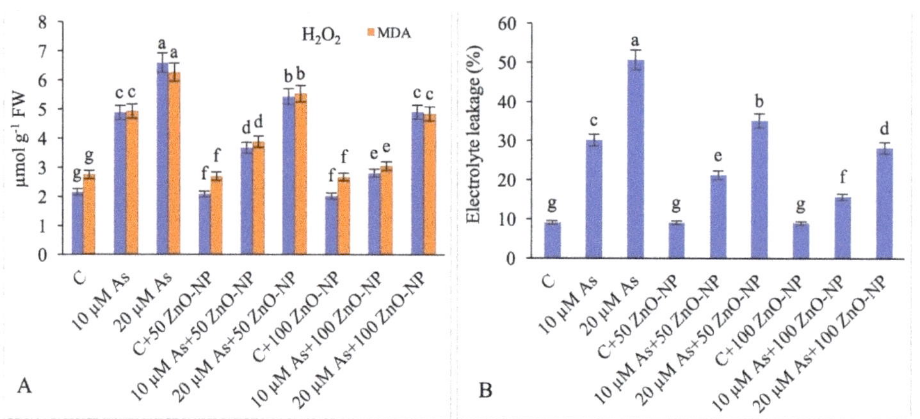

Figure 4. Effect of different concentrations of As and ZnO–NP on (**A**) accumulation of H_2O_2 and MDA content and (**B**) electrolyte leakage in soybean plants. Different letters indicate significant difference between the treatments. The data are the means ± SE ($n = 5$).

The MDA concentration increased by 126.97% and 206.51%, in the 10- and 20-µM-As-stressed plants, respectively, over those in the non-As-treated plants. The MDA content decreased by 21.13% in the 50-mg/L ZnO–NP-treated plants and 11.64% in the 100-mg/L ZnO–NP-treated plants under 10-µM As. The plants treated with 20-µM As when supplied with 50-mg/L and 100-mg/L ZnO–NP, the MDA concentration further declined by 38.0% and 22.64%, respectively, compared with those in the plants treated with As only (Figure 4A).

The As-stressed 10- and 20-µM plants showed increased electrolyte leakage by 230.98% and 456.81%, respectively relative to that in the non-As-treated plants. Exogenous application of 50-mg/L and 100-mg/L ZnO–NP declined the electrolyte leakage by 29.54% and 30.68%, respectively, in the

10-μM-As-stressed plants compared to those in the plants treated with As only. The 20-μM-As-stressed plants when provided ZnO–NP, showed reduced electrolyte leakage by 47.80% with 50-mg/L ZnO–NP and 44.36% with 100-mg/L ZnO–NP over those in the As-only-treated plants (Figure 4B).

2.8. Antioxidant Enzyme Activities and Ascorbate–Glutathione Cycle

The activities of SOD and CAT were found to be enhanced under 20-μM-As stress by 90.69% and 243.69%, respectively, over those in the non-As-fed plants. The activities of these enzymes were further enhanced when the 20-μM-treated plants were supplemented with 50-mg/L ZnO–NP (12.8% and 9.52%) and 100-mg/L ZnO–NP (24.8% and 22.85%) (Figure 5A).

The activities of the Asc–Glu cycle enzymes, APX and GR, increased under 20-μM-As stress by 90.69% and 243.69%, respectively, compared to those in the non-As-treated plants. The activities of these enzymes were further enhanced by 37.17% and 16.27%, respectively, in the 20-μM-As plants when they were supplemented with 50-mg/L ZnO–NP compared to those in the plants exposed to As only. The activities were further increased by 89.17% and 39.96%, respectively, with 100-mg/L ZnO–NP in the 20-μM-As-stressed plants (Figure 5B).

The activities of DHAR and MDHAR were found to be declined under 20-μM-As stress by 36.78% and 35.38%, respectively, compared to those of the non-As-treated plants. However, the activities of these enzymes increased markedly (8.29% and 12.25% relative to the controls) when supplemented with 50-mg/L ZnO–NP. On supplementation of 100-mg/L ZnO–NP to the 20-μM-As-stressed plants, the activity of DHAR further increased by 23.0% and that of MDHAR by 25.68% when compared with those of the plants exposed to As only (Figure 5C,D).

The amount of ascorbic acid (AsA) declined by 50% with 10-μM As and 20.83% by 20-μM-As stress over that in the non-As-treated plants. However, the levels of AsA in the As-stressed plants were found to be enhanced when they were supplemented with 50-mg/L and 100-mg/L ZnO–NP (Figure 5E).

The amount of glutathione (GSH) increased under 10- and 20-μM-As stress by 37.24% and 58.45%, respectively, compared to that in the non-As-treated plants. However, enhancement in the levels of GSH was observed with 50-mg/L and 100-mg/L ZnO–NP by 8.66% and 21.25%, respectively in the plants treated with 20-μM As compared to that in the plants exposed to As only (Figure 5F).

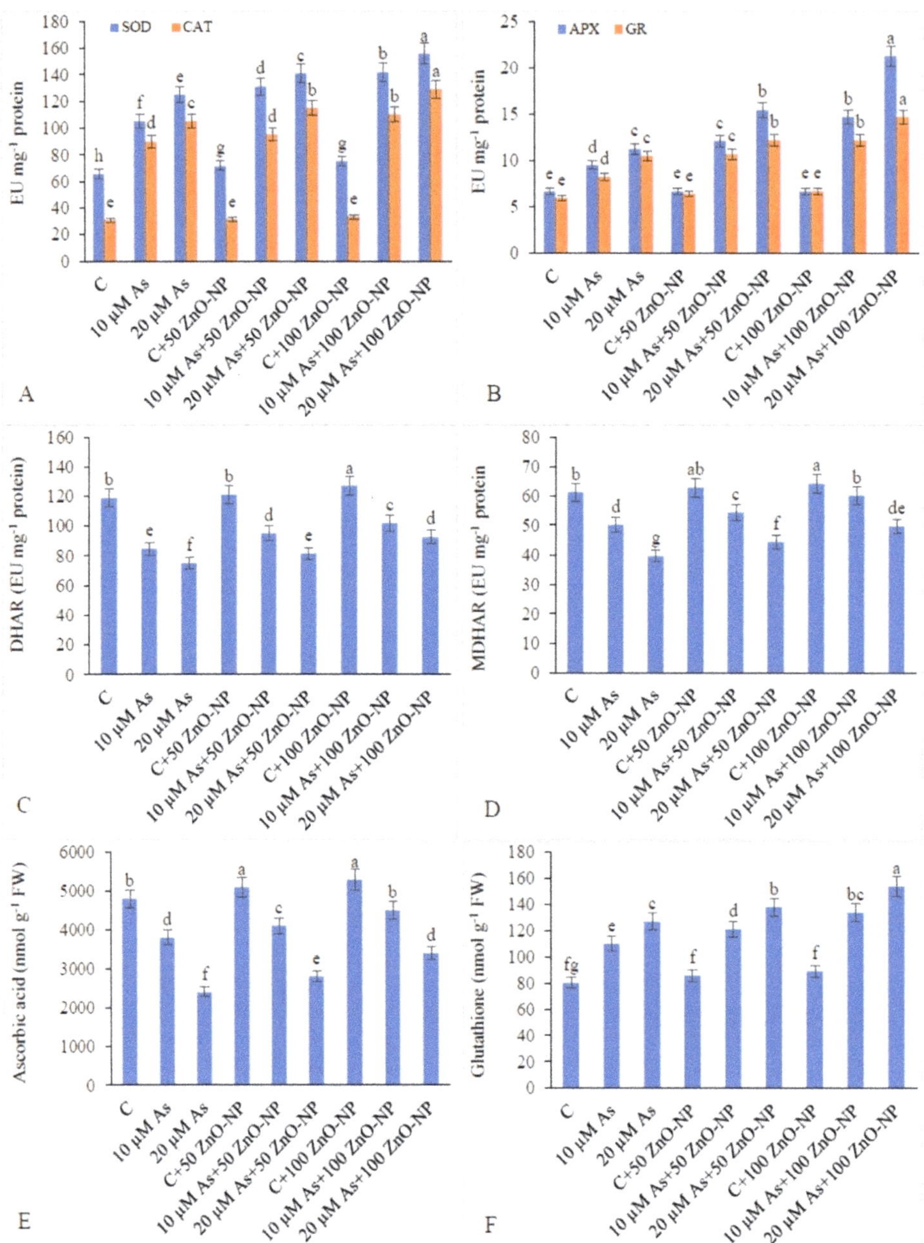

Figure 5. Effect of different concentrations of As and ZnO–NP on activity of (**A**) superoxide dismutase (SOD) and catalase (CAT), (**B**) ascorbate peroxidase (APX) and glutathione reductase (GR), (**C**) dehydroascorbate reductase (DHAR), (**D**) monodehydroascorbate reductase (MDHAR), (**E**) ascorbic acid and (**F**) glutathione content in soybean plants. Different letters indicate significant difference between the treatments. Data are means ± SE (n = 5).

2.9. Methyl Glyoxal (MG) and Glyoxalase (Gly) System in As-Stressed Plants

The As concentration of 10-μM and 20-μM increased the MG content by 76.90% and 144.02%, respectively. However, the levels of MG declined when 20-μM-As-stressed plants were supplemented with 50-mg/L (19.99%) and 100-mg/L ZnO–NP (28.15%) with respect to those in the plants exposed to As only (Figure 6A).

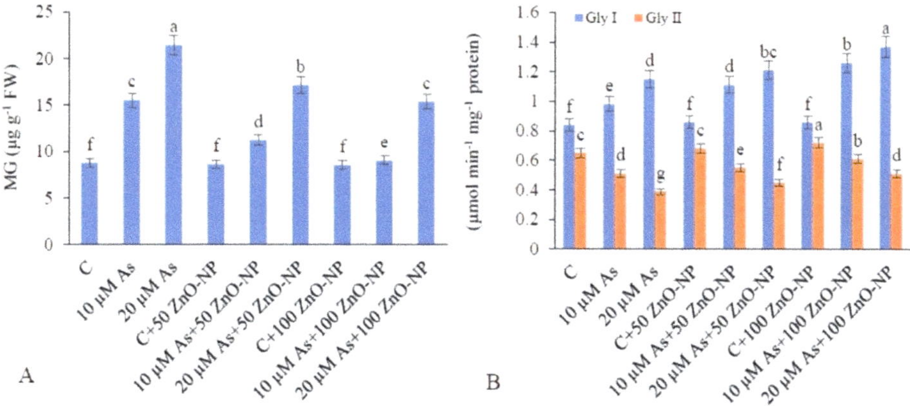

Figure 6. Effect of different concentrations of As and ZnO–NP on (**A**) MG content and (**B**) activity of GlyI and GlyII in soybean plants. Different letters indicate significant difference between the treatments. Fata are means ± SE (n = 5).

Application of 10-μM and 20-μM As stress increased the activity of GlyI by 16.66% and 36.90%, however, GlyII decreased by 21.53% and 40%, respectively, relative to those in the non-As-treated plants. The activity of GlyI increased by 5.21% and 19.13% when 20-μM-As-stressed plants were supplemented with 50-mg/L and 100-mg/L ZnO–NP, respectively. The activity of GlyII also increased by 15.38% with 50-mg/L and 30.76% with 100-mg/L ZnO–NP in the plants treated with 20-μM As over those in the plants exposed to As only (Figure 6B).

2.10. Effect of ZnO–NP and As on Accumulation of As in Shoot and Root

The 10-μM and 20-μM-As-stressed plants showed shoot As 395 μg g^{-1} DW and 500 μg g^{-1} DW, respectively, however, supplementation of 50-mg/L ZnO–NP decreased As accumulation in the shoots to 250 μg g^{-1} DW under 10-μM As and to 370 μg g^{-1} DW under 20-μM As. Accumulation of As in the shoots was further decreased to 175 μg g^{-1} DW and 220 μg g^{-1} DW, when the 10-μM and 20-μM-As-stressed plants were supplied with 100-mg/L ZnO–NP (Table 1).

In roots, the accumulation of 500-μg g^{-1} DW As under 10-μM As and 700-μg g^{-1} DW As under 20-μM As was recorded in the present study. However, application of 50-mg/L ZnO–NP reduced the accumulation of As to 420-μg g^{-1} DW and 600-μg g^{-1} DW in the 10-μM- and 20-μM-As-stressed plants, respectively. The accumulation of As in the roots was further decreased to 370-μg g^{-1} DW with 10-μM + 100-mg/L ZnO–NP and 400 with 20-μM + 100-mg/L ZnO–NP (Table 1).

Table 1. Effect of different concentrations of ZnO–NP on As accumulation in the leaf and root of soybean plants grown under As toxicity. Data presented are the means ± SE (*n* = 5). Different letters with mean values indicate significant difference at $p \leq 0.05$. (ND, not detected).

ZnO–NP (mg L^{-1})	As (μM)	Shoot As (μg g^{-1} DW)	Root As (μg g^{-1} DW)
	0	ND	ND
0	10	395 ± 25.45b	500 ± 33.25c
	20	500 ± 32.56a	700 ± 42.11a
	0	ND	ND
50	10	250 ± 15.77d	420 ± 29.67d
	20	370 ± 20.89c	600 ± 45.55b
	0	ND	ND
100	10	175 ± 10.23f	370 ± 21.17f
	20	220 ± 13.77e	400 ± 26.53e

3. Discussion

ZnO nanoparticles release Zn which regulates growth as it is an important element involved in the synthesis/accumulation of auxin, indole-3-acetic acid, which plays a vital role in cell division and cell expansion (Ali and Mahmoud, 2013). According to Hafeez et al. (2013) Zn application maintains membrane stability through binding to sulfhydryl groups and phospholipids under stress conditions. For example, Weisany et al. (2014) reported that Zn is essentially required for maintaining the structural integrity of membranes in soybean under salinity stress. In fact, Zn application boosts the uptake of micro- and macronutrients which are declined under stress conditions (Abd El-Hady 2007; Weisany et al. 2014). Furthermore, Zn supplementation was found to be beneficial in restoring the uptake of nutrients like K, Mg, Ca, Fe, P, thereby maintaining the structural and functional ability of different organelles like chloroplast, mitochondrion, etc. (Ahmad et al., 2018). Zinc helps in the uptake of Mg, one of the main constituents of the chlorophyll molecule, which can restore photosynthesis. Zinc is also reported to reduce ROS-induced oxidative stress, which may be attributed to the enhanced activity of enzymatic antioxidants (Ahmad et al., 2018), especially SOD, which contains different metal ions like copper and zinc (Cu/ZnSOD), manganese (MnSOD) and iron (FeSOD). However, in contrast, deficiency of Zn may lead to a variety of dysfunctions in plants. For example, Zn deficiency led to membrane disorganization in the thylakoids of sugar beet leaves (Henriques 2001). In other studies, Cakmak and Marschner (1988a, 1988b) reported that cotton roots deficient in Zn had enhanced potassium (K) leakage and it could be mitigated by the exogenous supplementation of Zn.

Arsenic toxicity affected the growth of soybean plants in terms of shoot length and shoot and root fresh and dry weights. These results endorse the findings of Jung, et al. [21] in As-stressed rice seedlings. Abedin, et al. [22] have also demonstrated decreased growth in rice due to As toxicity. Arsenic is believed to react with thiol groups of enzymes leading to overall inhibition of metabolism. Arsenic mediated growth inhibition was suggested to be attributable to cell cycle arrest and inhibition of DNA synthesis and repair mechanisms [23]. Many other heavy metals like cadmium, mercury, lead and chromium—as well as As—have also been shown to cause decreased growth and biomass yield in different plants [7,16,24,25]. This decreased growth and biomass yield of plants exposed to heavy metal stress may be due to restricted uptake of nutrient elements, reduced photosynthesis and other metabolic activities [7]. However, the supplementation of ZnO–NP in our study enhanced the growth and biomass yield of As-stressed soybean plants significantly. These results coincide with the findings of Rizwan, et al. [16], who also reported significantly increased growth of Cd-stressed maize plants with the application of ZnO–NP. ZnO–NPs have also been shown to enhance the growth and biomass yield of plants under Cd and Pb stress [26]. An interplay of various factors such as increase in mineral uptake, decrease in oxidative stress, triggering of biochemical pathways involved in biomass accumulation, etc., may play a significant role in such growth enhancement [26,27]. ZnO–NP application has been reported to decrease the oxidative stress induced by Cd and Pb in many other plants including *Leucaena leucocephala* [26]. Additionally, we also found that As hampered the growth of roots more than that of the shoots, possibly because retention of As in the roots was higher than that in the stem. Several studies have reported the accumulation of As being higher in roots than that in stem [5,28]. In contrast, some studies have also reported that the effect of As on root and shoot growth may vary depending on plant species, contamination level and plant tissue ability to uptake As [26].

All the photosynthetic parameters of the soybean plants were observed to be diminished under As stress. Arsenic is known to interfere with nitrogen metabolism leading to its reduced bioavailability [29]. Nitrogen is a core component of chlorophyll, so its reduced bioavailability due to arsenic stress diminishes the chlorophyll content of a plant [29]. This could be one of the main reasons for the decreased photosynthetic rate of plants in the presence of arsenic [29]. Our findings of reduced photosynthetic pigments and photosynthetic rate are in line with some previous studies on a variety of other plants exposed to arsenic stress [5,14]. Application of ZnO–NP considerably increased the amount of total chlorophyll in leaves, and it also had a positive effect on gas exchange characteristics including net assimilation rate, transpiration and stomatal conductance. These results endorse the findings of Faizan, et al. [30] which showed improved photosynthetic rates in rice plants on application of ZnO–NP. This may have been due to the ability of metal nanoparticles to enhance the production of chemical energy in photosynthetic systems, synthesis of photosynthetic pigments and improvement in quantum yield in plants [27,31–33]. Some earlier studies have highlighted the efficiency of NPs in enhancing photosynthesis in plants whose response varies in a dose-dependent and plant species-dependent manner [14,34]. Furthermore, our study showed that leaf relative water content (RWC) also decreased in the presence of As in the soybean plants. Similar results have been reported in other studies [35]. However, when arsenic-stressed soybean plants were supplemented with different concentrations of ZnO-NP, there was a considerable increase in water potential and RWC of the leaves. This may have been due to enhanced uptake of water and mineral elements in the presence of ZnO–NP.

Arsenic stress increased the content of essential osmolytes particularly of proline and glycinebetaine in the present study. These osmolytes are important in alleviating the stress-mediated damaging effects. Thus, an increase in their content under toxic conditions is a natural way of defense in plants [36,37]. Arsenic stress enhanced the proline accumulation in the soybean plants as has been reported by Choudhury, et al. [38] in *Oryza sativa* and Siddiqui, et al. [39] in *Withania somnifera*. The enhanced accumulation of proline and GB protects biochemical processes, promotes ROS scavenging, and maintains redox homeostasis and functions of different enzymes [40]. According to Hasanuzzaman, et al. [41], proline and GB help maintain the plant antioxidant and glyoxalase systems thereby enhancing tolerance against stress. Anjum, et al. [42] showed that GB and GSH help in metal chelation that reduces the noxious effects of metal stress. To decrease the stress or enhance the stress withstanding potential of plants, naturally occurring mechanisms could be made stronger by supplementation of agents such as metal oxides which can boost the antioxidant and glyoxalase systems. Thus, the use of ZnO–NP helps alleviate heavy metal toxicity, as has been found in this study. These findings are in line with other studies [18,43]). According to Sivakumar, et al. [44] the carboxylase activity of rubisco was protected by enhanced proline content which in turn impeded photoinhibition. An increase in photosynthesis under stress conditions correlates with the accumulation of GB [45]. ZnO–NP seems to upregulate the proline and GB biosynthetic pathways so as to attain maximum accumulation of these two osmolytes in plants, which could play a vital role in alleviation of As stress.

The increasing intensity of As stress in the soybean plants caused increased levels of H_2O_2 and MDA and electrolyte leakage (EL). Enhanced accumulation of H_2O_2 and MDA as well as increased EL has also been reported in other plants grown under different stresses ((Ahmad et al., 2018, 2020; Ahanger and Agarwal 2017 a,b; Kaya et al., 2020). The H_2O_2 accumulation needs to be maintained to a certain threshold level, otherwise it will lead to lipid peroxidation that in turn leads to electrolyte leakage. Supplementation of ZnO–NP in our study decreased the accumulation of H_2O_2 and MDA as well as EL in the present study, and our results coincide with the findings of some earlier reports with different plant species [16,18,34]. Enhanced activity of antioxidant enzymes by supplementation of ZnO–NPs may scavenge H_2O_2 and elevate mineral uptake, thereby reducing plant oxidative stress [16].

Plants are known to exhibit several detoxification mechanisms such as activation of different antioxidant enzymes such as superoxide dismutase (SOD), ascorbate peroxidase (APX) and catalase (CAT) and non-enzymatic antioxidants like glutathione, ascorbate, tocopherols, etc. against enhanced ROS accumulation [13,46]. Arsenic stress-induced enhanced activities of antioxidant enzymes corroborate with the findings of Talukdar [47], Kumar Yadav and Srivastava [48] and Siddiqui, et al. [49] in *Trigonella goenum-graecum, Zea mays* and *Ocimum tenuiflorum,* respectively. Lu, et al. [50] also reported enhanced activities of SOD, CAT, POD and APX in tartary buckwheat under Cd stress. Arsenic stress also decreased the components involved in the Asc–Glu cycle [50]. The activities of DHAR and MDHAR as well as the levels of AsA showed a decline with As toxicity in the present study and these results are in line with our earlier findings on *Vicia faba* [8]. Glutathione reductase (GR) has been reported as a rate limiting enzyme in the Asc–GSH cycle and it catalyzes the reaction of GSSG (oxidized form of GSH) to GSH with the electron donor NADPH [51,52]. The DHAR enzyme catalyzes the reduction of DHA to AsA [51,52]. According to Sharma [53] As binds to thiol groups of antioxidant enzymes which directly affect biochemical reactions thereby hampering overall plant growth. The enhanced activities of antioxidant enzymes against the oxidative stress have also been reported through the over-expression of SOD, APX and CAT isozymes [47]. Supplementation of As-stressed soybean plants with ZnO–NP further increased the activities of these defense enzymes, indicating a protective role of ZnO–NP in such adverse conditions. The ZnO–NP induced enhanced activities of SOD, CAT, APX and GR in the As-stressed soybean plants in the present study endorse the results of [26] in *Leucaena leucocephala* seedlings under Cd and Pb stresses. ZnO–NP-induced enhanced activities of SOD, CAT, APX and GR have been reported by many authors in different plants, e.g., Tripathi, et al. [43] in *Pisum sativum*, Hernandez-Viezcas, et al. [54] in *Prosopis juliflora*, Krishnaraj, et al. [55] in *Bacopa monnieri.* These enhanced activities of antioxidants helped the plants to scavenge the extra ROS produced under stressful cues.

Another detoxifying system known as the glyoxalase cycle which contains glyoxalase I and glyoxalase II (Gly I and Gly II) and their function is to remove the MG generated during stress [56]. MG accumulation under As stress is the main cause of cytotoxic and mutagenic effects and alteration of cellular ultrastructure and cell death [11]. However, Gly I and Gly II reduce the accumulation of MG as has been reported by earlier workers [8,57]. Supplementation of ZnO–NP enhanced the activity of Gly I and Gly II, thus providing an extra strength to minimize the production of MG. Little literature is available on the interaction of As and ZnO–NP in plants with respect to MG accumulation. Thus, further investigation at physiological and molecular levels needs to be carried out.

4. Materials and Methods

4.1. Plant Material and Experimental Design

Soybean seeds were decontaminated with sodium hypochlorite (0.5%, *v/v*) for 5 min and then soaked in distilled water for 10 min before germination. The germinated seeds (4-day-old seedlings) were shifted to pots containing vermicompost and sand (1:3 ratio) (sterilized) and kept for 3 weeks without any interruption. Thereafter, arsenic (AsIII, salt $NaAsO_2$, sodium arsenite) was added to each pot daily for 5 weeks (60-day-old plants). AsIII was used in this study because it is more soluble and mobile than other forms of As. Only full-strength Hoagland's nutrient solution was added to the experimental control. After 2 weeks of As treatment (39-day-old plants), three different concentrations (0, 50 and 100 mg L^{-1}) of ZnO–NP (Sigma Aldrich, Saint Louis, USA) were sprayed to plant foliage every alternate day for 2 weeks (60-day-old plants). After adding ZnO–NP in the pots, they were maintained in a growth chamber with the following conditions: 26/15 °C (day/night) temperature, 18 h/6 h (light/dark) photoperiod and 70%–75% relative humidity. Each treatment had five replications and each experiment was repeated 3 times. After growing the plants for 60-days, they were harvested and studied for different parameters.

4.2. Estimation of Length and Dry Mass of Shoot and Root

Shoot and root lengths were measured using a manual scale after uprooting the plants. The samples were measured for fresh weight and then subjected to an oven for drying at 70 °C for 24 h for dry weight estimation.

4.3. Evaluation of Photosynthetic Pigments

Leaf samples of 300 mg each were crushed with 80% acetone. The mixtures were centrifuged for 20 min at 3000× *g*. The optical density (OD) of the supernatant was estimated with a spectrophotometer at 480, 645 and 663 nm following Arnon [58].

4.4. Leaf Gas Exchange Parameters and Chlorophyll Fluorescence (Fv/Fm, ΦPSII, qP, NPQ)

Leaf gas exchange parameters including transpiration rate (*E*), net photosynthetic rate (*Pn*) and stomatal conductance (g_s) were estimated using the IRGA (LCA-4 model Analytical Development Company, Hoddesdon, UK) using fully expanded leaves from each replicate between 10:00 h and 12:00 h in full and bright sunlight.

For the determination of chlorophyll fluorescence parameters, Junior PAM Chlorophyll Fluorometer (H. Walz, Effeltrich, Germany) was used following the protocol of Li, et al. [59].

4.5. Determination of Leaf (LRWC)

For the assessment of relative water content of leaves, the method of Smart and Bingham [60] was used and its values were determined with the following formula:

$$\text{LRWC} = \frac{\text{FW} - \text{DW}}{\text{TW} - \text{DW}} \times 100 \tag{1}$$

4.6. Determination of Proline and Glycine Betaine (GB) Content

For determining the proline content, the protocol given by Bates, et al. [61] was used and the optical density (OD) was recorded at 520 nm with a spectrophotometer. For determining the glycine betaine (GB) content, the Grieve and Grattan [62] method was used, and the absorbance was taken at 365 nm.

4.7. Measurement of H_2O_2, Lipid Peroxidation (MDA) and Electrolyte Leakage (EL)

For measuring the H_2O_2 and MDA contents, the methods given by Velikova, et al. [63] and Madhava Rao and Sresty [64], respectively, were used. The OD was recorded at 390 nm for H_2O_2 and at 432 and 600 nm for MDA with a spectrophotometer. The method of Dionisio-Sese and Tobita [65] was used for measuring the electrolyte leakage, and its values were determined using the following formula:

$$\text{EL\%} = \frac{\text{EC1} - \text{EC0}}{\text{EC2} - \text{EC0}} \times 100 \tag{2}$$

4.8. Estimation of Activities of Antioxidant Enzymes and the Ascorbate–Glutathione Cycle

Fresh leaf sample (each 500 mg) was crushed in polyvinyl pyrrolidone solution (1%) and cold potassium phosphate buffer (pH 7.0, 100 mM) for extracting the antioxidant enzymes. The mixture was centrifuged at 4 °C at 12,000 *g* for 30 min. The supernatant was collected and used for determining the activities of SOD, CAT, APX and GR.

4.8.1. Superoxide Dismutase (SOD, EC1.15.1.1) Activity

The activity of SOD was determined by monitoring the diminishing of NBT (nitro blue tetrazolium) photochemically by the enzyme extract [66]. The OD was recorded at 560 nm and the SOD activity was presented as EU mg^{-1} protein.

4.8.2. Catalase (CAT, EC1.11.1.6) Activity

The CAT activity was evaluated by the protocol given by Aebi [67] and disappearance of H_2O_2 was recorded for 2 min at 240 nm. The CAT activity was presented as EU mg^{-1} protein.

4.8.3. Ascorbate Peroxidase (APX, EC1.11.1.1) Activity

APX activity was assayed by monitoring the H_2O_2-dependent oxidation of ascorbate at 290 nm for 2 min (Nakano and Asada 1981). The APX activity was presented as EU mg^{-1} protein.

4.8.4. Glutathione Reductase (GR, EC1.6.4.2) Activity

The GR activity was assayed by the procedure described by Foster and Hess [68]. The absorbance was taken at 340 nm for 3 min and the GR activity was presented as EU mg^{-1} protein.

4.8.5. Monodehydroascorbate Reductase (MDHAR, EC1.6.5.4) Activity

The protocol described by Miyake and Asada [69] was employed for the assay of MDHAR. The OD was recorded at 340 nm and the activity was presented as μmol NADPH oxidized (EU mg^{-1} protein).

4.8.6. Dehydroascorbate Reductase (DHAR, EC1.8.5.1) Activity

Nakano and Asada [70] protocol was exercised for the assay of DHAR activity and the OD was taken at 265 nm. The DHAR activity was presented as EU mg^{-1} protein.

4.8.7. Ascorbic Acid (AsA) and Glutathione

The methods described by Huang, et al. [71] and Yu, et al. [72] were followed for the assay of ascorbate and glutathione content, respectively.

4.9. Estimation of Methylglyoxal (MG), Gly I (EC4.4.1.5) and GlyII (EC3.1.2.6)

The protocol of Wild, et al. [73] was used for the determination of MG content and the OD was taken at 288 nm. For determining the activities of Gly 1 and Gly II, the protocols given by Hossain, et al. [74] and Mostofa and Fujita [75], respectively, were used. The OD was recorded at 240 nm and the enzymes' activities were expressed as μmol mg^{-1} min^{-1} protein.

4.10. Estimation of As in the Shoot and Root

The dry tissue (each 500 mg) was digested in an acid mixture and the concentration of As was quantified by an atomic absorption spectrophotometer (Perkin-Elmer Analyst Model 300, NJ, USA).

4.11. Statistical Analysis

The SPSS statistical software version 17.0 (SPSS, Inc., Chicago, USA) was used for analysis of variance of each data set and the mean values were compared using Duncan's multiple range test (DMRT) at $p \leq 0.05$. The data presented are mean± SE and $n = 5$.

5. Conclusions

This study marks that ZnO–NPs are efficient ameliorators of As toxicity in soybean plants at different concentrations. Thus, if used at certain specific doses, ZnO–NPs could prove to be an excellent foliar fertilizer in areas that are prone to arsenic exposure. The use of these nanoparticles can increase the efficiency of cropping systems by mitigating As stress and promoting the yield of plants.

Author Contributions: P.A., M.N.A. and M.A. drafted the experimental design and P.A. and M.N.A. performed the experiments. A.A.A.-H., M.A.A. and L.W. helped in writing the initial draft of the manuscript. A.A.A.-H., C.K. and A.B. analyzed the data. P.A., M.N.A. and M.A. revised the manuscript to the present form. All authors have read and approved this manuscript for publication.

Funding: Research Group No. RGP-231.

Acknowledgments: The authors also extend their appreciation to the Deanship of Scientific Research at King Saud University for supporting this work through Research Group No. RGP-231.

Conflicts of Interest: The authors declare no conflict of interest.

References

1. Bowell, R.J.; Alpers, C.N.; Jamieson, H.E.; Nordstrom, D.K.; Majzlan, J. The environmental geochemistry of arsenic—an overview. *Rev. Mineral. Geochem.* **2014**, *79*, 1–16. [CrossRef]
2. Sultana, R.; Kobayashi, K. Potential of barnyard grass to remediate arsenic contaminated soil. *Weed Biol. Manag.* **2011**, *11*, 12–17. [CrossRef]
3. Finnegan, P.M.; Chen, W. Arsenic toxicity: The effects on plant metabolism. *Front. Physiol.* **2012**, *3*, 182. [CrossRef] [PubMed]
4. Garg, N.; Singla, P. Arsenic toxicity in crop plants: Physiological effects and tolerance mechanisms. *Environ. Chem. Lett.* **2011**, *9*, 303–321. [CrossRef]
5. Stoeva, N.; Berova, M.; Zlatev, Z. Effect of arsenic on some physiological parameters in bean plants. *Biol. Plant.* **2005**, *49*, 293–296. [CrossRef]
6. Verbruggen, N.; Hermans, C.; Schat, H. Mechanisms to cope with arsenic or cadmium excess in plants. *Curr. Opin. Plant. Biol.* **2009**, *12*, 364–372. [CrossRef]
7. Ahmad, P.; Ahanger, M.A.; Alyemeni, M.N.; Wijaya, L.; Alam, P. Exogenous application of nitric oxide modulates osmolyte metabolism, antioxidants, enzymes of ascorbate-glutathione cycle and promotes growth under cadmium stress in tomato. *Protoplasma* **2018**, *255*, 79–93. [CrossRef]
8. Ahmad, P.; Alam, P.; Balawi, T.H.; Altalayan, F.H.; Ahanger, M.A.; Ashraf, M. Sodium nitroprusside (SNP) improves tolerance to arsenic (As) toxicity in *Vicia faba* through the modifications of biochemical attributes, antioxidants, ascorbate-glutathione cycle and glyoxalase cycle. *Chemosphere* **2020**, *244*, 125480. [CrossRef]
9. Yadav, S.K. Heavy metals toxicity in plants: An overview on the role of glutathione and phytochelatins in heavy metal stress tolerance of plants. *S. Afr. J. Bot.* **2010**, *76*, 167–179. [CrossRef]
10. Kumar, V.; Yadav, S.K. Proline and betaine provide protection to antioxidant and methylglyoxal detoxification systems during cold stress in *Camellia sinensis* (L.) O. Kuntze. *Acta Physiol. Plant.* **2009**, *31*, 261–269. [CrossRef]
11. Yadav, S.K.; Singla-Pareek, S.L.; Reddy, M.; Sopory, S. Methylglyoxal detoxification by glyoxalase system: A survival strategy during environmental stresses. *Physiol. Mol. Biol. Plants* **2005**, *11*, 1–11.
12. Ahmad, P.; Jaleel, C.A.; Salem, M.A.; Nabi, G.; Sharma, S. Roles of enzymatic and nonenzymatic antioxidants in plants during abiotic stress. *Crit. Rev. Biotechnol.* **2010**, *30*, 161–175. [CrossRef] [PubMed]
13. Ahmad, P.; Jaleel, C.A.; Sharma, S. Antioxidant defense system, lipid peroxidation, proline-metabolizing enzymes, and biochemical activities in two *Morus alba* genotypes subjected to NaCl stress. *Russ. J. Plant Physiol.* **2010**, *57*, 509–517. [CrossRef]
14. Rizwan, M.; Ali, S.; Hussain, A.; Ali, Q.; Shakoor, M.B.; Zia-ur-Rehman, M.; Farid, M.; Asma, M. Effect of zinc-lysine on growth, yield and cadmium uptake in wheat (*Triticum aestivum* L.) and health risk assessment. *Chemosphere* **2017**, *187*, 35–42. [CrossRef]
15. Dimkpa, C.O.; White, J.C.; Elmer, W.H.; Gardea-Torresdey, J. Nanoparticle and Ionic Zn Promote Nutrient Loading of Sorghum Grain under Low NPK Fertilization. *J. Agric. Food Chem.* **2017**, *65*, 8552–8559. [CrossRef]
16. Rizwan, M.; Ali, S.; Ali, B.; Adrees, M.; Arshad, M.; Hussain, A.; Zia ur Rehman, M.; Waris, A.A. Zinc and iron oxide nanoparticles improved the plant growth and reduced the oxidative stress and cadmium concentration in wheat. *Chemosphere* **2019**, *214*, 269–277. [CrossRef]

17. García-Gómez, C.; García, S.; Obrador, A.F.; González, D.; Babín, M.; Fernández, M.D. Effects of aged ZnO NPs and soil type on Zn availability, accumulation and toxicity to pea and beet in a greenhouse experiment. *Ecotoxicol. Environ. Saf.* **2018**, *160*, 222–230. [CrossRef]

18. Hussain, A.; Ali, S.; Rizwan, M.; Zia ur Rehman, M.; Javed, M.R.; Imran, M.; Chatha, S.A.S.; Nazir, R. Zinc oxide nanoparticles alter the wheat physiological response and reduce the cadmium uptake by plants. *Environ. Pollut.* **2018**, *242*, 1518–1526. [CrossRef]

19. Pullagurala, R.V.L.; Adisa, I.O.; Rawat, S.; Kim, B.; Barrios, A.C.; Medina-Velo, I.A.; Hernandez-Viezcas, J.A.; Peralta-Videa, J.R.; Gardea-Torresdey, J.L. Finding the conditions for the beneficial use of ZnO nanoparticles towards plants-A review. *Environ. Pollut.* **2018**, *241*, 1175–1181. [CrossRef]

20. Sturikova, H.; Krystofova, O.; Huska, D.; Adam, V. Zinc, zinc nanoparticles and plants. *J. Hazard. Mater.* **2018**, *349*, 101–110. [CrossRef]

21. Jung, H.-I.; Lee, J.; Chae, M.-J.; Kong, M.-S.; Lee, C.-H.; Kang, S.-S.; Kim, Y.-H. Growth-inhibition patterns and transfer-factor profiles in arsenic-stressed rice (*Oryza sativa* L.). *Environ. Monit. Assess.* **2017**, *189*, 638. [CrossRef] [PubMed]

22. Abedin, M.J.; Feldmann, J.; Meharg, A.A. Uptake kinetics of arsenic species in rice plants. *Plant Physiol.* **2002**, *128*, 1120–1128. [CrossRef] [PubMed]

23. Ghiani, A.; Fumagalli, P.; Nguyen Van, T.; Gentili, R.; Citterio, S. The combined toxic and genotoxic effects of Cd and As to plant bioindicator *Trifolium repens* L. *PLoS ONE* **2014**, *9*, e99239. [CrossRef] [PubMed]

24. Habiba, U.; Ali, S.; Rizwan, M.; Hussain, M.B.; Hussain, A.; Alam, P.; Alqarawi, A.A.; Hashem, A.; AbdAllah, E.F. The Ameliorative Role of 5-Aminolevulinic Acid (ALA) Under Cr Stress in Two Maize Cultivars Showing Differential Sensitivity to Cr Stress Tolerance. *J. Plant Growth Regul.* **2018**, *38*, 788–798. [CrossRef]

25. Kaya, C.; Okant, M.; Ugurlar, F.; Alyemeni, M.N.; Ashraf, M.; Ahmad, P. Melatonin-mediated nitric oxide improves tolerance to cadmium toxicity by reducing oxidative stress in wheat plants. *Chemosphere* **2019**, *225*, 627–638. [CrossRef] [PubMed]

26. Venkatachalam, P.; Jayaraj, M.; Manikandan, R.; Geetha, N.; Rene, E.R.; Sharma, N.C.; Sahi, S.V. Zinc oxide nanoparticles (ZnONPs) alleviate heavy metal-induced toxicity in *Leucaena leucocephala* seedlings: A physiochemical analysis. *Plant Physiol. Biochem.* **2017**, *110*, 59–69. [CrossRef] [PubMed]

27. Xie, Y.; Li, B.; Tao, G.; Zhang, Q.; Zhang, C. Effects of nano-silicon dioxide on photosynthetic fluorescence characteristics of *Indocalamus barbatus* McClure. *J. Nanjing For. Univ. (Nat. Sci. Ed.)* **2012**, *36*, 59–63.

28. Gomes, M.P.; Moreira Duarte, D.; Silva Miranda, P.L.; Carvalho Barreto, L.; Matheus, M.T.; Garcia, Q.S. The effects of arsenic on the growth and nutritional status of *Anadenanthera peregrina*, a Brazilian savanna tree. *J. Plant Nutr. Soil Sci.* **2012**, *175*, 466–473. [CrossRef]

29. Azizur Rahman, M.; Hasegawa, H.; Mahfuzur Rahman, M.; Nazrul Islam, M.; Majid Miah, M.A.; Tasmen, A. Effect of arsenic on photosynthesis, growth and yield of five widely cultivated rice (*Oryza sativa* L.) varieties in Bangladesh. *Chemosphere* **2007**, *67*, 1072–1079. [CrossRef]

30. Faizan, M.; Faraz, A.; Yusuf, M.; Khan, S.T.; Hayat, S. Zinc oxide nanoparticle-mediated changes in photosynthetic efficiency and antioxidant system of tomato plants. *Photosynthetica* **2018**, *56*, 678–686. [CrossRef]

31. Juhel, G.; Batisse, E.; Hugues, Q.; Daly, D.; van Pelt, F.N.A.M.; O'Halloran, J.; Jansen, M.A.K. Alumina nanoparticles enhance growth of *Lemna minor*. *Aquat. Toxicol.* **2011**, *105*, 328–336. [CrossRef] [PubMed]

32. Rico, C.M.; Peralta-Videa, J.R.; Gardea-Torresdey, J.L. Chemistry, Biochemistry of Nanoparticles, and Their Role in Antioxidant Defense System in Plants. In *Nanotechnology and Plant Sciences*; Springer International Publishing: Cham, Switzerland, 2015; pp. 1–17. [CrossRef]

33. Siddiqui, M.H.; Al-Whaibi, M.H.; Faisal, M.; Al Sahli, A.A. Nano-silicon dioxide mitigates the adverse effects of salt stress on *Cucurbita pepo* L. *Environ. Toxicol. Chem.* **2014**, *33*, 2429–2437. [CrossRef] [PubMed]

34. Venkatachalam, P.; Priyanka, N.; Manikandan, K.; Ganeshbabu, I.; Indiraarulselvi, P.; Geetha, N.; Muralikrishna, K.; Bhattacharya, R.C.; Tiwari, M.; Sharma, N.; et al. Enhanced plant growth promoting role of phycomolecules coated zinc oxide nanoparticles with P supplementation in cotton (*Gossypium hirsutum* L.). *Plant Physiol. Biochem.* **2017**, *110*, 118–127. [CrossRef] [PubMed]

35. Vezza, M.E.; Llanes, A.; Travaglia, C.; Agostini, E.; Talano, M.A. Arsenic stress effects on root water absorption in soybean plants: Physiological and morphological aspects. *Plant Physiol. Biochem.* **2018**, *123*, 8–17. [CrossRef] [PubMed]

36. Ahanger, M.A.; Agarwal, R. Potassium up-regulates antioxidant metabolism and alleviates growth inhibition under water and osmotic stress in wheat (*Triticum aestivum* L). *Protoplasma* **2017**, *254*, 1471–1486. [CrossRef] [PubMed]

37. Ahanger, M.A.; Agarwal, R.M. Salinity stress induced alterations in antioxidant metabolism and nitrogen assimilation in wheat (*Triticum aestivum* L) as influenced by potassium supplementation. *Plant Physiol. Biochem.* **2017**, *115*, 449–460. [CrossRef] [PubMed]

38. Choudhury, B.; Chowdhury, S.; Biswas, A.K. Regulation of growth and metabolism in rice (*Oryza sativa* L.) by arsenic and its possible reversal by phosphate. *J. Plant Interact.* **2011**, *6*, 15–24. [CrossRef]

39. Siddiqui, F.; Tandon, P.K.; Srivastava, S. Analysis of arsenic induced physiological and biochemical responses in a medicinal plant, *Withania somnifera*. *Physiol. Mol. Biol. Plants* **2015**, *21*, 61–69. [CrossRef]

40. Kaya, C.; Ashraf, M.; Alyemeni, M.N.; Ahmad, P. Responses of nitric oxide and hydrogen sulfide in regulating oxidative defence system in wheat plants grown under cadmium stress. *Physiol. Plant.* **2020**. [CrossRef]

41. Hasanuzzaman, M.; Alam, M.M.; Rahman, A.; Hasanuzzaman, M.; Nahar, K.; Fujita, M. Exogenous proline and glycine betaine mediated upregulation of antioxidant defense and glyoxalase systems provides better protection against salt-induced oxidative stress in two rice (*Oryza sativa* L.) varieties. *Biomed. Res. Int.* **2014**, *2014*, 757219. [CrossRef]

42. Anjum, N.A.; Hasanuzzaman, M.; Hossain, M.A.; Thangavel, P.; Roychoudhury, A.; Gill, S.S.; Rodrigo, M.A.M.; Adam, V.; Fujita, M.; Kizek, R.; et al. Jacks of metal/metalloid chelation trade in plants-an overview. *Front. Plant Sci.* **2015**, *6*, 192. [CrossRef] [PubMed]

43. Tripathi, D.K.; Singh, V.P.; Prasad, S.M.; Chauhan, D.K.; Dubey, N.K. Silicon nanoparticles (SiNp) alleviate chromium (VI) phytotoxicity in *Pisum sativum* (L.) seedlings. *Plant Physiol. Biochem.* **2015**, *96*, 189–198. [CrossRef] [PubMed]

44. Sivakumar, P.; Sharmila, P.; Pardha Saradhi, P. Proline Alleviates Salt-Stress-Induced Enhancement in Ribulose-1,5-Bisphosphate Oxygenase Activity. *Biochem. Biophys. Res. Commun.* **2000**, *279*, 512–515. [CrossRef] [PubMed]

45. Khan, M.I.R.; Asgher, M.; Khan, N.A. Alleviation of salt-induced photosynthesis and growth inhibition by salicylic acid involves glycinebetaine and ethylene in mungbean (*Vigna radiata* L.). *Plant Physiol. Biochem.* **2014**, *80*, 67–74. [CrossRef] [PubMed]

46. Sharma, I. Arsenic Stress in Plants: An Inside Story. In *Crop Improvement*; Springer: Boston, MA, USA, 2013; pp. 379–400. [CrossRef]

47. Talukdar, D. Arsenic-induced changes in growth and antioxidant metabolism of fenugreek. *Russ. J. Plant Physiol.* **2013**, *60*, 652–660. [CrossRef]

48. Kumar Yadav, R.; Srivastava, S.K. Effect of Arsenite and Arsenate on Lipid Peroxidation, Enzymatic and Non-Enzymatic Antioxidants in *Zea mays* Linn. *Biochem. Physiol.* **2015**, *4*. [CrossRef]

49. Siddiqui, F.; Tandon, P.K.; Srivastava, S. Arsenite and arsenate impact the oxidative status and antioxidant responses in *Ocimum tenuiflorum* L. *Physiol. Mol. Biol. Plant* **2015**, *21*, 453–458. [CrossRef]

50. Lu, Y.; Wang, Q.-F.; Li, J.; Xiong, J.; Zhou, L.-N.; He, S.-L.; Zhang, J.-Q.; Chen, Z.-A.; He, S.-G.; Liu, H. Effects of exogenous sulfur on alleviating cadmium stress in tartary buckwheat. *Sci. Rep.* **2019**, *9*, 7397. [CrossRef]

51. Liang, T.; Ding, H.; Wang, G.; Kang, J.; Pang, H.; Lv, J. Sulfur decreases cadmium translocation and enhances cadmium tolerance by promoting sulfur assimilation and glutathione metabolism in *Brassica chinensis* L. *Ecotoxicol. Environ. Saf.* **2016**, *124*, 129–137. [CrossRef]

52. Lou, L.; Kang, J.; Pang, H.; Li, Q.; Du, X.; Wu, W.; Chen, J.; Lv, J. Sulfur Protects Pakchoi (*Brassica chinensis* L.) Seedlings against Cadmium Stress by Regulating Ascorbate-Glutathione Metabolism. *Int. J. Mol. Sci.* **2017**, *18*, 1628. [CrossRef]

53. Sharma, I. Arsenic induced oxidative stress in plants. *Biologia* **2012**, *67*, 447–453. [CrossRef]

54. Hernandez-Viezcas, J.A.; Castillo-Michel, H.; Servin, A.D.; Peralta-Videa, J.R.; Gardea-Torresdey, J.L. Spectroscopic verification of zinc absorption and distribution in the desert plant *Prosopis juliflora-velutina* (velvet mesquite) treated with ZnO nanoparticles. *Chem. Eng. J.* **2011**, *170*, 346–352. [CrossRef] [PubMed]

55. Krishnaraj, C.; Jagan, E.G.; Ramachandran, R.; Abirami, S.M.; Mohan, N.; Kalaichelvan, P.T. Effect of biologically synthesized silver nanoparticles on *Bacopa monnieri* (Linn.) Wettst. plant growth metabolism. *Process. Biochem.* **2012**, *47*, 651–658. [CrossRef]

56. Ghosh, A.; Kushwaha, H.R.; Hasan, M.R.; Pareek, A.; Sopory, S.K.; Singla-Pareek, S.L. Presence of unique glyoxalase III proteins in plants indicates the existence of shorter route for methylglyoxal detoxification. *Sci. Rep.* **2016**, *6*, 18358. [CrossRef] [PubMed]

57. Hasanuzzaman, M.; Nahar, K.; Hossain, M.S.; Mahmud, J.A.; Rahman, A.; Inafuku, M.; Oku, H.; Fujita, M. Coordinated Actions of Glyoxalase and Antioxidant Defense Systems in Conferring Abiotic Stress Tolerance in Plants. *Int. J. Mol. Sci.* **2017**, *18*, 200. [CrossRef]

58. Arnon, D.I. Copper enzymes in isolated chloroplasts, polyphenoxidase in *Beta vulgaris*. *Plant Physiol.* **1949**, *24*, 1–15. [CrossRef]

59. Li, P.-M.; Cai, R.-G.; Gao, H.-Y.; Peng, T.; Wang, Z.-L. Partitioning of excitation energy in two wheat cultivars with different grain protein contents grown under three nitrogen applications in the field. *Physiol. Plant.* **2007**, *129*, 822–829. [CrossRef]

60. Smart, R.E.; Bingham, G.E. Rapid Estimates of Relative Water Content. *Plant Physiol.* **1974**, *53*, 258–260. [CrossRef]

61. Bates, L.S.; Waldren, R.P.; Teare, I.D. Rapid determination of free proline for water-stress studies. *Plant Soil* **1973**, *39*, 205–207. [CrossRef]

62. Grieve, C.M.; Grattan, S.R. Rapid assay for determination of water soluble quaternary ammonium compounds. *Plant Soil* **1983**, *70*, 303–307. [CrossRef]

63. Velikova, V.; Yordanov, I.; Edreva, A. Oxidative stress and some antioxidant systems in acid rain-treated bean plants. *Plant Sci.* **2000**, *151*, 59–66. [CrossRef]

64. Madhava Rao, K.V.; Sresty, T.V.S. Antioxidative parameters in the seedlings of pigeonpea (*Cajanus cajan* (L.) Millspaugh) in response to Zn and Ni stresses. *Plant Sci.* **2000**, *157*, 113–128. [CrossRef]

65. Dionisio-Sese, M.L.; Tobita, S. Antioxidant responses of rice seedlings to salinity stress. *Plant Sci.* **1998**, *135*, 1–9. [CrossRef]

66. Dhindsa, R.S.; Matowe, W. Drought Tolerance in Two Mosses: Correlated with Enzymatic Defence Against Lipid Peroxidation. *J. Exp. Bot.* **1981**, *32*, 79–91. [CrossRef]

67. Aebi, H. Catalase in vitro. In *Methods in Enzymology*; Colowick, S., Kaplan, N., Eds.; Elsevier: Florida, FL, USA, 1984; Volume 105, pp. 121–126.

68. Foster, J.G.; Hess, J.L. Responses of superoxide dismutase and glutathione reductase activities in cotton leaf tissue exposed to an atmosphere enriched in oxygen. *Plant Physiol.* **1980**, *66*, 482–487. [CrossRef]

69. Miyake, C.; Asada, K. Thylakoid-bound ascorbate peroxidase in spinach chloroplasts and photoreduction of its primary oxidation product monodehydroascorbate radicals in thylakoids. *Plant Cell Physiol.* **1992**, *33*, 541–553.

70. Nakano, Y.; Asada, K. Hydrogen peroxide is scavenged by ascorbate-specific peroxidase in spinach chloroplasts. *Plant Cell Physiol.* **1981**, *22*, 867–880.

71. Huang, C.; He, W.; Guo, J.; Chang, X.; Su, P.; Zhang, L. Increased sensitivity to salt stress in an ascorbate-deficient Arabidopsis mutant. *J. Exp. Bot.* **2005**, *56*, 3041–3049. [CrossRef]

72. Yu, C.-W.; Murphy, T.M.; Lin, C.-H. Hydrogen peroxide-induced chilling tolerance in mung beans mediated through ABA-independent glutathione accumulation. *Funct. Plant Biol.* **2003**, *30*, 955. [CrossRef]

73. Wild, R.; Ooi, L.; Srikanth, V.; Münch, G. A quick, convenient and economical method for the reliable determination of methylglyoxal in millimolar concentrations: The N-acetyl-l-cysteine assay. *Anal. Bioanal. Chem.* **2012**, *403*, 2577–2581. [CrossRef]

74. Hossain, M.A.; Hossain, M.Z.; Fujita, M. Stress-induced changes of methylglyoxal level and glyoxalase I activity in pumpkin seedlings and cDNA cloning of glyoxalase I gene. *Aust. J. Crop. Sci.* **2009**, *3*, 53.

75. Mostofa, M.G.; Fujita, M. Salicylic acid alleviates copper toxicity in rice (*Oryza sativa* L.) seedlings by up-regulating antioxidative and glyoxalase systems. *Ecotoxicology* **2013**, *22*, 959–973. [CrossRef] [PubMed]

MDPI
St. Alban-Anlage 66
4052 Basel
Switzerland
Tel. +41 61 683 77 34
Fax +41 61 302 89 18
www.mdpi.com

Plants Editorial Office
E-mail: plants@mdpi.com
www.mdpi.com/journal/plants